Mathematical Engineering

Series Editors

Jörg Schröder, Institute of Mechanics, University of Duisburg-Essen, Essen, Germany

Bernhard Weigand, Institute of Aerospace Thermodynamics, University of Stuttgart, Stuttgart, Germany

Today, the development of high-tech systems is unthinkable without mathematical modeling and analysis of system behavior. As such, many fields in the modern engineering sciences (e.g. control engineering, communications engineering, mechanical engineering, and robotics) call for sophisticated mathematical methods in order to solve the tasks at hand.

The series Mathematical Engineering presents new or heretofore little-known methods to support engineers in finding suitable answers to their questions, presenting those methods in such manner as to make them ideally comprehensible and applicable in practice.

Therefore, the primary focus is—without neglecting mathematical accuracy—on comprehensibility and real-world applicability.

To submit a proposal or request further information, please use the PDF Proposal Form or contact directly: Dr. Thomas Ditzinger (thomas.ditzinger@springer.com)

Indexed by SCOPUS, zbMATH, SCImago.

More information about this series at https://link.springer.com/bookseries/8445

Sandro G. Longo

Principles and Applications of Dimensional Analysis and Similarity

Springer

Sandro G. Longo
Department of Engineering and Architecture
University of Parma
Parma, Italy

ISSN 2192-4732   ISSN 2192-4740   (electronic)
Mathematical Engineering
ISBN 978-3-030-79219-0   ISBN 978-3-030-79217-6   (eBook)
https://doi.org/10.1007/978-3-030-79217-6

This Springer imprint is published by the registered company Springer Nature Switzerland AG
The registered company address is: Gewerbestrasse 11, 6330 Cham, Switzerland

To my parents Vincenzo and Gilda,

who have never left me alone

Preface

Dimensional analysis is a research tool that covers all research sectors. Much has already been written to survey the general concepts or to deepen some subtle aspects, and it is difficult to add something new or different. A list of the contributions would inevitably be incomplete.

It is a common opinion that dimensional analysis allows us to clarify facts already known and appears not very effective for the analysis of new processes and for progress; see also the comment by Gibbings.[1] It is worthwhile to remember that the search for greater clarity in the analysis of physical processes inevitably leads to greater knowledge. The criteria of analogy between distinct processes, introduced by dimensional analysis and then developed by the theory of similarity, are very effective for improving the average level of scientific arguments.

The discussion of the principles and applications of dimensional analysis leaves many questions on relevant issues still unanswered, although the discussions on the fundamentals are commonly neglected while taking care of the more applicative aspects. For example, the question of the number and nature of fundamental dimensions, as well as the validity of reducing their number on the basis of new physical relationships or of their increase by discriminating them by giving different roles to the same dimension, is still unanswered. This is not a trivial matter since, as demonstrated in the book, an increase in the number of fundamental quantities leads to a reduction in the number of dimensionless groups and, ultimately, to a simplification of the structure of physical equations—all this only on the basis of choices unjustified and not justifiable a priori. This is one of the reservations and is perhaps the most relevant. However, we remind ourselves that classical mechanics is based on principles and axioms (homogeneity and isotropy of the space; stationarity; see Landau & Lifshitz 1978[2]), assumed to be the basis of a complete theory that, ultimately, has undergone experimental validation.

---

[1] J. C. Gibbings. (1980). On dimensional analysis. *Journal of Physics A: Mathematical and General* *13*(1), 75.

[2] L. D. Landau & E. M. Lifshitz. (1960). *Course of Theoretical Physics* (Vol. 1: Mechanics). Pergamon Press.

Practically nonexistent are reservations on the theory of similarity and models, the second important topic discussed here. Current thinking on the scientific representation of reality assumes that the resources useful for such representation are linguistic, considering mathematics as a specific language (Giere 2004[3]): the language of science has its own syntax, a semantic aspect and a pragmatic one. In physical modelling, it is the pragmatic aspect that is emphasised, with syntax and semantic aspects adapted accordingly.

Leaving aside the issues of foundation, which are here only briefly mentioned, the purpose of the book is to provide the necessary and appropriate tools for an accurate and consistent interpretation of physical processes, both through mathematical analysis and through physical modelling. The first chapters address the definitions, with few dimensional analysis theorems and similarity criteria. There is also the analysis of self-similarity, both of first and second kind, with a couple of completely solved problems, framed within the group theory. I have deliberately not dealt with the theory of renormalised groups, which I consider more specialised.

From Chap. 5 onward, the focus is on applications in some of the engineering sectors. The number of topics is necessarily limited, but, almost always, there are details, calculations and treatment of assumptions. To avoid making the book too broad, I have left out the description of many experimental devices used for physical modelling, such as channels and tanks with wave generators and rotary platforms for geophysical models. I have included only a brief description of the centrifuge of the shaking table, although an entire chapter is devoted to wind tunnel technology. I have also omitted the treatment of measurement techniques and instrumentation, available, for example, in Doebelin, 2008[4] and in several other books focused on different branches of research. Some more specific notions, required by the context, are reported in the Appendix, where appears also the description of numerous dimensionless groups, all of engineering interest, but with the exclusion of many others related to physical processes of electrical nature or physics of particles. I prefer to repeatedly explain the symbols next to the formulas that use them, rather than listing them in an Appendix. As far as I could, I tried to make them uniform: a specific symbol corresponds to the same variable throughout the book. The glossary lists the meaning of some specific terms used in the book.

The book is addressed to those trained in engineering disciplines and natural and physical sciences and to graduate students and fellow researchers interested in a clarification of the methodology, of dimensional analysis and theory of similarity aims and purposes.

Much of the work involved in writing this book was done during my stay on sabbatical leave in Granada, Spain, in the Spring-Summer of 2010, hosted by Prof. Miguel A. Losada, Emeritus of the University of Granada and at that time Director of the Centro Andaluz de Medio Ambiente (CEAMA), a joint research structure between the Junta de Andalucia and the Universidad de Granada. At CEAMA, in addition

---

[3] R. N. Giere. (2004). How models are used to represent reality, *Philosophy of Science, 71,* 742–752.

[4] E. O. Doebelin. (1990). *Measurement systems: Application and design.* McGraw-Hill Book Co.

to the availability of the laboratory and all the equipment to carry out my experimental research program, co-monitored by Luca Chiapponi, Mara Tonelli, Simona Bramato and Christian Mans, I had the complete availability of what was necessary for bibliographic investigations and to meditate and to write. I had a fantastic time in Granada, where numerous young researchers and students were deeply involved in wind tunnel and sea gravity waves and related phenomena research.

The English language version, revised and extended with new sections, was completed during the COVID-19 pandemic in 2020, which is about to be finally eradicated thanks to biomedical research, with several vaccines being developed very quickly. Thanks are due to all the researchers involved in that field and in memory of those who did not survive.

Parma, Italy   Sandro G. Longo
April 2021

Acknowledgements

Much of the expertise acquired prior to the drafting of this monograph, is the result of decades of scientific collaboration with Miguel A. Losada, Emeritus of the University of Granada (Spain), Marius Ungarish, Emeritus of Technion (Israel), and Vittorio Di Federico, Full Professor at the University of Bologna (Italy). I would like to thank them and their research groups for sharing their knowledge on some fascinating phenomena of Fluid Mechanics.

I gratefully acknowledge critical reading of some chapters by colleagues; Andrea Maranzoni critically commented on Chap. 3, Andrea Spagnoli commented on Chap. 7, and Diana Petrolo revised and commented on all the chapters. I thank Luca Chiapponi, who created some of the most complex figures with great accuracy.

Introduction

...avvengaché quello che noi ci immaginiamo bisogna che sia o una delle cose già vedute, o un composto di cose o di parti delle cose altra volta vedute; ché tali sono le sfingi, le sirene, le chimere, i centauri, etc.

(*... for that which we imagine must be either something already seen or a composite of things and parts of things seen at different times; such are sphinxes, sirens, chimeras, centaurs, etc.*).
Galileo Galilei

In the introductory speech of the Conference "Models in Technique", Various authors (1956), organised in 1955 by the Accademia dei Lincei in Venice, an anecdote taken from Winston Churchill's biography was reported. During the last world war, Winston Churchill was visiting a British naval base, where models of warships were docked and used for camouflage. Those models were very well made and served very well the purpose of their realisation. However, Churchill recognised from afar one of those fake ships, saying that there were no seagulls around it and, therefore, there was no crew. No one had thought of perfecting the model by throwing some garbage overboard, as always there is for an uninhabited vessel. This anecdote shows us how important details are in the realisation of the models. They are not only aesthetic details, of course, which are often of negligible importance in physical model making; they are the details of the design of the model, of the conditions of partial similarity, which are of greatest interest and must be carefully considered by the scientist.

Surely, a model designer must be aware of dimensional analysis and the consequent rules of mechanical and physical similarity. However, this ends up being very little in the face of the multiform knowledge and intuition that may require the claim to experiment "at the small scale", or "by analogy", complex phenomena.

The need for great experience, also intended as experimental practise, to accompany dimensional analysis, has been widely stressed by many other eminent scholars, such as Bridgman (1922), who, in his treatise, described the intrinsic limitations of the mathematical theory of dimensional analysis: *"Problem cannot be solved by the philosopher in his armchair, but the knowledge involved was gathered only by someone at some time soiling his hands with direct contact."* He specifically referred to the validation of a set of governing variables to adequately define a governed variable, and the concept expressed by Bridgman can be extended to all the steps necessary to make workable dimensional analysis and theory of similarity.

For itself only, the dimensional analysis reflects the fact that the measurement of each physical quantity is agreed upon in such a way as to attribute an intrinsic value to the relative measurement, that is, the ratio between "the measurements of two quantities of the same species".

It is amazing that such a simple concept has generated a multitude of schemes of reasoning and methods of dealing with physical problems and has brought order to the approach to the description of problems and their solution in all areas of science.

In the next section, we outline a brief history of dimensional analysis and similarity, with more details and curiosities that can be retrieved in Macagno (1971) and Sterrett (2017).

Brief History of Dimensional Analysis: The Different Aims and Approaches

An interesting publication by Macagno (1971) describes the origins of dimensional analysis, when it was not yet definable with the same terminology adopted in our days, at a time when it was born from flashes of ideas of great mathematicians of the past, such as Euclid (IV–III centuries b.c.), Pappus (c.350–c.290 b.c.), and Ptolemaeus of Alexandria (c.170–c.100 b.c.).

Egyptians and Romans developed the concept of size and measure, both for fiscal reasons and for the design and realisation of their great constructions. However, the ideas were quite confused, and many of the operations that have fallen into current practise, such as the combination of various dimensions to define a new dimension, were considered aberrations.

It was in the Middle Age that the idea of numbers to represent quantities took shape, perhaps not yet understood as the ratio between the quantity to be measured and a reference quantity (not necessarily fundamental, in today's meaning). In the early Renaissance, the first thinkers and carriers of ideas about the nature of dimension and measure were born, which then led to Galileo and Descartes, both responsible in many ways for a reversal of perspective in scientific and philosophical observation, a continuous contrast between matter and spirit. Leonardo (1452–1519) was, in addition to many other roles, one of the first hydraulic and coastal engineers, a keen observer of hydraulic phenomena and an avant-garde experimenter, aware that physical modelling was of great help in describing nature. Subsequently, Newton (1642–1726) in his *Principia* (Newton 1687), Leibniz (1642–1716) (letter of October 16, 1707 Fleckstein (1957)), and Euler (1707–1783) in his *Mechanica* (1736) began to deal with dimensional problems on a permanent basis, first as support for the mathematical models proposed for the study of the motion of falling bodies and motion in general, and then as a separate analysis within the framework of the development of physical-mathematical knowledge.

However, even before Newton, Galilei (1564–1642), following some visits to the Arsenal in Venice, in his *Dialogues of the New Sciences*[5] (Galilei 1638, 1718) in the *Giornata Prima: Scienzia nuova prima, intorno alla resistenza de i corpi solidi all'essere spezzati*, wrote:

"Why did they make so much more apparatus of supports, reinforcements and other shelters and fortifications around that great Galeazza[6] that had to be launched, which is not done around smaller vessels?"

This consideration is accompanied by numerous others in which, for example, Salviati introduces Sagredo to the rudiments of knowledge that explain why structures with different geometric scales do not have the same proportion of resistance. In practise, emphasis is placed on the concept of scaling relationships between variables of similar processes, with several other discussions involving, for example, the relationship between the period of oscillation of two pendulums of different lengths. In these discussions, the seed of similarity appears, including the discrimination between variables of interest (length and period of the pendulums) and irrelevant variables (amplitude of oscillations and mass of the pendulums). Beyond the conceptual outline, it is relevant that Galileo applies the results to calculate the length of a pendulum by measuring its period of oscillation and knowing the period of oscillation and length of another similar pendulum. In fact, Galileo analysed a physical model, taking advantage of the method he proposed to perform an indirect estimate of a variable. This is exactly what we do today when we perform experiments, for example, on the physical hydraulic model of a dam.

Euler in particular posed the problem of units of measurement and homogeneity (Euler 1765). Euler, a prolific writer of 866 scientific papers[7], was more interested in the mathematical aspect of physical processes than in the process itself. From this point of view, the notion that a set of variables characterise the state of a system is illuminating, and as a consequence, similarity involves multiple relationships. It should not be assumed, however, that Euler did not participate in the many discussions on the physical implications of similarity criteria, as did Lagrange and many other scientists of the time (Hepburn 2007).

As often happens, reading today those notes and chapters in light of modern knowledge reveals numerous errors and many inaccuracies. However, it is undeniable the great conceptual effort of scientists in trying to give shape and norms to concepts that in many ways are still vague.

---

[5] The book is presented as a series of discussions among two philosophers and a layman; Salviati (named after Galileo's friend Filippo Salviati 1582–1614) presents some of Galileo's views directly; Sagredo (named after Galileo's friend Giovanni Francesco Sagredo 1571–1620) is an intelligent layman; Simplicio (most likely a pun on Simplicius of Cilicia, a sixth-century commentator on Aristotle and "un simpliciotto", "a simple minded" in Italian), a follower of opponents of Galileo's.

[6] Galeazza represented the zenith of shipping technology of the 16th and 17th century, with superior maneuverability and deadly firepower (Ercole 2011).

[7] According to Gustav Eneström, a Swedish mathematician who completed a comprehensive survey of Euler's work.

The concepts introduced by Euler were summarily used by Lagrange and Laplace but without being considered truly enlightening. It was instead Fourier (1768–1830) who established the principles of dimensional analysis in his *Analytical Theory of Heat* (Fourier, 1822). For the first time, he described the exponents of a dimension, and he started with the introduction of dimensional tables, even if incomplete since they were strictly related to a research topic that involved only length, time and temperature (what today we call a subset of dimensional space). He established that equations representing a physical process (in his case, heat propagation) must be independent of units of measurement and derived the principle of dimensional homogeneity. He missed the last steps, that is, (i) the identification of the potential structure of the relationship between dimensional variables, which he had already grouped into nondimensional groups; (ii) the similarity in connection with the dimensionless groups.

Undeniably, Bertrand's (1822–1900) great contribution (Bertrand 1848) marked an important step toward the application of Newton's results. His most valuable Theorem states that *One can ignore the constraint reactions even in many questions of mechanical similarity, as long as their virtual work is identically zero.* Bertrand was the first to promote the practical applications of dimensional analysis.

In the following decades, there were numerous discussions on the concepts of similarity, with a remarkable paper by Maxwell (1831–1879) *On the mathematical classification of physical quantities* (Maxwell, 1869), where the symbols for mass, length and time *M, L, T* were introduced for the first time; the seeds sown by Fourier finally took firm root in Lord Rayleigh's (1842–1919) *Theory of Sound*, 1877 (Strutt 1877), with a chapter entitled *Method of Dimensions*.

Rayleigh later wrote numerous publications applying the method, introducing numerous nondimensional groups such as the Reynolds number and the Weber number. Meanwhile, in France, Carvallo (1856–1945) (Carvallo 1892) and Vaschy (1857–1899) (Vaschy 1892) aimed to formulate a more general Theorem about dimensions, and Vaschy was the first in a race involving other eminent French scientists. Vaschy published a book, *Théorie de l'Électricité* (Vaschy 1896), and the first chapter deals with units of measurement and dimensions.

However, Vaschy's Theorem long remained almost unknown, so much so that in 1911 Riabouchinsky (1882–1962) formulated a Theorem with practically identical results (Riabouchinsky 1911), without knowledge of Carvallo and Vaschy's work. Riabouchinsky worked extensively on dimensional analysis problems, with numerous theoretical and practical applications. In particular, when commenting on a passage by Rayleigh in an article published in Nature (Strutt 1915) concerning Boussinesq's problem of heat transmission, Riabouchinsky highlighted a paradox of dimensional analysis: the selection of truly independent quantities is not unambiguous and, as a consequence, the number of nondimensional groups that arise in the description of the physical process is also not unique. Incidentally, a paradox in a fuzzy field such as dimensional analysis is not particularly disturbing if one considers the numerous paradoxes listed by Birkhoff (1884–1944) (Birkhoff 1950) in the field of fluid mechanics, where the fundamentals are very strong and deep-rooted. Apart from the interesting examination of paradoxes, mainly attributed to a lack of rigour, we recognise that Birkhoff has managed to fill the gap between pure mathematics and physics, as he himself hopes at the end of his book, by linking group theory to fluid mechanics.

The discussion of the "Rayleigh-Riabouchinsky problem" was later deepened by Sedov (1907–1999) (Sedov 1959), who began a process of rationalisation in the choice of the quantities to be considered fundamental, which involves a critical analysis of the physical process under investigation: the selection of the quantities must have a physical ground, initially dictated by intuition or experience alone, then to be verified experimentally. For example, there is no single "pressure": pressure can be a thermodynamic variable and, as such, involves temperature; or, for incompressible fluids, it is a mechanical variable (defined, for example, as one-third of the stress tensor trace) and, as such, involves purely mechanical quantities. A clear interpretation of the Rayleigh-Riabouchinsky problem was outlined in Gibbings (1980) and then fully explained in Gibbings (2011).

In general, some relations not participating in the process should be ignored. In other circumstances, it may be convenient to look at more fundamental quantities than is considered strictly necessary (and we remark "considered" necessary because there are no unambiguous selection criteria). This is how Bridgman (1882–1961) (Bridgman 1922) discussed Stokes' problem of the sphere falling into a fluid. Bridgman started from the evidence that the problem of a falling sphere is a static problem if the analysis is carried out in steady state at constant speed and, as so, in the absence of acceleration. In this respect, the standard approach considering mass, length and time as fundamental quantities does not make sense since the process does not involve acceleration. The introduction of force as a fourth independent variable makes it possible to elucidate the effect of all the variables on speed (the governed variable), which does not happen if we proceed considering only three fundamental quantities and the force as a derived quantity. Of course, we could further discuss Bridgman's position, for example, by pointing out that if the sphere is in motion with a turbulent regime, turbulent fluctuations induce accelerations that are zero on average only and not uniformly zero. Therefore, neglecting a priori Newton's principles in formulating a dimensional analysis scheme where force is an independent variable could be arbitrary. This possible discussion, like many others that have animated the early days of dimensional analysis up to the present day, is part of a vision of science that transcends the philosophical vision of the World and Nature, against which the scientist can only seek confirmation in Galilean empiricism: a scheme consistent with experimental evidence must be assumed to be correct.

In the first decades of the past century, Buckingham (1867–1940) (Buckingham 1914) wrote numerous publications on the topic, being perfectly aware of Riabouchinsky's work. Today, the Π-Theorem is often named Buckingham's Theorem, but it should actually be defined as Carvallo-Vaschy's Theorem, if not only Vaschy's Theorem. In fact, it is likely to be referred to as the Π-Theorem to avoid a precise attribution to a scientist, and *it would not be wrong to name it as 'I-Theorem', the Invariant Theorem* (Signorini A., in Various authors 1956) since it is based on a reformulation of the equations describing a physical process, with the introduction of invariants. The analysis in terms of invariants is particularly attractive and illuminates the most recondite aspects of physical reality. And yet, as Bridgman well pointed out, reality is not unequivocally revealed by the criteria of dimensional analysis but requires an interpretive effort such that "... *dimensional analysis is essentially of the nature of an analysis of an analysis.*"

Bridgman deserves a historical review of his contribution, which was among the most important of the many scientists involved. In a review paper by Macagno (Macagno 1971), Bridgman is attributed a second Theorem that, in fact, should precede the Π-Theorem. In essence, regardless of the system of units of measurement, the ratio between the measurements of two quantities of the same nature is invariant. This is an apparently simple and trivial concept, but it is the foundation that justifies any variety of system of units of measurement: whatever the choice of the system of units of measurement, it does not change the intrinsic value of the quantities. If this were not the case, we would be faced with a physics that behaves according to the observer and the conventions he or she has chosen.

In the second part of the past century, the topic has been extensively treated by Langhaar (1909–1992) (Langhaar 1951), Corrsin (1920–1986) (Corrsin 1951), with numerous disputes and discussions. From the patchwork of new approaches, often aimed at giving rigorous proofs of theorems, inspectional analysis was then developed by Ehrenfest-Afanasjeva (1876–1964) (Ehrenfest 1915) and by Ruark (1899–1979), Ruark (1935), Focken (1901–1978), Focken (1953), and Ipsen (1921–?), Ipsen (1960). A corollary to Buckingham's Theorem (or a generalisation, according to Sonin's script) is due to Sonin (1937–2010) (Sonin 2004), for the case when some variables are kept constant. Numerous books have been published on dimensional analysis, each of which has made inspiring contributions on the subtle aspects of the topic, also highlighting the closeness between science and philosophy when dealing with the subject. A very clear book full of enlightening details was recently published by Gibbings (Gibbings 2011), who has also actively written numerous articles and discussed several controversial aspects of dimensional analysis. A precious contribution is due to Tan (Tan 2011), with his book containing several handled cases with rigorous mathematics and practical indications. A *sui generis* contribution was made by Barenblatt, who has long developed the analysis of self-similar solutions with a clear framework within group theory. His books on fundamentals (Barenblatt 2003), and on specific applications (Barenblatt 2014; Barenblatt et al. 1989) are a continuous source of indications, clarifications and new ideas for those who wish to handle dimensional analysis and similarity theory, particularly when applied to mathematical physics.

To date, discussions on methods continue, albeit in apparently less passionate tones than in the past. Books on the subject are flourishing, but innovations are limited. However, the international standardisation of unit systems is still a slow process. A disaster investigation board reported that NASA's Mars Climate Orbiter burned up in the Martian atmosphere in 1999 because engineers failed to convert units from

English to metric, and the space probe costing \$125 million was lost.[8] The problem was in the software controlling the orbiter's thrusters because the software calculated the force the thrusters needed to exert in pounds of force, whereas a separate piece of software took in the data assuming it was newtons.

The engineering applications of similarity are among the most numerous and flourishing, having often represented, especially in the past, the only "calculation" tool available for the design of large works.

In the structural sector, we recall the contributions by Thomson (1822–1922) (brother of William Thomson (1824–1907) Lord Kelvin). Thomson deepened the criteria of similarity, differentiating them according to the objective of the analysis, which could be related to strength and deformation, or stability.

In the field of hydraulics, Froude (1810–1879), who, prompted by the need to identify a criterion for the design of large warships, developed the criterion of similarity that still bears his name. The problem of transposing the results of a model to a prototype had already been addressed by other scientists, including Russell (1808–1882), without finding a solution. In contrast, it was the perseverance and great competence of Froude, aided by the availability of an experimental tank, that led to the genesis of his criterion of similarity. The criterion was extended by Froude to be able to compare forces, not only lengths and times, finally becoming a criterion of dynamic similarity. Froude undoubtedly benefited from the knowledge developed on the subject by Reech (1805–1884), although his decisive impulse is undeniable. Figure 1 shows the most relevant milestones in dimensional analysis and similarity.

Overview of Methods and Application Scenarios

The methods for studying and applying dimensional analysis and similarity criteria have also developed as a result of the numerous heated disputes that have animated the scientific community in the last century. Among these methods, Rayleigh's method is based on the assumption that equality between two power functions representing physical quantities necessarily requires equality of the exponents of the fundamental quantities involved in the description of the two quantities. The result is an immediate relation that has the limit, however, of relegating the functional relation to the monomials only, a limit later overcome by Rayleigh himself assuming that a series expansion of the generic power of the monomials can reproduce any function. However, not all series are convergent, and some assumptions are requested to expand a function.

The best known method is that of Vaschy, although it is usually better known as Buckingham's method, based on the Π-Theorem. After identifying the relevant

---

[8] Stephenson A. G., La Piana L. S., Mulville D. R., Rutledge P. J., Bauer F. H., Folta D., Dukeman G. A., Sackheim R., Norvig P. (1999, November 10). *Mars Climate Orbiter Mishap Investigation Board Phase I Report.* NASA.

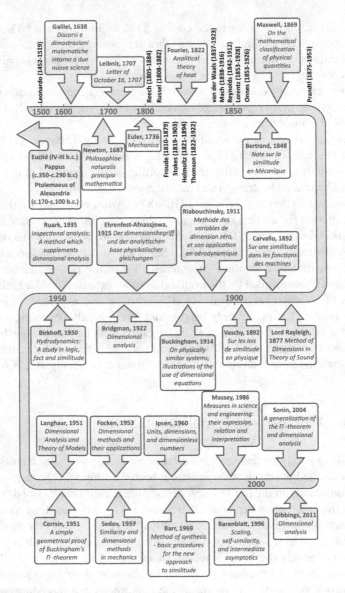

Fig. 1 Milestones in dimensional analysis and similarity

variables involved in a physical process, it is necessary to select a set of fundamental variables and construct a number of dimensionless groups. The method has
an adequate level of generality, even excessive in reality, since it is not possible to
identify a priori which groups must be constructed and which function correlates
them. Physical sensitivity of the process being analysed and the experience of the
scientist do help. The experiments, then, allow the construction of representations

of the function of the dimensionless groups, which is of practical use. However, the number of experiments required rises prohibitively if numerous dimensionless groups are involved.

An indication of the most relevant groups to be used with Vaschy's method was suggested by Barr. The method by Barr, which can be considered a variant of Vaschy's method, proposes a reorganisation of the variables involved to obtain monomials or power functions characterised by being dimensional and having the dimension of a length. The criterion is defined as *linear proportionality* and can be generalised by combining the variables into monomials all having the dimension of a time, or a mass. The apparent complication, compared to Vaschy's method, has some advantages: the aim is to reduce the number of groups in which the dependent variable appears (the optimum would be to have only one); the presence of a greater number of dimensional groups facilitates the selection in order to achieve the goal.

For the remaining methods, some techniques have been explored in depth without additional methods. The matrix technique, for example, facilitates calculations in the case of a large number of variables involved in the process and can be applied both to the selection of dimensionless groups and to dimensional groups (e.g., having the size of a length, as suggested by Barr 1969).

Of some interest are other techniques for reducing the number of dimensionless groups based on vectorisation or discrimination. Reducing the number of groups facilitates the analysis of the structure of the functional relationship between the groups and the interpretation of the experimental results, if any. As a general notation, some techniques have been strongly criticised, such as vectorisation; see, e.g., Massey 1978, 1986, and Gibbings 1981.

The application scenarios mainly belong to model analysis in similarity. Figure 2 shows a road map of the topics treated in the present book.

Perspectives and Further Developments

In little more than a century, dimensional analysis and the criteria of similarity have passed from epistolary exchanges between scientists to specialised texts to scholastic texts and are now well-established knowledge with practical application. It is difficult to predict what further evolution we can expect and what possible upheavals might occur, at least as far as basic concepts and fundamentals are concerned. It is very probable that we will find further future links between the criteria of dimensional analysis and other areas of mathematics, as happened with group theory. Interesting perspectives are emerging for the analysis of differential problems describing physical processes, today the object of intense activity for the identification of solutions and their experimental validation.

Applications of dimensional analysis have extended to the humanities, social sciences, and economics, and we expect that they can permeate all areas of knowledge.

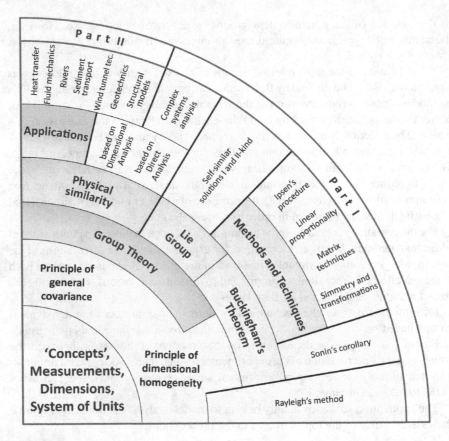

Fig. 2 Road map of the topics treated in the book

For the more applicative aspects, model theory is becoming increasingly relevant, with a pressing demand for approximations such that scale effects can be reduced or well quantified. In this respect, we recall that for interactions in complex systems, such as geofluid dynamics and turbulence, excessive approximations can lead to results that are only remotely linked to reality. Hence, further investigation is needed to elaborate similarity criteria that better reproduce real phenomena.

References

Barenblatt, G. I. (1996). *Scaling, self-similarity, and intermediate asymptotics.* Cambridge University Press.

Barenblatt, G. I. (2003). *Scaling.* Cambridge University Press.

Barenblatt, G. I. (2014). *Flow, deformation and fracture: Lectures on fluid mechanics and the mechanics of deformable solids for mathematicians and physicists* (Vol. 49). Cambridge University Press.

Barenblatt, G. I., Entov, V. M., & Ryzhik, V. M. (1989). *Theory of fluid flows through natural rocks.* Kluwer Academic Publishers.

Barr, D. I. H. (1969). Method of synthesis–basic procedures for the new approach to similitude. *Water Power*, *21*(4), 148–153.

Bertrand, J. (1848). Note sur la similitude en Mécanique. *Journal de l'École Polythechnique* XIX, Cahier XX XII, 189–197.

Birkhoff, G. (1950). *Hydrodynamics: A study in logic, fact and similitude*. Dover.

Bridgman, P. W. (1922). *Dimensional Analysis*. Yale University Press.

Buckingham, E. (1914). On physically similar systems; illustrations of the use of dimensional equations. *Physical Review*, *4*(4), 345–376.

Carvallo, E. (1892). Sur une similitude dans les fonctions des machines. *Journal de Physique Théorique et Appliquée*, *1*(1), 209–212.

Corrsin, S. (1951). A simple geometrical proof of Buckingham's Π-theorem. *American Journal of Physics*, *19*(3), 180–181.

Ehrenfest-Afanassjewa, T. (1915). Der Dimensionsbegriff und der analytischen Base physikalischer Gleichungen. *Mathematische Annalen*, *77*, 259–276.

Ercole, G. (2011). *Vascelli e fregate della Serenissima: navi di linea della Marina veneziana 1652–1797 (in Italian)*. Gruppo modellistico trentino di studio e ricerca storica.

Euler, L. (1765). *Theoria motus corporum solidorum seu rigidorum*. A. F. Röse.

Fleckstein, G. O. (1957). *Leonhardi Euleri opera omnia*, vol. 5(II). Orell Füssli, Lausanne.

Focken, C. M. (1953). *Dimensional methods and their applications*. Arnold.

Fourier, J. B. (1822). *Theorie analytique de la chaleur, par M. Fourier*. Chez Firmin Didot, père et fils, reprinted by Cambridge University Press, 2009.

Galilei, G. (1638). *Discorsi e dimostrazioni matematiche intorno a due nuove scienze attinenti la meccanica e i movimenti locali (only first 4 days + appendix, in Italian and in Latin)*. Elzeviri.

Galilei, G. (1718). *Discorsi e dimostrazioni matematiche intorno a due nuove scienze attinenti la meccanica e i movimenti locali (fifth and sixth days, in Italian and in Latin)*. Tartini and Franchi.

Gibbings, J. C. (1980). On dimensional analysis. *Journal of Physics A: Mathematical and General*, *13*(1), 75.

Gibbings, J. C. (1981). Directional attributes of length in dimensional analysis. *International Journal of Mechanical Engineering Education*, *8*(3), 263–272.

Gibbings, J. C. (2011). *Dimensional analysis*. Springer Science & Business Media.

Hepburn, B. S. (2007). *Equilibrium and explanation in 18th century mechanics*. PhD thesis, University of Pittsburgh.

Ipsen, D. C. (1960). *Units, Dimensions, and Dimensionless Numbers*. McGraw-Hill Book Co.

Langhaar, H. L. (1951). *Dimensional Analysis and Theory of Models*. Wiley.

Macagno, E. O. (1971). Historico-critical review of dimensional analysis. *Journal of the Franklin Institute*, *292*(6), 391–402.

Massey, B. S. (1978). Directional analysis? *International Journal of Mechanical Engineering Education*, *6*(1), 33–36.

Massey, B. S. (1986). *Measures in science and engineering: Their expression, relation and interpretation*. Ellis Horwood Ltd.

Maxwell, C. J. (1869). Remarks on the mathematical classification of physical quantities. *Proceedings of the London Mathematical Society*, *1*(1), 224–233.

Newton, I. (1687). *Philosophiae naturalis principia mathematica (in Latin)*. Societatis Regiae.

Riabouchinsky, D. (1911). Methode des variables de dimension zéro, et son application en aérodynamique. *L'Aérophile*, *1*, 407–408.

Ruark, A. E. (1935). Inspectional analysis: A method which supplements dimensional analysis. *Journal of the Elisha Mitchell Scientific Society*, *51*(1), 127–133.

Sedov, L. I. (1959). *Similarity and Dimensional Methods in Mechanics*. Academic Press.

Sonin, A. A. (2004). A generalization of the Π-theorem and dimensional analysis. *Proceedings of the National Academy of Sciences*, *101*(23), 8525–8526.

Sterrett, S. G. (2017). Physically Similar Systems–A History of the Concept. In Bertolotti Magnani (Eds.), *Springer Handbook of model-based science* (377–411). Springer.

Strutt, J. W. (Lord Rayleigh) (1877). *The theory of sound*. Macmillan Publishers Ltd.

Strutt, J. W. (Lord Rayleigh) (1915). The principle of similitude. *Nature*, *95*, 66.

Tan, Q.-M. (2011). *Dimensional Analysis: with case studies in mechanics.* Springer Science & Business Media.

Various authors (1956). *I modelli nella tecnica. Vol. I–II.* Accademia Nazionale dei Lincei (in Italian).

Vaschy, A. (1892). Sur les lois de similitude en physique. *Annales Télégraphiques, 19,* 25–28.

Vaschy, A. (1896). *Théorie de l'électricité: Exposé des phénomènes électriques et magnétiques fondé uniquement sur l'expérience et le raisonnement.* Librairie Polytechnique, Baudry et cie.

Contents

About the Author

Sandro G. Longo, PhD, is full professor in Hydraulics at the University of Parma, Italy. His research interests include sea gravity waves, turbulence, sediment transport, flows of non-Newtonian fluids, gravity currents. He currently teaches Hydraulics to undergraduates and Environmental and Coastal Hydraulics to graduates, and has written books on Hydraulic Measurements and Controls and educational textbooks. He has consulted extensively in the field of Maritime Hydraulics, designing or participating in the realization of physical models of hydraulic and hydraulic-maritime works. He was a visiting researcher in Wallingford (UK) and an academic visitor in Granada (Spain) and Cambridge (UK).

Part I
The Methods

Chapter 1
Dimensional Analysis

The opposite of a correct statement is a false statement. But the opposite of a profound truth may well be another profound truth.

Niels Bohr

The expression of the equations describing the physical phenomena requires the respect of some formal rules and the knowledge of some basic principles that guarantee correctness and logical coherence. In particular, the structure of an equation that connects physical quantities must respect the principle of dimensional homogeneity. The comprehensive description of this principle, first formalised by Fourier (1822), requires the *classification of physical quantities* and the definition of *systems of units of measurement*. This is a principle that today seems self-evident and does not require any specific discussion but has nevertheless stimulated numerous discussions and diatribes over the past centuries. The need for uniformity in the choice of unit systems, for a writing that intrinsically accounts for and respects the dimensional nature of variables, has required a conceptual effort in a framework in which theorems are few (perhaps only one) but rules are numerous. Furthermore, the change of unit systems, whether belonging to the same or different classes, has been rationalised.

1.1 Classification of Physical Quantities

A *physical quantity*, which can be defined as an element, a class of entities that pertain in physical processes, to each of which a measure can be assigned, can be classified as a *constant*, a *parameter* or a *variable*.

© The Author(s), under exclusive license to Springer Nature Switzerland AG 2021
S. G. Longo, *Principles and Applications of Dimensional Analysis and Similarity*,
Mathematical Engineering, https://doi.org/10.1007/978-3-030-79217-6_1

We observe that for practical applications, physical quantities are closely related to their measurement, while in fact, they can be defined as concepts independent of the measurement operations. In this regard, there are different points of view: Kelvin considered concepts and their measurement closely related: *"When you cannot express it in numbers, your knowledge is of a meagre and unsatisfactory kind."* Thomson (1883) and Fourier (1822), however, defined concepts and measurement independently of each other; see the discussion in Gibbings (2011), who also clearly depicted the sequence of concepts, dimensions and units. Gibbings also defines four 'primary' concepts as being independent of each other, making a distinction between 'extension', 'time', with a measurement that has a relative linear scale (no absolute origin, but measurement as difference of); and 'force', 'quantity', with an absolute linear scale, with a zero. Linear scale means that their measure (which, according to Gibbing, is not part of the definition) is obtained by the addition of unit quantities.

A *constant* is a physical quantity that does not vary in space or time. For example, the speed of light in vacuum and Planck's constant are constants.

Constants derive from the knowledge and descriptive modes of the physical world and can be known with infinite precision if they are implicitly defined on the basis of a physical process or with limited precision if, although their origin and invariant nature are known, they must be measured. Other constants, such as the ratio $\pi = 3.141\,5\ldots$ between the length of a circle circumference and its diameter in a plane space, or the number of Euler $e = 2.717\,8\ldots$, are irrational numbers, independent of the nature of the Universe, and known only with approximation.

It is conceivable that in the future, as knowledge advances, some constants may no longer be so, and new constants will have to be introduced.

A *parameter* is a physical quantity that can vary but which, in the context of a physical problem or process, takes on a constant value. Thus, for example, the acceleration of gravity is a parameter in many processes. The same is true for Young's modulus, Poisson's ratio, and specific heat.

A *variable* is a physical quantity whose numerical value can change during a physical process involving it. If the variation has the nature of *effect*, the variable is called *dependent* or *governed*; if it has the nature of *condition*, the variable is called *independent* or *governing*. The *condition* is distinct from the *cause*, which is usually expressed through variables external to the process but able to influence the dependent variables that characterise its internal behaviour.

The *causes* are generally variables that do not depend on the process (except for stability problems).

Many dependent and independent variables can coexist in a physical process or phenomenon. It is appropriate to formalise the process analytically to highlight a single dependent variable, but this is not always possible, as is the case, for example, in the coupled problems of flow, heat, and transport.

All physical quantities, whether constants, parameters or variables, can be *dimensional* or *dimensionless*. A *dimensional* quantity has a measure with value that depends on the specific unit measurement system. A *dimensionless* quantity has a measure with value independent of the unit system. An exception is represented by the plane angle, which, although dimensionless (in fact, it is the ratio between the length of the arc of a circle and its radius), has a measure that varies according to the chosen system: in the sexagesimal system of measuring angles, a radian is equal to $57°17'44.8''$; in the sexadecimal system it is equal to $57°.2958$, and in the centesimal system it is equal to $63.661\ 977\ 2$ centesimal degrees.

We observe that a dimensional constant changes its numerical value in two different systems of units, but this does not affect its nature as a constant.

Quantities can be *extensive* or *intensive*. The measure of an extensive quantity depends on the size of the system (e.g., mass, volume, surface area). The measure of an intensive quantity does not depend on the size of the system: temperature, specific heat, and dynamic viscosity are intensive quantities. The ratio of two extensive quantities is an intensive quantity if the two extensive quantities refer to the same dimensions of the system.

The quantities can be *scalar*, if their measure is sufficient to specify them completely, or *vectorial* if, in addition to the measure, it is also necessary to indicate a line of action and a direction. *Applied vectorial* quantities also require the indication of their application point. *Tensors* are mathematical objects that generalise algebraic structures based on a vector space: a tensor of order 0 is named a scalar, and a tensor of order 1 is named a vector.

Vectorial quantities are defined as *pseudo-vectors* or *axial vectors* if their orientation requires a convention; otherwise, they are defined as *true vectors* or *polar vectors*. Pseudo-vector quantities are, for example, all those involving rotation and, in general, all vectors resulting from a vectorial product (*bivectors*): the angular momentum is a pseudo-vector, since it is necessary to specify a convention to determine whether the rotation associated with it is clockwise or counter-clockwise (typically, the right-hand rule is adopted). In common applications, it is not necessary to distinguish pseudo-vectors from true vectors, but in some applications (for instance, in particle physics), the distinction is essential since, for example, pseudo-vectors do not satisfy the symmetry for a reflection (the *parity*), and on reflection, they require the change of orientation (see Fig. 1.1). This limits their use in the expression of the physical laws that are believed to satisfy parity. It is also possible to define pseudo-scalar and pseudo-tensor quantities.

Finally, we define *coefficient* as a multiplicative factor of a term of an equation, usually dimensionless or in any case not a function of any variable of the equation in which it appears.

1.2 The Systems of Units

The measurement of a variable of a physical process requires, as a first step, the identification of the entity and the selection of a sample to be used as a reference unit. We then proceed to relate the quantity to be measured to the sample and obtain a number that defines the modulus. The modulus represents the ratio of the physical quantity to the unit of measurement of the sample.

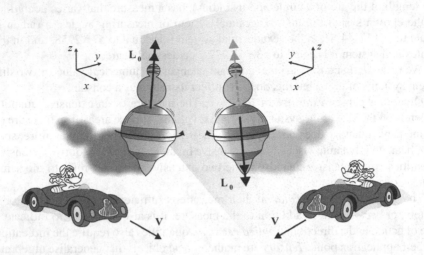

Fig. 1.1 The angular momentum L_0 is a *pseudo-vector*: for a specular reflection, it requires transformation to remain consistent with the convention (in the example shown, the right-hand rule applies). The dashed vector is the reflected vector, which must be inverted to be consistent with the convention. The velocity V, in contrast, is a *true vector*

Therefore, for example, to measure the volume of a tank, we can take a bottle b as the standard quantity, and for comparison, we can estimate how many sample bottles (or fractions of bottles) are needed to fill the tank volume W. If n is a positive real number indicating this value, we will say that the tank measures $W = n\,b$ bottles. From this example, the need to ensure the stability of the sampling unit over time immediately emerges, since the use of different bottles would result in different n numbers for the same tank volume.

The choice of a bottle as a sampling unit, although appropriate in principle, leads to inconsistent measurements in individuals who choose different bottles, with the consequence that a large number of conversion factors have to be introduced in the relationships linking the various quantities. There is a need to refer to commonly agreed sampling units. This operation results in a coherent system of measurement, such as the International System of Units (SI, abbreviated from the French Système International (d'Unités)).

The International System of Units was codified in 1960 at the XI General Conference on Weights and Measures (XI GFCM), although the works of the conference date back to 1948. Another measuring system still in use, especially in Anglo-Saxon countries and in the US, is the British Imperial system.

In any system of units of measurement, there are *fundamental quantities* and *derived quantities*. Fundamental quantities are the basic quantities that, independent of each other, are in total necessary and sufficient to describe any other physical quantity; all the other quantities are named *derived quantities*, which can be derived from fundamental quantities by means of laws or definitions in elementary physics.

For example, the speed U of a particle is the ratio between a spatial quantity s (the particle displacement) and a temporal quantity t (the duration of the displacement):

$$U = \frac{s}{t}. \tag{1.1}$$

If we choose the *metre* (symbol is m) as the fundamental quantity for space and the *second* (symbol is s) for the time t, it follows that the velocity is a derived quantity; see Eq. (1.1) and its unit of measure in the SI is the *metre per second*, symbolically m/s or m s^{-1}, or $\frac{m}{s}$.

In this regard, we observe that it is not strictly necessary to define a minimal system in which there is no unit redundancy. For example, one can define a system in which the unit of length is the metre, the unit of time is the second, and the unit of speed is the node (1 knot = 1 nautical mile per hour, ≈ 0.5 m s^{-1}). Of course, other fundamental quantities can be chosen, such as the second s for t and U for U; in that case, the distance is a quantity derived through the relationship (1.1) and is measured in U s or with a derived unit of measure with its own name and with dimension $U\,T$. This assumption is contrary to the spirit of SI but can be effective in obtaining functional links between quantities, which are very useful in physical applications.

The previous observation requires the identification of the minimum number of fundamental quantities, a value that depends on the nature of the problem at hand. For example, in fluid mechanics, temperature never appears explicitly, so three fundamental quantities are needed to write the dimensions, which in the SI are *mass*, *length* and *time*.

Two systems of units of measurement that adopt the same set of fundamental quantities belong to the same *class* of systems of units of measurement: the CGS (acronym for centimetre, gram, second, from the units of measurement of the fundamental quantities) and the MKS system (metre, kilogram, second) belong to the same class $M\,L\,T$ (mass, length, time), as they share the choice of such fundamental quantities.

As already mentioned, the number of fundamental quantities does not necessarily coincide with the minimum number required. Thus, temperature is a fundamental quantity, but it can also be expressed as a function of velocity and mass because, by definition, temperature is a measure of the average kinetic energy of atoms or molecules and can be described as a function of a force and a length (there are numerous controversial arguments about this choice, which should be corroborated by physical insights into the process). However, it is more practical to consider the temperature as fundamental, also because it is so close to everyday experience that it would not be suitable for definition in terms of combinations of other quantities.

In addition to systems with a finite number of fundamental quantities, we are also free to choose *monodimensional* systems, with a single fundamental quantity, and *omnidimensional* systems, without derived quantities, in which all quantities are fundamental. The choice has several practical implications.

1.2.1 Monodimensional Systems

In a monodimensional system, once a single fundamental quantity has been established, all the other quantities are expressed through it and through constants. For example, if we wish to use a monodimensional system having the length as its sole fundamental quantity, naming it a *system of units of measure L*, time can be expressed as a function of the speed of propagation of light in vacuum c, which is assumed to be constant. Hence,

$$\begin{cases} [t] = \dfrac{1}{c} L \equiv L, \\ t = AL, \end{cases} \tag{1.2}$$

where A is a constant dimensionless coefficient in the unit system L; in other unit systems, A is a constant coefficient (by convention, square brackets indicate the dimension of the argument and are omitted if the argument is a fundamental quantity in the used system). A time of 1 m in the L monodimensional system is equivalent to approximately 3.3 billionths of seconds in the International System. The coefficient A becomes a conversion factor between systems of units of measurement of different classes, with a numerical value that depends on the units of measurements chosen in the two systems.

The velocity

$$[U] = \frac{L}{[t]} \tag{1.3}$$

is dimensionless in the monodimensional L-system, and the acceleration

$$[a] = \frac{[U]}{[t]} \equiv L^{-1} \tag{1.4}$$

has the dimension of the inverse of a length. The force can be expressed with respect to the gravitational attraction between two masses, with dimension equal to

$$[F] = \frac{[k][M]^2}{L^2}, \tag{1.5}$$

where k is the gravitational constant, that is, for Newton's second law

$$[F] = [M][a]. \tag{1.6}$$

Equating these yields

$$[M] = \frac{[a]L^2}{[k]} \equiv L, \tag{1.7}$$

Table 1.1 Dimensions of some quantities in a monodimensional system L

Quantity	Symbol	Dimension
Time	t	L
Velocity	U	L^0
Acceleration	a	L^{-1}
Mass	M	L
Force	F	L^0
Energy	E	L
Power	W	L^0
Temperature	Θ	L^0
Dynamic viscosity	μ	L^{-1}
Density	ρ	L^{-2}

since the gravitational constant is dimensionless in the L-system. Therefore, the force is dimensionless, and the conversion factor with the International System is

$$\frac{c^4}{k} \text{ N}. \tag{1.8}$$

A force of magnitude 1.0 in the monodimensional L-system is equivalent to a force of magnitude

$$1.0 \times \frac{c^4}{k} = 1.0 \times \frac{(299.792\ 458 \cdot 10^6)^4}{6.674\ 28 \cdot 10^{-11}} = 1.21 \cdot 10^{44} \text{ N} \qquad \therefore \tag{1.9}$$

in the International System.

In a similar manner, we can calculate the dimensions of all quantities in the L-system. Table 1.1 lists the dimensions of some of them. Generalising the procedure, it is possible to derive the dimensions of any quantity in any monodimensional system, regardless of the selected fundamental quantity.

However, a monodimensional measuring system is useless for the application of the criteria of dimensional analysis. For example, suppose we wish to investigate an elastic mass-spring system to calculate the period of oscillation in a three-dimensional system of the class $M\,L\,T$. Assuming that the period depends on the mass and on the stiffness of the spring, we can write

$$t = C_1\, m^\alpha\, k^\beta, \tag{1.10}$$

where t is the period, m is the mass, k is the elastic constant of the spring, and C_1 is a dimensionless coefficient. Applying the principle of dimensional homogeneity (see Sect. 1.3), we calculate $\alpha = 1/2$ and $\beta = -1/2$, yielding

$$t = C_1 \sqrt{\frac{m}{k}}. \tag{1.11}$$

If we wish to apply the same analysis in a monodimensional L-system, we obtain

$$L = L^\alpha L^{-\beta}, \tag{1.12}$$

leading to the equation $(\alpha - \beta) = 1$. This equation admits infinite solutions and is of no help in solving the problem.

1.2.2 Omnidimensional Systems

In contrast to monodimensional systems, we consider omnidimensional systems, in which all quantities are fundamental. In this case, according to the principle of dimensional homogeneity, one or more dimensional constants must appear in all physical equations. For example, for the simple physical law

$$F = m\,a, \tag{1.13}$$

we should write

$$F = [C_1]\,m\,a, \tag{1.14}$$

which indicates that C_1 is a coefficient with the following dimensions:

$$[C_1] = \frac{F}{m\,a}. \tag{1.15}$$

Omnidimensional systems, as with monodimensional systems, have no practical application. Reconsidering the example of the mass-spring system reported in Sect. 1.2.1, in an omnidimensional system in which t, m and k (all the quantities that appear in the physical process) are fundamental quantities, the constant has the expression

$$[C_1] = \frac{t}{m^\alpha k^\beta}, \tag{1.16}$$

and can always accommodate any value of α and β in order to satisfy the principle of dimensional homogeneity. Unfortunately, this procedure does not provide any indication of the value of the two exponents α and β.

1.2.3 Multidimensional Systems

The category of multidimensional systems includes universally adopted systems, including SI. In SI, there are 7 fundamental quantities, all listed in Table 1.2.

The choice of a set of fundamental quantities is dictated by many factors, not least the ease of reproduction of unit samples, and is largely arbitrary. In many ways, a

Table 1.2 Fundamental quantities and units of measurements in the International System (SI)

Fundamental quantities	Symbol of the dimension	Denomination of the unit	Symbol of the unit
Length	L	metre	m
Mass	M	kilogram	kg
Time	T	second	s
Thermodynamic temperature	Θ	kelvin	K
Electric current	I	ampere	A
Luminous intensity	C	candela	cd
Amount of substance	N	mole	mol

set of fundamental quantities is equivalent to the basis of a vector space. In fact, in vector algebra, a basis is defined as a set of unit vectors that are linearly independent and for which a linear combination allows any vector in the space to be represented.

The basic units of the fundamental quantities are fixed conventionally but with reference elements, as far as possible, that are not linked to the time or place of measurement.

In 2019, the fundamental units in SI were redefined in terms of seven dimensional constants:

- the caesium hyperfine frequency $\Delta\nu_{Cs}$;
- the speed of light in vacuum c;
- Planck constant h;
- elementary charge e;
- Boltzmann constant k;
- Avogadro constant N_A;
- the luminous efficacy of defined visible radiation K_{cd}.

The conversions have been introduced once the requested accuracy in measuring these constants has been achieved. Although the seven dimensional constants could be adopted as fundamental, the past seven fundamental units are still in use, but they are now defined in terms of the new constants. Figure 1.2 shows the logo used to disseminate the new standard by the *Bureau International des Poids et Mesures* (BIPM). The BIPM is an international organisation established by the Metre Convention, through which member states act together on matters related to measurement science and measurement standards.

Units derived by means of the basic units must have a numerical unit coefficient, and the multiples and submultiples of the units of measurement must be expressed as integer exponent powers of ten; see Table 1.3. There are also some derived units with their own names, and those of interest in fluid mechanics are listed in Table 1.4.

There are some derived units not defined in the SI that are permanently allowed; see Table 1.5.

Fig. 1.2 Logo of the SI
constants (internal ring), in
force since 20 May 2019 to
define the fundamental units
(external ring)

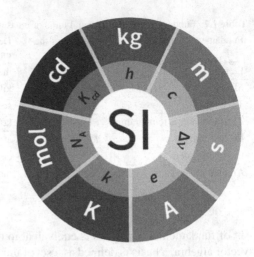

Table 1.3 Multiples and
submultiples in SI

Coefficient	Name	Symbol
10^{24}	yotta	Y
10^{21}	zetta	Z
10^{18}	exa	E
10^{15}	peta	P
10^{12}	tera	T
10^{9}	giga	G
10^{6}	mega	M
10^{3}	kilo	k
10^{2}	etto	h
10^{1}	deca	da
10^{-1}	deci	d
10^{-2}	centi	c
10^{-3}	milli	m
10^{-6}	micro	μ
10^{-9}	nano	n
10^{-12}	pico	p
10^{-15}	femto	f
10^{-18}	atto	a
10^{-21}	zepto	z
10^{-24}	yocto	y

Table 1.4 Some derived units with names

Quantity	Name	Symbol	Expression in derived SI units	Expression in fundamental SI units
Frequency	hertz	Hz		s^{-1}
Force	newton	N		$m\,kg\,s^{-2}$
Pressure	pascal	Pa	N/m^2	$m^{-1}\,kg\,s^{-2}$
Energy, work	joule	J	N m	$m^2\,kg\,s^{-2}$
Power, energy flux	watt	W	J/s	$m^2\,kg\,s^{-3}$

Table 1.5 Non-SI units accepted for use with SI

Quantity	Name	Symbol	Expression in SI units
Time	minute	min	60 s
	hour	h	3600 s
	day	d	86 400 s
	year	a	31 536 000 s
Angle	degree of arc	deg	$\pi/180$ rad
	minute of arc	′	$\pi/10\,800$ rad
	second of arc	″	$\pi/648\,000$ rad
Area	hectare	ha	10 000 m^2
	acre	ac	4046.872 61 m^2
Volume	litre	l, L	0.001 m^3
Mass	tonne (metric)	ton	1000 kg
Linear density	tex	tex	$10^{-6}\,kg\,m^{-1}$
Energy	electronvolt	eV	$1.602\,177\,33 \cdot 10^{-9}$ J
Atomic mass	unit of atomic mass	u	$1.660\,640\,2 \cdot 10^{-27}$ kg
Length	astronomic unit	au	$1.495\,979 \cdot 10^{11}$ m
	light-year	ly	$9.460\,528\,405 \cdot 10^{15}$ m
	parsec	pc	$3.085\,678\,186 \cdot 10^{16}$ m

If there is no limit to the value assumed by the exponents, then the number of derived units is infinite. If we assume that the absolute value of the exponent (integer) is p, including also the zero, there are $(2p+1)$ values (characterised by a positive and negative sign).

The number of derived quantities is equal to the number of permutations with repetition of the $(2p+1)$ possible exponents in the number of classes n represented by the number of fundamental quantities:

$$N_d = D'_{2p+1,\,n} \equiv (2p+1)^n. \tag{1.17}$$

Eliminating the cases of null exponents and of exponents all null except one, unitary (which would lead back to a fundamental quantity), we calculate a number of derived quantities equal to

$$N_d = (2p+1)^n - n - 1. \tag{1.18}$$

1.2.4 The Dimension of a Physical Quantity and the Transformation of the Units of Measurement

The *measurement* of an object, also called the *measuring procedure*, meaning a property, a characteristic of any material entity, involves the assignment of a range of values to that object. The measurement method is the set of theoretical and practical operations, expressed in general terms, that are used in the execution of a particular measurement. These operations, which are always conventional, must be clearly described so that the result of the measurement can be shared, reproduced and used.

Recalling some concepts expressed in the introduction of the systems of units of measurement, to measure an object, it is necessary (i) to identify its physical dimension, (ii) to choose a unit of measurement for that dimension; and (iii) to estimate the ratio between the object and the unit of measurement. All this entails the choice of a system of units since changing from one system of units to another even within the same class implies a change in the numerical value of the ratio between the object and the new unit of measurement; this guarantees the invariance of the intrinsic value of the object and, therefore, of the ratios between different objects of the same dimension. We observe that the *true value* of the measure of the object is inaccessible and that the measure of the object, expressed by assigning a range of values and not a single value, is the best estimate of the true value.

We consider, for example, the *average height of the European male population*, equal to 1.78 m in the International System and 5 ft 10 in in the British Imperial System (ft and in are the symbols of the units for foot and inch, respectively). We also consider the *average height of the European female population*, equal to 1.69 m in the International System and 5 ft 6 $\frac{1}{2}$ in in the British Imperial System. The ratio between the measurement of the two objects in the International System is given by

$$\frac{\text{average height of European females}}{\text{average height of European males}} = \frac{1.69}{1.78} = 0.95. \tag{1.19}$$

This ratio must also be equal to 0.95 in the British Imperial System and in any other system:

$$\frac{\text{average height of European females}}{\text{average height of European males}} = \frac{5 \text{ ft } 6\frac{1}{2} \text{ in}}{5 \text{ ft } 10 \text{ in}} = 0.95. \tag{1.20}$$

In general, the expression of a derived quantity as a function of the fundamental quantities must have a structure that ensures the *objective value of the ratios*.

We define the *dimension of a physical quantity* as that function that determines the measure of the quantity in different systems of units of measurement belonging to the same class.

For example, let us consider the physical quantity *speed*; its dimension in systems of the class $M\,L\,T$ is $L\,T^{-1}$, and the measure of the speed is expressed as follows:

$$U = \frac{\text{space}}{\text{time}} = \frac{\text{measure of space}}{\text{measure of time}}. \tag{1.21}$$

The space measurement is the ratio between the space travelled and the unit of space measurement in the chosen system, for example, the metre in SI; the time measurement is the ratio between the travel time and the unit of time measurement, the second in SI. For any other system $M\,L\,T$, for instance the British Imperial System, the measurement of space is the ratio of the space travelled to the unit of measurement of space in the British Imperial System, the yard; the measurement of time is the ratio of the travel time to the unit of measurement of time, again the second. Different numerical values of the measurement of the same speed in the two systems must correspond to an intrinsic value of speed that is calculated only on the basis of the dimension of speed.

As already mentioned, by convention, the dimension of a quantity G is indicated in square brackets, $[G]$; the dimension of the fundamental quantities is an exception and is indicated without square brackets. For example, the dynamic viscosity dimension μ in each system of the class $M\,L\,T$ is $[\mu] = M\,L^{-1}\,T^{-1}$. Other authors have used a different style to indicate the dimension, such as the symbol "Δ" by Gibbings (2011) $(F \Delta M\,L\,T^{-2})$ and "$\triangleq$" by Duncan (1953) $(F \triangleq M\,L\,T^{-2})$. If the class of the system of units changes, the form of the dimensional function changes.

Thus, while the intrinsic value of a physical quantity does not depend on the chosen system of units of measurement, the measure of the quantity is closely related to the unit of measurement of the system. What makes it possible to preserve the intrinsic value unchanged is the dimension of the quantity.

When the measure of a quantity is the same for all systems of the same class, the quantity is *dimensionless*. We bear in mind the particular case of the plane angle (see Sect. 1.1).

To identify the general structure of the dimension of a quantity, we suppose that G is a quantity function of mass, length and time, and we express its dimension as

$$[G] = f(M,\,L,\,T). \tag{1.22}$$

In this book, this equation is named the *typical equation*.

We indicate with g_1 the measure of G in *system 1; g_1* depends on the units of measure chosen for the fundamental quantities, defined as m_1, l_1 and t_1:

$$g_1 = f(m_1,\,l_1,\,t_1). \tag{1.23}$$

If we indicate with *system 2* a new system of units of measurement of the same class, varying only the units but not the fundamental quantities (which remain mass, length and time), the result is

$$g_2 = f(m_2, l_2, t_2). \tag{1.24}$$

The same class of systems of units of measurement of the two systems implies the functional identity of f. Consequently, we can write

$$\frac{g_2}{g_1} = \frac{f(m_2, l_2, t_2)}{f(m_1, l_1, t_1)}. \tag{1.25}$$

In the fundamental hypothesis that within the same class of systems of units, all systems are equivalent, that is, that there is no preferential system distinguishable from all the others, *system 2* can be obtained from *system 1*, multiplying the units of measure of *system 1* by the ratios m_2/m_1, l_2/l_1 and t_2/t_1, i.e.,

$$\frac{g_2}{g_1} = f\left(\frac{m_2}{m_1}, \frac{l_2}{l_1}, \frac{t_2}{t_1}\right), \tag{1.26}$$

or

$$r_G = f(r_M, r_L, r_T), \tag{1.27}$$

where $r_G = g_2/g_1$ is the ratio between the G measurement in the new system and in the old system and where r_M, r_L, r_T are the ratios between the units of measurement for the fundamental quantities in the two systems. We can assume that the transformation from *system 1* to *system 2* occurs via an intermediate system of the same class, which we refer to as *system 3*. The transition from *system 1* to *system 3* must satisfy

$$r'_G = f\left(r'_M, r'_L, r'_T\right), \tag{1.28}$$

where r'_G, r'_M, r'_L, r'_T are the ratios between the measure of G and the three fundamental quantities in *system 3* and in *system 1*. The transition from *system 3* to *system 2* requires that

$$r''_G = f\left(r''_M, r''_L, r''_T\right), \tag{1.29}$$

where r''_G, r''_M, r''_L, r''_T are the ratios of the measure of G and of the three fundamental quantities in *system 2* and in *system 3*. Consequently,

$$r_G = r'_G \, r''_G; \tag{1.30}$$

hence,

$$f(r_M,\ r_L,\ r_T) \equiv f\left(r_M' r_M'',\ r_L' r_L'',\ r_T' r_T''\right)$$

$$= f\left(r_M',\ r_L',\ r_T'\right) \cdot f\left(r_M'',\ r_L'',\ r_T''\right). \tag{1.31}$$

Applying the logarithm and defining the operator

$$\tilde{f}(\ln r_M,\ \ln r_L,\ \ln r_T) \equiv f(r_M,\ r_L,\ r_T), \tag{1.32}$$

yields

$$\ln \tilde{f}\left(\ln r_M' + \ln r_M'',\ \ln r_L' + \ln r_L'',\ \ln r_T' + \ln r_T''\right)$$
$$= \ln \tilde{f}\left(\ln r_M',\ \ln r_L',\ \ln r_T'\right) + \ln \tilde{f}\left(\ln r_M'',\ \ln r_L'',\ \ln r_T''\right). \tag{1.33}$$

This last relation requires that the operator $\ln \tilde{f}$ is linear; thus, the relation can be expressed as

$$\ln \tilde{f}(\ln r_M,\ \ln r_L,\ \ln r_T) \equiv \ln r_G = \alpha \ln r_M + \beta \ln r_L + \gamma \ln r_T. \tag{1.34}$$

Applying the properties of logarithm yields

$$r_G = r_M^\alpha r_L^\beta r_T^\gamma; \tag{1.35}$$

hence,

$$[G] = M^\alpha L^\beta T^\gamma. \tag{1.36}$$

In summary, the dimension of a quantity must be a power function. A quantity is dimensionless in an assigned system of units of measure if all the exponents of the power function are null.

Equation (1.35) allows us to quickly calculate the new measurement of the G quantity if a change in the units of measurements of mass, length and time occurs and r_G is a unit-conversion factor.

Example 1.1 Given an acceleration a with a measure equal to 350 ft min^{-2}, calculate its measure a' in a system in which the units are the inch and the second.

The two systems belong to the same class $L\,T$; the dimension of a is equal to

$$[a] = L\,T^{-2}. \tag{1.37}$$

Using Eq. (1.35) yields

$$r_a \equiv \frac{a'}{a} = r_L\, r_T^{-2}; \tag{1.38}$$

hence,

$$a' = a\, r_L\, r_T^{-2} = a\,\left(\frac{L'}{L}\right)\left(\frac{T'}{T}\right)^{-2} \rightarrow$$

$$a' = 350 \times \frac{12}{1} \times \left(\frac{60}{1}\right)^{-2} = 1.166 \text{ in s}^{-2}, \quad \therefore \quad (1.39)$$

where the ratios $12/1 = $ in/ft and $60/1 = $ s/min.

Example 1.2 Given the measure of a pressure p equal to 1 psi (pounds per square inch), calculate the measure p' of the same pressure in pascals.

Pressure has the dimensions of a force per unit surface area:

$$[p] \equiv [F]L^{-2}. \tag{1.40}$$

Using Eq. (1.35) yields

$$r_p = \frac{p'}{p} = r_F\, r_L^{-2}; \tag{1.41}$$

hence,

$$p' = p\, r_F\, r_L^{-2} = p\left(\frac{F'}{F}\right)\left(\frac{L'}{L}\right)^{-2} \rightarrow$$

$$p' = 1 \times \left(\frac{0.453 \times 9.806}{1}\right) \times \left(\frac{0.0254}{1}\right)^{-2} = 6885.3 \text{ Pa}, \quad \therefore \quad (1.42)$$

with the ratios $(0.453 \times 9.806)/1 = $ newton/pound and $0.0254/1 = $ metre/inch.

We observe that the measure of a quantity and the quantity are often indicated ambiguously by the same symbol.

1.2.5   Some Writing Rules

In technical writing, it is advisable to follow some basic rules in order to be clear and avoid errors.

Units of measurement expressed in symbolic form always begin with a lowercase letter, unless they are derived from the name of a person. For example: 1 s and not 1 S; 12 A (from the name of André-Marie Ampère) and not 12 a. In addition, a space between the number and the symbol (23 m and not 23m) is always required, and symbols should never be indicated in italic or bold: 1 s and not 1 *s* or 1 **s**.

If we need to write the unit of measurement in full in the text, we always use lowercase characters, even if the unit is derived from a person name: two amperes and not two Amperes, one newton and not one Newton.

A unit of measurement symbol consisting of the product of two or more units can be written either by interposing a point or by leaving a space: 131.8 N · m or 131.8 N m.

In the case of the quotient between two units of measurement, we can write, for example, 6.9 m/s, 6.9 m s^{-1}, 6.9 m · s^{-1}, or 6.9 $\frac{m}{s}$. The second form is the advisable one.

For multiple or submultiple prefixes, only those greater than 10^6 are shown with a capital letter; therefore: 1.5 MJ and not 1.5 mJ; 252 kg and not 252 Kg. We observe, in this regard, that the prefix "m" (milli-) indicates 10^{-3}, while the prefix "M" (mega-) indicates 10^6. Again, the multiple or submultiple symbol is placed next to the unit of measurement symbol, without a space: 16.2 mW and not 16.2 m W.

In the scientific notation, it is necessary that the units are the basic ones: we write, then, 3.2×10^5 m and not 3.2×10^2 km. In the notation with the prefixes, it is also advisable to choose the prefix so that the number is between 0.1 and 1000: 7.8 MJ and not 7800 kJ. Double prefixes are not allowed, so 1.2 μF (microfarad) is used and not 1.2 mmF (millimillifarad).

When writing numbers containing more than 4 digits in sequence, spacing is appropriate, grouping the digits in groups of 3 to the left and to the right of the decimal point: 12 130 and not 12130, and 33.224 38 and not 33.22438.

Finally, we observe that the International Organization for Standardization (ISO) suggests the comma as the decimal separator, while in English-speaking countries, the comma is the separator of the thousands and the point is the decimal separator. Therefore, to avoid confusion between the International System notation and the Anglo-Saxon notation, it is not advisable to use a point or comma to separate the thousands. Since 2003, the use of the decimal point has also been allowed in English texts.

As a last point: how many figures should be used in writing a measure of a variable?

The expression of measurements is a way to communicate the uncertainty of the value as well, and the number of significant figures in a measurement is related to the overall uncertainty of the entire procedure. More accuracy in measurements requires a greater number of significant figures. As a rule, zeros do not contribute to the number of significant figures unless they are between nonzero numbers or unless they are underlined. 0.000 123 and 452 and 121.0 have three significant figures, but 0.010 204 and 12010, and 303.12, and 862.0000 have five significant figures. The scientific notation helps avoid ambiguities: the number 121.0, with three significant figures, can be written as 1.21×10^2; 862.0000 can be written as 8.620×10^2; and 0.001 204 can be written as 1.204×10^{-3}. The number before the power of ten should be preferably between 1 and 10 and should contain all the significant figures, without the need to underline zeros that are significant.

In combining values of measurements or data with different uncertainties, follow these rules:

– For multiplication and division, the number of significant digits in the result can be no greater than the number of significant digits in the least precise measured value;

Table 1.6 Notations for significant figures

Number	Significant figures	
3.651	4	There are no zeros, and all numbers are significant
1010.56	6	The two zeros are significant here because they occur between other significant figures
0.2198	4	The first zero is only a placeholder for the decimal point and is not significant
0.000 044 2	3	The first five zeros are placeholders needed to report the data to the hundred-thousandths place
33.100	3	With no underlining or scientific notation, the last two zeros are placeholders and are not significant
11 891 000	7	The two underlined zeros are significant, while the last zero is not, as it is not underlined
5.457×10^{13}	4	In scientific notation, all numbers reported before the power of ten are significant
6.520×10^{-23}	4	In scientific notation, all numbers reported before the power of ten, including zeros, are significant
0.320×10^{-2}	3	In scientific notation, all numbers reported before the power of ten, including zeros (but not before decimal point), are significant

– For addition and subtractions, the result should have the same number places (tens place, ones place, tenths place, etc.) as the least-precise starting value: if you have 1.010 kg of salt (uncertainty to 1 thousandth of a kilogram) and then buy 5.4 kg of salt (uncertainty 0.1 kg), you have 1.010 + 5.4 = 6.4 kg of salt.

Table 1.6 lists a series of numbers with different numbers of figures and notations.

1.3   Principle of Dimensional Homogeneity

According to the principle of dimensional homogeneity,

All the terms of an equation representing a physical process must have the same dimension.

We consider the motion of an oscillating mass linked to a spring neglecting gravity and friction, which is described by the following equation:

$$\underbrace{m\,\frac{\mathrm{d}^2 x}{\mathrm{d}t^2}}_{I} + \underbrace{k(x - x_0)}_{II} = 0, \tag{1.43}$$

where m is the mass, x is the abscissa, x_0 is the abscissa of the equilibrium position, t is the time, and k is the stiffness of the spring.

The formalisation of this equation does not require the selection of a system of units of measurement, but only the use of the same units of measurement for all the quantities of the same nature that appear there (the lengths, the mass and the time); here, for example, x and x_0 have the dimensions of a length and shall both be expressed in metres, or both in inches. Based on the principle of dimensional homogeneity, the dimension of term I must coincide with the dimension of term II. In addition, regardless of what the two terms represent (inertia is the first, force exerted by the spring the second), on the basis of the equation that connects them, they have the same physical role. This applies to each pair of terms contained in an equation, with the logical consequence that in physical equations with more than two terms, all terms must have the same dimension.

If we add viscous damping and gravity, Eq. (1.43) becomes

$$m \underbrace{\frac{d^2x}{dt^2}}_{I} + \underbrace{k(x - x_0)}_{II} + \underbrace{\beta \frac{dx}{dt}}_{III} = \underbrace{- mg}_{IV}, \tag{1.44}$$

where β is the viscous damping coefficient and x is positive upward, where the terms III and VI have the same dimension of the other two terms, with β having the dimension of mass per unit time.

Thus, if a physical process is expressed by an equation such as:

$$A = B \tanh C + D e^F - \frac{G_1 + G_2}{H} + N, \tag{1.45}$$

the principle of dimensional homogeneity requires that C and F be dimensionless, that G_1 and G_2 have the same dimension as $A \cdot H$, and that B, D, N and A have the same dimensions.

The principle of dimensional homogeneity allows us to check the dimensional correctness of an equation and implies, as a consequence, that the argument of trigonometric or transcendent functions appearing in the physical equations must be necessarily dimensionless.

In fact, let us suppose that in an equation appears a term of the type $\cos x$; expanding as a Taylor series in the vicinity of the origin yields:

$$\cos x = 1 - \frac{x^2}{2!} + \frac{x^4}{4!} + \cdots . \tag{1.46}$$

Since all terms have to be dimensionally homogeneous, we conclude that x must be dimensionless. If we consider the logarithm, we recall that the logarithm is also defined as

$$\int_{x_0}^{x} \frac{dx'}{x'} = \ln x - \ln x_0 \equiv \ln \frac{x}{x_0}. \tag{1.47}$$

The principle of dimensional homogeneity ensures that the integral is dimensionless, being a sum of infinitesimal dimensionless terms (whatever dimension has x'); hence, $\ln x - \ln x_0$ must be dimensionless, and reporting only $\ln x$ is substantially wrong and misleading unless x is dimensionless.

The enunciation of the principle at the beginning of the paragraph actually refers to a particular case of a more general principle. We can formulate here a stricter and more extensive definition:

An equation is said to be dimensionally homogeneous if its form does not depend on the units of measure chosen.

Given the function

$$y = f(x_1, x_2, \ldots, x_n), \tag{1.48}$$

if, as a consequence of a change in the system of units, the dependent variable y and the independent variables $x_1, x_2, \ldots, x_n$ have the new values $y', x_1', x_2', \ldots, x_n'$, then we have

$$y' = f\left(x_1', x_2', \ldots, x_n'\right), \tag{1.49}$$

where f is coincident with the function in Eq. (1.48).

We have already shown that any quantity G must be expressed dimensionally as a power function of the fundamental quantities:

$$[G] = M^\alpha L^\beta T^\gamma, \tag{1.50}$$

and that the change of units of measurement, within the same class of systems, allows us to calculate the new measure of G as a function of the previous measure, according to the relation:

$$G' = G\, r_M^\alpha r_L^\beta r_T^\gamma. \tag{1.51}$$

That means that

$$y = M^\alpha L^\beta T^\gamma, \tag{1.52}$$

and if

$$[x_i] = M^{\alpha_i} L^{\beta_i} T^{\gamma_i}, \quad (i = 1, 2, \ldots, n), \tag{1.53}$$

then it also turns out that

$$\begin{cases} y' = y\, r_M^\alpha r_L^\beta r_T^\gamma \equiv y\,k, \\ x_i' = x_i\, r_M^{\alpha_i} r_L^{\beta_i} r_T^{\gamma_i} \equiv x_i\,k_i, \quad (i = 1, 2, \ldots, n), \end{cases} \tag{1.54}$$

where k and k_i are dimensionless coefficients. Equation (1.48) becomes

$$k\,y = k\,f(x_1, x_2, \ldots, x_n) = f(x_1 k_1, x_2 k_2, \ldots, x_n k_n), \tag{1.55}$$

and the condition of dimensional homogeneity requires that the identity (1.55), in the variables x_i, ($i = 1, 2, \ldots, n$) and (r_M, r_L, r_T) (defining k and k_i), is satisfied.

In particular, if the f function is a sum (or generally a linear combination of terms),

$$y = f(x_1, x_2, \ldots, x_n) \equiv x_1 + x_2 + \ldots + x_n, \tag{1.56}$$

the condition of dimensional homogeneity requires that

$$k \, y = k \, (x_1 + x_2 + \ldots + x_n) \equiv k_1 \, x_1 + k_2 \, x_2 + \ldots + k_n \, x_n; \tag{1.57}$$

hence,

$$k \equiv k_1 \equiv k_2 \equiv \cdots \equiv k_n. \tag{1.58}$$

These identities are satisfied iff all terms x_1, x_2, $\ldots$, x_n and y have the same dimension.

If f is a power function

$$y = f(x_1, x_2, \ldots, x_n) = x_1^{\delta_1} x_2^{\delta_2} \cdots x_n^{\delta_n}, \tag{1.59}$$

the condition of dimensional homogeneity yields the expression

$$k = k_1^{\delta_1} k_2^{\delta_2} \cdots k_n^{\delta_n}, \tag{1.60}$$

hence,

$$\begin{cases} \alpha_1 \, \delta_1 + \alpha_2 \, \delta_2 + \ldots + \alpha_n \, \delta_n = \alpha, \\ \beta_1 \, \delta_1 + \beta_2 \, \delta_2 + \ldots + \beta_n \, \delta_n = \beta, \\ \gamma_1 \, \delta_1 + \gamma_2 \, \delta_2 + \ldots + \gamma_n \, \delta_n = \gamma. \end{cases} \tag{1.61}$$

This is a linear system of equations in the unknown exponents δ_1, δ_2, $\ldots$, δ_n. Here, it is highlighted that dimensional variables must be defined as positive since power functions containing negative terms with fractional exponents would be imaginary.

1.3.1 The Arithmetic of Dimensional Calculus

In dimensional calculus, it is advantageous to use some rules with an elementary demonstration based on simple algebraic criteria and definitions that are briefly summarised here.

Product rule. The dimension of the product (or quotient) of the dimensions of two or more quantities is equal to the dimension of the product variable (or quotient). Having adopted square brackets to indicate the dimension of a quantity (except when

the quantity is fundamental in the selected system of units of measurement), this rule
has the following formalisation:

$$[A_1][A_2]\cdots[A_n] = [A_1 A_2 \cdots A_n]. \tag{1.62}$$

The associative rule. This is the associative property extended to dimensional calculus: if the product is grouped differently between pairs of terms, the product of the
dimension of more quantities does not change:

$$[A_1]\,([A_2][A_3]) = ([A_1][A_2])\,[A_3]. \tag{1.63}$$

Rule of exponents. The dimension of the power of a quantity is equal to the power
of the dimension of the quantity itself:

$$[A^n] = [A]^n. \tag{1.64}$$

Simple derivative rule. The dimension of the derivative of a quantity is equal to the
ratio between the dimension of the quantity and the dimension of the increment:

$$\left[\frac{dA}{dx}\right] = \frac{[A]}{[x]}. \tag{1.65}$$

The demonstration is immediate by using the product rule and recalling that the
derivative is the limit of an incremental ratio.

Rule of the nth-order derivative. The dimension of the derivative of order n of a
quantity is equal to the ratio between the dimension of the quantity and the nth power
of the dimension of the increment:

$$\left[\frac{d^n A}{dx^n}\right] = \frac{[A]}{[x]^n}. \tag{1.66}$$

For the demonstration, it is sufficient to calculate the derivative as a derivative $(n-1)$
times and repeatedly apply the simple derivative rule.

Rule of the simple integral. The dimension of the integral of a quantity is equal to the
product between the dimension of the quantity and the dimension of the increment:

$$\left[\int A\,dx\right] = [A][x]. \tag{1.67}$$

The demonstration is achieved by considering that the integral is defined as the limit
of a summation function of elementary areas, applying the rule of product dimension
and the principle of dimensional homogeneity.

Multiple integral rule. The dimension of the multiple integral of a quantity is equal to the product between the dimension of the quantity and the product of the dimensions of the increments:

$$\left[\int \int \cdots \int A \, dx_1 \, dx_2 \cdots dx_n \right] = [A][x_1][x_2] \cdots [x_n]. \qquad (1.68)$$

Again, the demonstration is obtained by starting from the definition of multiple integrals and repeatedly applying the rule of dimension of a simple integral.

The dimension of the Dirac function. The dimension of the Dirac function is the inverse of the dimension of its argument. The Dirac function must satisfy the constraint

$$\int_{-\infty}^{+\infty} \delta(x) \, dx = 1; \qquad (1.69)$$

hence, applying the rule of the simple integral results in $[\delta(x)] = 1/[x]$.

We recall here that it is permissible to perform numerical operations on quantities only when they belong to the same system of units and are expressed with consistent values of the base units or multiples or submultiples. Particular attention must be paid to preserving the physical meaning of the coefficients in the case of product and division operations. For example, kinetic energy can be considered the product of a mass by the square of a velocity, i.e., of two scalar coefficients, of which the second has no physical meaning. If, instead, we consider it as the product of the momentum and the velocity, two collinear vectors in the field of the dynamics of the particles, the kinetic energy preserves its vectorial connotation that is of great help in the interpretation, for instance, of the kinetic energy of a continuous body. Indeed, it is precisely when integration in space is required that the vector connotation of quantities that appear scalar is of greater help. See the discussion in Maxwell (1869).

1.4 The Structure of the Typical Equation on the Basis of Dimensional Analysis

Some useful information on the combination of variables involved in a physical process, when known, can be obtained by applying the criteria of dimensional analysis and following one of the two most known procedures, the *method of Rayleigh* and the *method of Buckingham*. Other procedures, such as linear proportionality, are just modifications of these basic procedures.

1.4.1 The Method of Rayleigh

Rayleigh's method is based on evidence that in an equality between two dimensionally homogeneous power functions, the exponent of each dimension of the power function to the left-hand side must be equal to the sum of the exponents of the corresponding dimension in the right-hand power function. An example clarifies the method.

We consider the resistance to motion F of a sphere of diameter D, which is moving with speed U in an incompressible fluid of density ρ and dynamic viscosity μ; see Fig. 1.3. The physical process is described by the following typical equation:

$$F = f(D,\ U,\ \rho,\ \mu). \tag{1.70}$$

A simple expression of the f function is as a power function; hence,

$$F = C_1\, D^a\, U^b\, \rho^c\, \mu^d, \tag{1.71}$$

where C_1 is a dimensionless coefficient.

Expressing all the variables in function of M, L and T, selected as fundamental quantities, results in

$$M\,L\,T^{-2} = L^a\, L^b\, T^{-b}\, M^c\, L^{-3c}\, M^d\, L^{-d}\, T^{-d}. \tag{1.72}$$

Imposing the equality among the exponents of an equal dimension, we obtain a system of 3 equations in the 4 unknowns a, b, c, d

$$\begin{cases} 1 = c + d, \\ 1 = a + b - 3c - d, \\ -2 = -b - d, \end{cases} \tag{1.73}$$

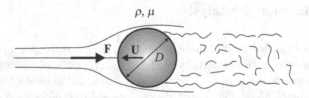

Fig. 1.3 Schematic for calculating the drag on a sphere moving in an incompressible fluid. The pattern of the streamlines refers to $\mathrm{Re} > 2 \cdot 10^5$, for instance, the Reynolds number of a grapefruit with $D = 8$ cm shot at 2.5 m s^{-1} in air

which admits ∞^1 solutions. The solution for a, b and c, as a function of d, is

$$\begin{cases} a = 2 - d, \\ b = 2 - d, \\ c = 1 - d; \end{cases} \tag{1.74}$$

hence,

$$F = C_1\, \rho\, D^2\, U^2 \left(\frac{\mu}{\rho\, U\, D} \right)^d. \tag{1.75}$$

In general, if the dependent variable depends on n variables and if k is the number of fundamental quantities, we can write k equations in the unknown exponents, admitting ∞^{n-k} solutions for arbitrary values of the $(n - k)$ exponents.

We observe that Eq. (1.75) has an overly constrained structure to represent the general phenomenon, as it forces a power function expression for F. To overcome that limit, the Rayleigh method suggests that, given the arbitrariness of the exponent d in the equation, we can express the dependent variable as a linear combination of power functions having the same structure, that is,

$$F = C_1\, \rho\, D^2\, U^2 \sum_{n=0}^{\infty} a_n \left(\frac{\mu}{\rho\, U\, D} \right)^n, \tag{1.76}$$

with a_n dimensionless coefficients. Since the summation can be considered a series expansion of the function f in an appropriate functional space, Eq. (1.76) can be re-written as:

$$F = C_1\, \rho\, D^2\, U^2\, f \left(\frac{\mu}{\rho\, U\, D} \right), \tag{1.77}$$

or

$$\underbrace{\frac{F}{\rho\, D^2\, U^2}}_{\text{Newton number}} = C_1\, f \left(\underbrace{\frac{\rho\, U\, D}{\mu}}_{\text{Reynolds number}} \right) \equiv C_1\, f(\mathrm{Re}). \tag{1.78}$$

The dimensionless group on the l.h.s. of Eq. (1.78) is named the *Newton number*, while the dimensionless group argument of the function, is named the *Reynolds number* (see Sect. 5.1).

Example 1.3 We consider the speed of propagation of surface gravity waves in a liquid. We wish to evaluate the structure of the typical equation.

We assume that the waves propagate over infinite depth, that the stabilising component is gravity and that, ultimately, the physical process depends on the wavelength l, the acceleration of gravity g and the density of the fluid ρ. Applying Rayleigh's method, we can write

$$c = C_1\, l^a\, g^b\, \rho^d, \tag{1.79}$$

where C_1 is a dimensionless coefficient and a, b and d are exponents. Expressing the variables as a function of the fundamental quantities M, L and T yields

$$LT^{-1} = L^a L^b T^{-2b} M^d L^{-3d}; \qquad (1.80)$$

hence,

$$\begin{cases} 1 = a + b - 3d, \\ -1 = -2b, \\ 0 = d, \end{cases} \longrightarrow \begin{cases} a = \dfrac{1}{2}, \\ b = \dfrac{1}{2}, \\ d = 0, \end{cases} \qquad (1.81)$$

and

$$c = C_1 \sqrt{g l}, \qquad (1.82)$$

equivalent to

$$\frac{c^2}{g l} = \text{const.} \qquad (1.83)$$

The value of the coefficient C_1 is obtained from the theoretical analysis of the physical process and is equal to $1/(2\pi)$.

If gravity waves propagate at a finite water depth equal to h, we can assume that this last variable also appears; hence,

$$c = C_1 l^a g^b \rho^d h^e, \qquad (1.84)$$

and

$$\begin{cases} 1 = a + b - 3d + e, \\ -1 = -2b, \\ 0 = d, \end{cases} \longrightarrow \begin{cases} a = \dfrac{1}{2} - e, \\ b = \dfrac{1}{2}, \\ d = 0. \end{cases} \qquad (1.85)$$

The solution is parametric in function of e. The wave speed can be expressed as

$$c = C_1 \sqrt{g l} \left(\frac{h}{l}\right)^e; \qquad (1.86)$$

hence,

$$\frac{c^2}{g l} = f\left(\frac{h}{l}\right). \qquad (1.87)$$

in general. The theoretical approach brings the following expression of the function f:

$$f\left(\frac{h}{l}\right) \equiv \frac{1}{2\pi} \tanh\left(\frac{2\pi h}{l}\right). \qquad \therefore \quad (1.88)$$

We observe that even though density is included in the list of variables, it does not appear in the final expression because it is dimensionally irrelevant (see Sect. 2.1).

1.4.2 The Method of Buckingham (Π-Theorem, Buckingham's Theorem)

Buckingham's method, which was first formulated by Vaschy (1892) and then taken up by Buckingham (1914), can be formally reduced to the following statement:

Assigning a physical process dependent on n quantities, we can always express it with a function of only (n–k) dimensionless groups, where k is the number of fundamental quantities.

Several authors have provided different demonstrations of this Theorem. Buckingham's demonstration (Buckingham 1914) is based on MacLaurin's series expansion of the functional relationship between the quantities involved in the process. Other demonstrations, such as in Bridgman (1922), require the differentiability of the functional relationship and are based on the conditions necessary for the resolution of a system of partial derivative linear equations of the first order. Duncan's demonstration (Duncan 1953) adopts Euler's theory of homogeneous functions and Euler's Theorem. Other demonstrations are algebraic in their nature or make use of Lie group theory (Bluman and Kumei 1989).

The simplest proof of Buckingham's Theorem is by *reductio ad absurdum*, is formalised in a more rigorous way in Sedov (1959) and reduced to the invariance of the function f by Birkhoff (1950). See also Sect. 1.4.4, where group theory is explained and invoked to demonstrate Buckingham's Theorem.

Let us assume that a physical process depends on n variables and is expressed by a function such as

$$f(a_1, a_2, \ldots, a_n) = 0, \qquad (1.89)$$

where $(a_1, a_2, \ldots, a_n)$ are the variables involved in the process.

If there exists a group $(a_1, a_2, \ldots, a_k)$ of k independent variables, it is possible to express the other variables $(a_{k+1}, a_{k+2}, \ldots, a_n)$ as a function of $(a_1, a_2, \ldots, a_k)$, and we define $(a_1, a_2, \ldots, a_k)$ as a *basis* (see Sect. 1.4.2.1).

For the generic variable that is not an element of the basis, we may write its dimension as

$$a_{k+i} = a_1^{\alpha_{k+i}} a_2^{\beta_{k+i}} \cdots a_k^{\delta_{k+i}}, \quad (i = 1, 2, \ldots, n-k), \qquad (1.90)$$

where $\alpha_{k+i}, \beta_{k+i}, \ldots, \delta_{k+i}$ are the dimensional exponents of a_{k+i} with respect to $a_1, a_2, \ldots, a_k$.

The process described by Eq. (1.89) can be expressed with a new different function $\tilde{f}$ of the new set of variables as follows:

$$\tilde{f}\left(a_1, a_2, \ldots, a_k, \frac{a_{k+1}}{a_1^{\alpha_{k+1}} a_2^{\beta_{k+1}} \cdots a_k^{\delta_{k+1}}}, \right.$$
$$\left. \frac{a_{k+2}}{a_1^{\alpha_{k+2}} a_2^{\beta_{k+2}} \cdots a_k^{\delta_{k+2}}}, \ldots, \frac{a_n}{a_1^{\alpha_n} a_2^{\beta_n} \cdots a_k^{\delta_n}}\right) = 0. \qquad (1.91)$$

The power function arguments of $\tilde{f}$ with the expression

$$\Pi_i = \frac{a_{k+i}}{a_1^{\alpha_{k+i}} a_2^{\beta_{k+i}} \cdots a_k^{\delta_{k+i}}}, \quad (i = 1, 2, \ldots, n-k), \qquad (1.92)$$

are dimensionless and are indicated by the symbol Π, which, in mathematics, indicates the multiple product operation.

We observe, in particular, that when swapping to a new system of units by changing only the units and keeping the basis unchanged (the new system, therefore, belongs to the same *class*), the dimensionless terms do not change their numerical value, unlike the dimensional terms $(a_1, a_2, \ldots, a_k)$. In this case, the new $\tilde{f}$ function must depend only on dimensionless terms, and Eq. (1.91) simplifies to:

$$\tilde{f}\left(\frac{a_{k+1}}{a_1^{\alpha_{k+1}} a_2^{\beta_{k+1}} \cdots a_k^{\delta_{k+1}}}, \frac{a_{k+2}}{a_1^{\alpha_{k+2}} a_2^{\beta_{k+2}} \cdots a_k^{\delta_{k+2}}}, \right.$$
$$\left. \ldots, \frac{a_n}{a_1^{\alpha_n} a_2^{\beta_n} \cdots a_k^{\delta_n}}\right) \equiv \tilde{f}(\Pi_i) = 0, \quad (i = 1, 2, \ldots, n-k). \qquad (1.93)$$

If this is not true, that is, if the function $\tilde{f}$ is still dependent on dimensional variables such as $(a_1, a_2, \ldots, a_k)$, changing the system of units changes the analytical dependence between the variables, which is obviously impossible, given that the physical process cannot depend on the chosen system of measurement. This concludes the proof.

A more rigorous formulation of the Theorem also requires verification of the independence and completeness of the set of dimensionless groups. According to Van Driest (1946) the statement of the Theorem should be amended as follows:

Given a physical process dependent on n variables, it is always possible to express it as a function of only (n−r) dimensionless groups, where the number r represents the maximum number of quantities that cannot form any dimensionless group and that are, therefore, really independent.

It can be shown that the r number represents the rank of the dimensional matrix of the quantities involved, with $r \leq k$ (the rank of a matrix is equal to the maximum order of the nonzero determinant minors extracted from the matrix itself).

The demonstration of this generalised Theorem of Buckingham (or Vaschy method) guarantees the completeness of the set of dimensionless groups that are obtained from the variables once the basis is chosen:

The set of dimensionless groups is complete if each dimensionless group is independent from the other and each other dimensionless group obtained from the variables involved is a power function of the dimensionless groups of the same set.

Ultimately, the notion of completeness, analysed in detail in Sect. 1.4.2.2, reflects the potential of the set of dimensionless groups to fully represent the space variables that appear in the physical process.

It can be demonstrated (Langhaar 1951) that the dimension of the complete set of dimensionless groups is equal to $(n - r)$, with r rank of the dimensional matrix of the involved quantities.

1.4.2.1 The Definition of a Dimensional Basis

To check if 3 quantities can be used as a dimensional basis, we express them in terms of definitely independent quantities, such as M, L and T:

$$\begin{cases} [a_1] = M^{\alpha_1} L^{\beta_1} T^{\gamma_1}, \\ [a_2] = M^{\alpha_2} L^{\beta_2} T^{\gamma_2}, \\ [a_3] = M^{\alpha_3} L^{\beta_3} T^{\gamma_3}. \end{cases} \tag{1.94}$$

Applying the logarithm yields

$$\begin{cases} \alpha_1 \ln M + \beta_1 \ln L + \gamma_1 \ln T = \ln a_1, \\ \alpha_2 \ln M + \beta_2 \ln L + \gamma_2 \ln T = \ln a_2, \\ \alpha_3 \ln M + \beta_3 \ln L + \gamma_3 \ln T = \ln a_3. \end{cases} \tag{1.95}$$

The system of equations in the three unknowns $\ln M$, $\ln L$, $\ln T$ admits a single solution if, and only if, it results in

$$\det \begin{bmatrix} \alpha_1 & \beta_1 & \gamma_1 \\ \alpha_2 & \beta_2 & \gamma_2 \\ \alpha_3 & \beta_3 & \gamma_3 \end{bmatrix} \neq 0. \tag{1.96}$$

This condition ensures dimensional independence between the variables of the triad. The second requirement is that the elements of the triad must be in sufficient number to describe the entire dimensional space: if temperature is also involved, in addition to mass, length and time, our triad is not a basis, and a further independent quantity is required. The extension to the case of k dimensions is immediate. Ultimately, it turns out that:

A set of quantities is a basis if their dimensional matrix, expressed in terms of another set of quantities that certainly represents a basis, has a nonzero determinant. In addition, the set of quantities should permit the expression of any quantity of the dimensional space.

The same result is achieved with a functional approach. If a system of definitely independent quantities is known, for example, M, L and T, any other quantity that pertains to the physical process is a function of M, L and T and can be expressed as

$$\begin{cases} a_1 = f_1(M, L, T), \\ a_2 = f_2(M, L, T), \\ a_3 = f_3(M, L, T), \end{cases} \tag{1.97}$$

a necessary and sufficient condition for the 3 new quantities to be independent is that the Jacobian has a nonzero determinant. That is,

$$\left| \frac{\partial(a_1, a_2, a_3)}{\partial(M, L, T)} \right| = \begin{vmatrix} \dfrac{\partial a_1}{\partial M} & \dfrac{\partial a_1}{\partial L} & \dfrac{\partial a_1}{\partial T} \\ \dfrac{\partial a_2}{\partial M} & \dfrac{\partial a_2}{\partial L} & \dfrac{\partial a_2}{\partial T} \\ \dfrac{\partial a_3}{\partial M} & \dfrac{\partial a_3}{\partial L} & \dfrac{\partial a_3}{\partial T} \end{vmatrix} \neq 0. \tag{1.98}$$

1.4.2.2 The Completeness of the Set of Dimensionless Groups

Let us assume that we have identified the following set of dimensionless groups:

$$\begin{cases} \Pi_1 = a_1^{\alpha_1} a_2^{\beta_1} \cdots a_k^{\delta_1}, \\ \Pi_2 = a_1^{\alpha_2} a_2^{\beta_2} \cdots a_k^{\delta_2}, \\ \cdots, \\ \Pi_{n-k} = a_1^{\alpha_{n-k}} a_2^{\beta_{n-k}} \cdots a_k^{\delta_{n-k}}, \end{cases} \tag{1.99}$$

with the dimensional matrix

	a_1	a_2	$\ldots$	a_k
Π_1	α_1	β_1	$\ldots$	δ_1
Π_2	α_2	β_2	$\ldots$	δ_2 .
$\ldots$	$\ldots$	$\ldots\ldots$	$\ldots$	$\ldots$
Π_{n-k}	α_{n-k}	β_{n-k}	$\ldots$	δ_{n-k}

(1.100)

The definition of completeness of the set of dimensionless groups requires the following product of powers:

$$\Pi_1^{h_1} \Pi_2^{h_2} \cdots \Pi_{n-k}^{h_{n-k}}, \tag{1.101}$$

assumes a unit value only if the exponents $(h_1, h_2, \ldots h_{n-k})$ are all null. In fact, if this were not the case, we could express a dimensionless group, for example, the first, as a function of the others as

$$\Pi_1 = \frac{1}{\Pi_2^{h_2/h_1} \cdots \Pi_{n-k}^{h_{n-k}/h_1}}, \tag{1.102}$$

and dimensionless groups would no longer be independent.

We prove the following:

A necessary and sufficient condition for a set of dimensionless groups to be independent is that the rows of their dimensional matrix (see (1.100)) are linearly independent.

The condition is necessary. In fact, let us assume that the groups are independent and that the rows are linearly dependent. From the definition of linear dependency, there is a set of exponents $(h_1, h_2, \ldots h_{n-k})$ that are not all null, such that

$$\begin{cases} h_1 \alpha_1 + h_2 \beta_1 + \ldots + h_{n-k} \delta_1 = 0, \\ h_1 \alpha_2 + h_2 \beta_2 + \ldots + h_{n-k} \delta_2 = 0, \\ \ldots, \\ h_1 \alpha_{n-k} + h_2 \beta_{n-k} + \ldots + h_{n-k} \delta_{n-k} = 0; \end{cases} \tag{1.103}$$

hence,

$$\Pi_1^{h_1} \Pi_2^{h_2} \cdots \Pi_{n-k}^{h_{n-k}} = $$
$$a_1^{(h_1 \alpha_1 + h_2 \beta_1 + \ldots + h_{n-k} \delta_1)} a_2^{(h_1 \alpha_2 + h_2 \beta_2 + \ldots + h_{n-k} \delta_2)} \cdots$$
$$a_k^{(h_1 \alpha_{n-k} + h_2 \beta_{n-k} + \ldots + h_{n-k} \delta_{n-k})} = a_1^0 a_2^0 \cdots a_k^0 = 1, \tag{1.104}$$

contrary to the hypothesis.

The condition is also sufficient. Let us assume that the rows of the dimensional matrix (1.100) are linearly independent. If there exists a set of exponents $(h_1, h_2, \ldots h_{n-k})$ not all null, such that

$$\Pi_1^{h_1} \Pi_2^{h_2} \cdots \Pi_{n-k}^{h_{n-k}} = 1, \tag{1.105}$$

then it must hold that

$$a_1^{(h_1\,\alpha_1+h_2\,\beta_1+\ldots+h_{n-k}\,\delta_1)}\ a_2^{(h_1\,\alpha_2+h_2\,\beta_2+\ldots+h_{n-k}\,\delta_2)}\cdots$$

$$a_k^{(h_1\,\alpha_{n-k}+h_2\,\beta_{n-k}+\ldots+h_{n-k}\,\delta_{n-k})} = a_1^0\,a_2^0\cdots a_k^0 = 1, \qquad (1.106)$$

which is satisfied by imposing that the system (1.103) in the unknowns $(h_1,\ h_2,\ \ldots\ h_{n-k})$ admits a solution. Such a homogeneous system admits a non-trivial solution only if the determinant of the matrix of coefficients is null, that is, if the rows are linearly dependent, against the hypothesis.

1.4.3   A Further Demonstration of Buckingham's Theorem

The proof by *reductio ad absurdum* of Buckingham's Theorem has been accompanied by several other demonstrations, such as the demonstration reported by Duncan (1953) which requires a preliminary reference to some notions on homogeneous functions, listed in Appendix A.

Bearing in mind the assumptions that lead to the formulation of the physical process with Eq. (1.89), we suppose that there are k fundamental quantities $(a_1,\ a_2,\ \ldots,\ a_k)$. Buckingham's Theorem states that the starting relationship between the n variables that define the process can be expressed as a function of $m = (n - k)$ new quantities:

$$f(\phi_1,\ \phi_2,\ \ldots,\ \phi_m) = 0 \qquad (1.107)$$

and that these quantities are dimensionless and in *minimum number* (the minimum number here has a mathematical meaning and does not contradict the fact that, experimentally, the physical process can be described using even fewer dimensionless groups than this minimum number).

These quantities are power functions of the fundamental quantities:

$$\phi_r = a_1^{\beta_{r1}}\,a_2^{\beta_{r2}}\cdots a_k^{\beta_{rk}}, \qquad (r = 1,\ 2,\ \ldots,\ m). \qquad (1.108)$$

We wish to show that the exponents are all null and that, therefore, the quantities are dimensionless.

We modify the units of measure of the k fundamental quantities so that the new units are equal to $(1/a_1)$ for the first, $(1/a_2)$ for the second, and $(1/a_k)$ for the k-th. For the first quantity,

$$\phi_1' = \left(\frac{1}{a_1}\right)^{\beta_{11}}\left(\frac{1}{a_2}\right)^{\beta_{12}}\cdots\left(\frac{1}{a_k}\right)^{\beta_{1k}}\ \phi_1 \to \phi_1 = \phi_1'\,a_1^{\beta_{11}}\,a_2^{\beta_{12}}\cdots a_k^{\beta_{1k}}. \qquad (1.109)$$

Similar results are obtained for the other $(m - 1)$ quantities involved. Equation (1.107) becomes

$$f\left(\phi_1' a_1^{\beta_{11}} \cdots a_k^{\beta_{1k}}, \; \phi_2' a_1^{\beta_{21}} \cdots a_k^{\beta_{2k}}, \; \ldots, \; \phi_m' a_1^{\beta_{m1}} \cdots a_k^{\beta_{mk}}\right) = 0, \qquad (1.110)$$

and must be valid for any numerical value assumed by $(a_1, a_2, \ldots, a_k)$.

Differentiating Eq. (1.110) with respect to a_1 yields

$$\beta_{11} \, \phi_1' \, a_1^{\beta_{11}-1} \, a_2^{\beta_{12}} \cdots a_k^{\beta_{1k}} \, \frac{\partial f}{\partial \phi_1} +$$

$$\beta_{21} \, \phi_2' \, a_1^{\beta_{21}-1} \, a_2^{\beta_{22}} \cdots a_k^{\beta_{2k}} \, \frac{\partial f}{\partial \phi_2} + \ldots$$

$$+ \beta_{m1} \, \phi_m' \, a_1^{\beta_{m1}-1} \, a_2^{\beta_{m2}} \cdots a_k^{\beta_{mk}} \, \frac{\partial f}{\partial \phi_m} = 0, \qquad (1.111)$$

which can be written as

$$\beta_{11} \frac{\phi_1}{a_1} \frac{\partial f}{\partial \phi_1} + \beta_{21} \frac{\phi_2}{a_1} \frac{\partial f}{\partial \phi_2} + \ldots + \beta_{m1} \frac{\phi_m}{a_1} \frac{\partial f}{\partial \phi_m} = 0. \qquad (1.112)$$

If we set $a_1 = 1$, Eq. (1.112) becomes

$$\beta_{11} \, \phi_1 \, \frac{\partial f}{\partial \phi_1} + \beta_{21} \, \phi_2 \, \frac{\partial f}{\partial \phi_2} + \ldots + \beta_{m1} \, \phi_m \, \frac{\partial f}{\partial \phi_m} = 0. \qquad (1.113)$$

By introducing the new variables

$$\psi_1 = \phi_1^{\frac{1}{\beta_{11}}}, \; \psi_2 = \phi_2^{\frac{1}{\beta_{21}}}, \; \ldots, \; \psi_m = \phi_m^{\frac{1}{\beta_{m1}}}, \qquad (1.114)$$

Equation (1.113) becomes

$$\psi_1 \frac{\partial f}{\partial \psi_1} + \psi_2 \frac{\partial f}{\partial \psi_2} + \ldots + \psi_m \frac{\partial f}{\partial \psi_m} = 0, \qquad (1.115)$$

which, as a consequence of Euler's Theorem (see Appendix A), has a homogeneous function of $f(\psi_1, \psi_2, \ldots, \psi_m)$ for the solution.

For one of the properties of homogeneous functions (see Appendix A), equation $f(\psi_1, \psi_2, \ldots, \psi_m) = 0$ is equivalent to a relationship between only $(m-1)$ variables, but this contradicts the hypothesis that m is the minimum number of variables. Therefore, all the exponents β_{ri}, $r = 1, 2, \ldots, m$, $i = 1, 2, \ldots, k$ must be null, and the variables $(\phi_1, \phi_2, \ldots, \phi_m)$ are dimensionless.

Example 1.4 We consider a physical process that can be expressed by the typical equation

$$f(U, \, l, \, F, \, \rho, \, \mu, \, g) = 0, \qquad (1.116)$$

where U is the speed, l is the length, F is the force, ρ is the density, μ is the dynamic viscosity, and g is the acceleration of gravity. The dimensional matrix with M, L and T fundamental quantities is

$$
\begin{array}{c|cccccc}
 & U & l & F & \rho & \mu & g \\
\hline
M & 0 & 0 & 1 & 1 & 1 & 0 \\
L & 1 & 1 & 1 & -3 & -1 & 1 \\
T & -1 & 0 & -2 & 0 & -1 & -2
\end{array}
\tag{1.117}
$$

and has rank 3. In fact, the extracted square matrix

$$
\begin{array}{c|ccc}
 & \rho & \mu & g \\
\hline
M & 1 & 1 & 0 \\
L & -3 & -1 & 1 \\
T & 0 & -1 & -2
\end{array}
\tag{1.118}
$$

has a nonzero determinant.

According to Buckingham's Theorem, it is possible to express the physical process as a function of $(6 - 3) = 3$ dimensionless groups.

Buckingham's Theorem does not give any indication of the form of the $\tilde{f}$ function or the structure of dimensionless groups, since the product of two or more dimensionless groups, or the power of one dimensionless group, is still dimensionless. The indications may come from the experimental investigation accompanied by careful processing of the results or from some symmetry properties of the process (see Chap. 3). However, it is advisable and appropriate that the experimental investigation be preceded by a theoretical analysis for the identification of dimensionless groups that, in light of targeted experiments, can assume the dignity of groups representative of the physical process.

To facilitate the choice of groups with a physical meaning, according to some authors (for example, Woisin 1992) it is necessary and appropriate to distinguish between *variables in input* or *governing*, which control a given physical process, and *variables of response* or *governed*, which represent the response of the physical process to the action of the governing variables; these are the input and output variables well known in the analysis of systems. In this sense, the initial functional relationship should be nonhomogeneous and should express the response variables as a function of the governing variables. Then, instead of writing

$$
f(a_1, \ a_2, \ \ldots, \ a_n) = 0,
\tag{1.119}
$$

it would be appropriate to write

$$
a_1 = \tilde{f}(a_2, \ \ldots, \ a_n),
\tag{1.120}
$$

assuming that a_1 is the governed variable.

These are well-founded and logical arguments that, however, are of limited help in advance, when the variables involved in a given physical process are being preliminarily identified, and become logical afterwards, possibly after completion of the necessary experiments.

Other authors suggest that governing variables should be selected from those whose value can be more easily modified in the laboratory (generally, they are variables representative of extensive quantities). In fact, although dimensional analysis does not require an experimental support activity, in practise, it often derives from this last force and applicative importance, such that an appropriate choice of the governing variables can result in a greater simplicity in the programming and in the execution of the experiments.

It is also experimentation that suggests the negligibility of some dimensionless groups or of some parameters considered relevant in the preliminary analysis, such as, for instance, the greater significance of power function combinations of some of the variables involved instead of the individual variables. Thus, in many physical processes, the dynamic viscosity μ and the density ρ appear with their ratio $\mu/\rho = \nu$, that is, the kinematic viscosity. This leads to a reduction in the number of variables and in the number of dimensionless groups. In this sense, Buckingham's Theorem indicates the maximum number of dimensionless groups and dimensionless parameters needed to describe a given process, but only some of these groups can be sufficient.

One of the great advantages of applying Buckingham's Theorem is in reducing the dimensions of the domain of the unknown function, with a consequent reduction in the number of experimental tests to be performed.

Example 1.5 We demonstrate the Pythagoras Theorem using the criteria of dimensional analysis (Barenblatt 2003).

In Fig. 1.4, the area A_c of the surface of the largest triangle depends on the length of the diagonal and the angle θ. The typical equation is

$$A_c = f(\theta, \, c). \tag{1.121}$$

Fig. 1.4 A proof of the Pythagoras Theorem using the criteria of dimensional analysis

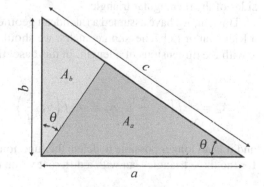

The problem is only geometric: the angle is already dimensionless, we select the length of the hypotenuse c as the fundamental quantity, and we can write the expression

$$\frac{A_c}{c^2} = \tilde{f}(\theta).$$ (1.122)

Similarly, for the two triangles contained in the larger rectangular triangle, we can write:

$$\frac{A_a}{a^2} = \tilde{f}(\theta),$$ (1.123)

$$\frac{A_b}{b^2} = \tilde{f}(\theta).$$ (1.124)

In these three equations, the function $\tilde{f}$ is the same since the triangles are similar. Therefore, we have

$$A_c = A_a + A_b \rightarrow c^2\,\tilde{f}(\theta) = a^2\,\tilde{f}(\theta) + b^2\,\tilde{f}(\theta),$$ (1.125)

and by eliminating the function $\tilde{f}$ (or assuming that it is unitary), the Pythagoras Theorem is obtained:

$$c^2 = a^2 + b^2.$$ (1.126)

The Pythagoras Theorem applies not only to squares but also to all similar polygons that can be constructed on three sides. For example, it is easy to show that the sum of the areas of the equilateral triangles built on the two catheti is equal to the area of the equilateral triangle built on the hypotenuse:

$$a^2\,\frac{\sqrt{3}}{4} + b^2\,\frac{\sqrt{3}}{4} = c^2\,\frac{\sqrt{3}}{4}.$$ (1.127)

This is tantamount to imposing that the function $\tilde{f}$ in Eq. (1.125) assumes a non-unitary value and depends on the geometry of similar polygons constructed on the sides of the rectangular triangle.

Thus far, we have assumed a Euclidean geometry. In a different geometry, such as a Riemann or Lobachevskii geometry, we should introduce an additional parameter λ with the dimensions of a length. In this case, the result is

$$A_c = c^2\,\tilde{f}\!\left(\theta,\ \frac{\lambda}{c}\right),\ A_a = a^2\,\tilde{f}\!\left(\theta,\ \frac{\lambda}{a}\right),\ A_b = b^2\,\tilde{f}\!\left(\theta,\ \frac{\lambda}{b}\right) \quad \therefore \ (1.128)$$

and it is no longer possible to delete the function $\tilde{f}$, which has an identical structure for the three triangles but with a different value of one of the two arguments.

Fig. 1.5 Conical fracture generated by a punch applying a load on a block of brittle material

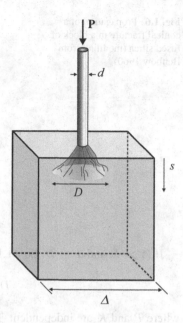

Example 1.6 Using the criteria of dimensional analysis, we wish to derive the relationship between the diameter of a conical fracture in a brittle material and the load (Barenblatt 2003).

The conical fracture, as shown in Fig. 1.5, is in a condition of variable equilibrium, in the sense that an increase in load results in an extension of the fracture zone.

From the theory of elasticity, it is known that the normal stress in an elastic medium beneath a point load decreases proportionally to the square root of the distance,

$$\sigma \propto \frac{N}{\sqrt{s}}, \tag{1.129}$$

where N is a constant and s is the distance.

In the hypothesis that the elastic field, under the same load conditions, is independent of the nature of the material, the factor of intensity N of the normal stress depends exclusively on the characteristics of the material:

$$N = \frac{K}{\pi}, \tag{1.130}$$

where K is called the *fracture resistance coefficient*.

In the general case, the diameter D of the base of the fracture cone depends on the load P, the fracture resistance coefficient K, Poisson's ratio v, the specimen size Δ and the punch diameter d:

$$D = f(P, \ K, \ v, \ \Delta, \ d). \tag{1.131}$$

Fig. 1.6 Propagation of a conical fracture in a block of fused silica (modified from Benbow 1960)

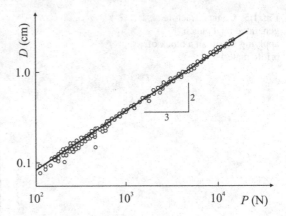

If $\Delta \gg D$ and $d \ll D$, Eq. (1.131) simplifies to

$$D = f(P, \; K, \; \nu), \qquad (1.132)$$

where P and K are independent. The dimensional matrix is

$$
\begin{array}{c|cccc}
 & D & P & K & \nu \\
\hline
F & 0 & 1 & 1 & 0 \\
L & 1 & 0 & -3/2 & 0
\end{array}
\qquad (1.133)
$$

and has rank 2. Applying Rayleigh's method, we can write:

$$\frac{D}{P^{2/3}\,K^{-2/3}} = \Phi(\nu), \qquad (1.134)$$

or

$$D = \left(\frac{P}{K}\right)^{2/3} \Phi(\nu). \qquad \therefore \;\; (1.135)$$

Figure 1.6 shows some experimental tests validating this result.

Example 1.7 Using the criteria of dimensional analysis, we wish to determine the evolution of the shock wave front generated by an explosion.

The rigorous analysis of the physical process involves the classic equations of mass conservation, momentum balance and energy balance. In addition, it is necessary to indicate the initial condition and the boundary conditions at the wave front. These boundary conditions, an instant after the explosion, are a source of the high analytical complexity of the phenomenon: speed, density and pressure are essentially unknown and cannot be assigned values (Fig. 1.7).

A brilliant approach to solving the problem was developed by Taylor (1950), who supposed that the explosion was comparable to an instantaneous release of energy

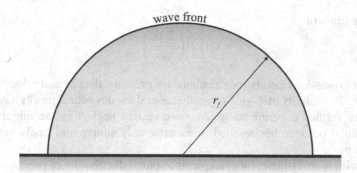

Fig. 1.7 Schematic for the analysis of the shock wave generated by an explosion in an ambient fluid at rest

from an ideal point source with a null radius r_0. This implies that the initial conditions for density, velocity and pressure in the $r < r_0$ domain become unessential. Taylor also narrowed the analysis to a stage where the pressure at the wave front is much greater than the ambient pressure p_0; this implies that the pressure p_0 disappears both from the initial conditions and from the boundary conditions.

The expansion process can be expressed with the typical equation

$$r_f = f(E, t, \rho_0, \gamma), \tag{1.136}$$

where r_f is the radius of the wave front, E is the initial energy of the explosion, t is time, ρ_0 is the initial density, and γ is the exponent of transformation, which is assumed to be adiabatic.

The dimensional matrix is

$$
\begin{array}{c|ccccc}
 & r_f & E & t & \rho_0 & \gamma \\
\hline
M & 0 & 1 & 0 & 1 & 0 \\
L & 1 & 2 & 0 & -3 & 0 \\
T & 0 & -2 & 1 & 0 & 0
\end{array}
\tag{1.137}
$$

and has rank 3. If we choose E, t and ρ_0 as fundamental quantities (they are independent), applying Buckingham's Theorem results in 2 dimensionless groups, one of which is the exponent of the adiabatic transformation. A second possible dimensionless group is

$$\frac{r_f \, \rho_0^{1/5}}{E^{1/5} \, t^{2/5}}, \tag{1.138}$$

and the typical equation becomes

$$\tilde{f}\left(\frac{r_f \, \rho_0^{1/5}}{E^{1/5} \, t^{2/5}}, \, \gamma\right) = 0. \tag{1.139}$$

Hence, we can write

$$r_f = \Phi(\gamma) \left(\frac{E t^2}{\rho_0} \right)^{1/5}. \qquad (1.140)$$

Similar expressions can also be calculated for pressure, density and velocity at the wave front. The validity of Taylor's hypotheses and the theoretical results have been extensively verified experimentally. We observe that neglecting the initial radius and the initial pressure has resulted in an extremely simple and easily verifiable relationship.

If we wish to investigate, for example, the spatial distribution of pressure within the wave front, we can hypothesise spherical symmetry, and we can introduce as a further variable the distance r from the centre of the explosion. In this case the following solution is obtained:

$$\rho = \rho_f \, f_r \left(\frac{r}{r_f}, \, \gamma \right). \qquad \therefore \quad (1.141)$$

This phenomenon enjoys the property of self-similarity, and the ratio of ρ/ρ_f does not depend on time but only on the dimensionless distance r/r_f and γ. This is a lucky case of self-similarity, not the general situation.

Example 1.8 We wish to study the normal oscillating modes of a star (Strutt 1915). A star can be represented as a fluid body that can oscillate, also assuming symmetrical shapes with respect to an axis. Some of the oscillating modes are visible in Fig. 1.8. Viscosity, unless it is very high, does not affect the oscillating modes and is neglected here. Let us assume that the density is constant and homogeneous and that the frequency of a natural mode of oscillation n depends only on the diameter D, the density ρ and the gravitational constant k:

$$n = f(D, \, \rho, \, k). \qquad (1.142)$$

Fig. 1.8 Three normal oscillating modes of a star

The dimensional matrix is

$$
\begin{array}{c|cccc}
 & n & D & \rho & k \\
\hline
M & 0 & 0 & 1 & -1 \\
L & 0 & 1 & -3 & 3 \\
T & -1 & 0 & 0 & -2
\end{array}
\qquad (1.143)
$$

and has rank 3. If we select as fundamental quantities D, ρ and k, they are independent, but D is dimensionally irrelevant (see Sect. 2.1) since its elimination reduces the rank of the dimensional matrix from 3 to 2. Hence, only n, ρ and k are involved, with a dimensional matrix of rank 2, and a single dimensionless group is obtained, namely,

$$
\tilde{f}\left(\frac{n^2}{k\rho}\right) = 0, \qquad (1.144)
$$

or

$$
n = C_1 \sqrt{k\rho}, \qquad \therefore \ (1.145)
$$

where C_1 is a dimensionless coefficient (see Sect. 2.4.1).

Example 1.9 We wish to study the volumetric flow rate Q of a centrifugal pump (see Fig. 1.9), assuming that it depends on the density of the fluid ρ, the rotational rate n and the impeller diameter D, the pressure p and the dynamic viscosity of the fluid μ:

$$
Q = f(\rho, \ n, \ D, \ p, \ \mu). \qquad (1.146)
$$

Three of the 6 variables are independent and can be selected as fundamental. In fact, the following dimensional matrix is expressed as a function of 3 independent quantities, such as mass, length and time:

$$
\begin{array}{c|cccccc}
 & Q & \rho & n & D & p & \mu \\
\hline
M & 0 & 1 & 0 & 0 & 1 & 1 \\
L & 3 & -3 & 0 & 1 & -1 & -1 \\
T & -1 & 0 & -1 & 0 & -2 & -1
\end{array}
\qquad (1.147)
$$

which has rank 3. For example, ρ, n and D are 3 independent quantities, and Eq. (1.146) can be re-written in terms of 3 dimensionless groups. A set of possible groups is obtained by solving the 3 dimensional equations:

$$
\begin{cases}
Q = \rho^{\alpha_Q} n^{\beta_Q} D^{\gamma_Q}, \\
p = \rho^{\alpha_p} n^{\beta_p} D^{\gamma_p}, \\
\mu = \rho^{\alpha_\mu} n^{\beta_\mu} D^{\gamma_\mu},
\end{cases}
\longrightarrow
\begin{cases}
L^3 T^{-1} = M^{\alpha_Q} L^{-3\alpha_Q} T^{-\beta_Q} L^{\gamma_Q}, \\
M L^{-1} T^{-2} = M^{\alpha_p} L^{-3\alpha_p} T^{-\beta_p} L^{\gamma_p}, \\
M L^{-1} T^{-1} = M^{\alpha_\mu} L^{-3\alpha_\mu} T^{-\beta_\mu} L^{\gamma_\mu}.
\end{cases}
\qquad (1.148)
$$

Fig. 1.9 Schematic and variables of interest for the analysis of the flow rate of a centrifugal pump

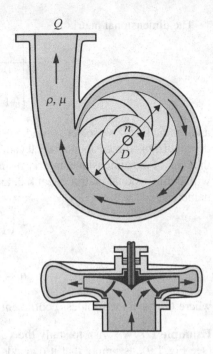

Equating exponents of the same quantity, for the first equation, we obtain the following system of linear equations:

$$\begin{cases} 0 = \alpha_Q, \\ -1 = -\beta_Q, \\ 3 = -3\alpha_Q + \gamma_Q, \end{cases} \longrightarrow \begin{cases} \alpha_Q = 0, \\ \beta_Q = 1, \\ \gamma_Q = 3, \end{cases} \tag{1.149}$$

and the first dimensionless group is

$$\frac{Q}{n\,D^3}. \tag{1.150}$$

Similarly, for the second and third equations,

$$\begin{cases} 1 = \alpha_p, \\ -2 = -\beta_p, \\ -1 = -3\alpha_p + \gamma_p, \end{cases} \longrightarrow \begin{cases} \alpha_p = 1, \\ \beta_p = 2, \\ \gamma_p = 2, \end{cases} \tag{1.151}$$

$$\begin{cases} 1 = \alpha_\mu, \\ -1 = -\beta_\mu, \\ -1 = -3\alpha_\mu + \gamma_\mu, \end{cases} \longrightarrow \begin{cases} \alpha_\mu = 1, \\ \beta_\mu = 1, \\ \gamma_\mu = 2, \end{cases} \tag{1.152}$$

and other 2 possible dimensionless groups are

$$\frac{p}{\rho\, n^2\, D^2}, \quad \frac{\mu}{\rho\, n\, D^2}. \tag{1.153}$$

The physical process can be described with the new function

$$\frac{Q}{n\, D^3} = \tilde{f}\left(\frac{p}{\rho\, n^2\, D^2}, \frac{\mu}{\rho\, n\, D^2}\right). \tag{1.154}$$

It is always advisable to select dimensionless groups with a physical meaning that is closely related to the specific nature of the problem. In the present case, the pressure can be conveniently expressed as $\rho\, g\, H$, where H is the energy per unit weight. The second dimensionless group in parentheses is the Reynolds number of the impeller. Equation (1.154) can be written as:

$$Q = n\, D^3 \tilde{f}_1\left(\frac{g\, H}{n^2\, D^2}, \text{Re}\right). \tag{1.155}$$

Experiments indicate that if the flow is turbulent, viscosity plays no role, and the Reynolds number can be dropped (this is named asymptotic independence of the turbulence and is typical of all turbulent flows, both internal and external). This justifies the assumption that, for $\text{Re} \rightarrow \infty$, Eq. (1.155) reduces to

$$Q = n\, D^3 \tilde{f}_2\left(\frac{g\, H}{n^2\, D^2}\right). \qquad \therefore \tag{1.156}$$

In some ways, Buckingham's method, compared to Rayleigh's method, leads to a more general formulation of the structure of physical equations. Buckingham's method, in fact, does not impose any constraint on the structure of the function, while Rayleigh's method leads at least to a homogeneous binomial structure of the function, although Rayleigh himself suggested that this limitation could be formally overcome.

1.4.4 Interpretation Through Group Theory

Dimensional analysis is framed within group theory, which can help to further extend some concepts that the principles of dimensional analysis do not clarify.

Given a set of transformations for the n variables with r parameters:

$$a_i' = g_i(a_1, a_2, \ldots, a_n; A_1, A_2, \ldots, A_r), \quad (i = 1, 2, \ldots, n), \tag{1.157}$$

where g_i are smooth functions, we define (1.157) as a r-parameter group of transformations if:

- it contains identity transformation (identity property);
- the inverse of each transformation belongs to (1.157) (invertibility property);
- the product of two transformations still belongs to (1.157); that is, the transformation of a transformation still belongs to (1.157) and can be obtained directly (associativity property).

A Lie group is a group and a differentiable manifold, a type of manifold that is locally similar to a linear space and allows one to do calculus (Warner 2013).

Let us consider the transformation from one system of units to another system of units belonging to the same class, where the fundamental variables (common to both systems) transform as

$$
\begin{cases}
a_1' = A_1\, a_1, \\
a_2' = A_2\, a_2, \\
\ldots, \\
a_k' = A_k\, a_k,
\end{cases}
\tag{1.158}
$$

where the dependent variables transform as

$$
\begin{cases}
a_{k+1}' = \left(A_1^{\alpha_{k+1}} A_2^{\beta_{k+1}} \cdots A_k^{\delta_{k+1}} \right) a_{k+1}, \\
a_{k+2}' = \left(A_1^{\alpha_{k+2}} A_2^{\beta_{k+2}} \cdots A_k^{\delta_{k+2}} \right) a_{k+2}, \\
\ldots, \\
a_n' = \left(A_1^{\alpha_n} A_2^{\beta_n} \cdots A_k^{\delta_n} \right) a_n,
\end{cases}
\tag{1.159}
$$

and where the k parameters $A_1, A_2, \ldots, A_k$ are the coefficients of the transformation; in the present case, the class of system is $a_1, a_2, \ldots, a_k$; hence, $A_1 = a_1'/a_1 \equiv r_{a_1}$, $A_2 = a_2'/a_2 \equiv r_{a_2}, \ldots, A_k = a_k'/a_k \equiv r_{a_k}$.

In compact form, the fundamental variables transform as

$$
a_i' = r_{a_i}\, a_i, \quad i = 1, 2, \ldots, k,
\tag{1.160}
$$

and the generic variable transforms as

$$
a_i' = \left(r_{a_1}^{\alpha_i} r_{a_2}^{\beta_i} \cdots r_{a_k}^{\delta_i} \right) a_i.
\tag{1.161}
$$

Identity, invertibility and associativity are satisfied, and (1.160) is a k-parameter group transformation.

Buckingham's Theorem is a consequence of the general covariance principle: if a physical process is invariant upon a transformation, it can be expressed as a function of variables that are invariant within the same transformation. The dimensionless groups $\Pi_1, \Pi_2, \ldots, \Pi_{n-k}$ in Eq. (1.91) are invariant with respect to the transformation

(1.160), whereas the elements of the basis $a_1, a_2, \ldots, a_k$ are not. Therefore, the functional relationship describing the process cannot depend on the variables of the basis but only on the $n - k$ dimensionless groups, as shown in Eq. (1.93).

We can check the invariance of the dimensionless groups considering, for example, the Reynolds number, $\mathrm{Re} = UD/v$, where U is a velocity scale, D is a length scale, and v is the kinematic viscosity. The transformation of velocity, length, kinematic viscosity between two systems of units belonging to the M, L, T class is

$$
\begin{cases}
U' = \dfrac{r_L}{r_T}U \rightarrow U' = r_U U, \\[2mm]
D' = r_L D \rightarrow D' = r_D D, \\[2mm]
v' = \dfrac{r_L^2}{r_T}v \rightarrow v' = r_v v;
\end{cases}
\tag{1.162}
$$

the transformation of the Reynolds number is

$$
\mathrm{Re}' = \frac{r_U\, r_D}{r_v}\mathrm{Re} \rightarrow \mathrm{Re}' = \frac{\dfrac{r_L}{r_T}\,r_L}{\dfrac{r_L^2}{r_T}}\mathrm{Re} \rightarrow \mathrm{Re}' = \mathrm{Re}.
\tag{1.163}
$$

This is equivalent to $r_{\mathrm{Re}} = 1$, and this result is valid for all dimensionless groups; for the dimensional variables: at least one of the conversion factors (or parameters of the group) of the fundamental variables $r_{a_i} \neq 1$, $i = 1, 2, \ldots, k$, hence results in $r_{a_j} \neq 1$, $j = k + 1, k + 2, \ldots, n$, and as a consequence dimensional variables are generally not invariant.

1.4.5   The Generalised Proof of the Π-Theorem: Ipsen's Procedure Leading to Gibbing's Demonstration

Ipsen (1960) introduced a procedure of sequential cancellation of the dimensions, obtaining a typical equation reduced to a minimum number of arguments, all dimensionless. Gibbings (2011) suggested that this was not simply a procedure, but was a demonstration of Buckingham's Theorem and extended it in a general form.

Following the methodology suggested by Ipsen, we consider the physical process of the energy dissipation of a fluid flowing in a circular-cross-section pipe, described by the following typical equation:

$$
J = f(g,\ V,\ D,\ \mu,\ \rho,\ \varepsilon),
\tag{1.164}
$$

where J is the energy dissipation per unit weight and per unit length, g is gravity acceleration, V is the cross-section-averaged velocity of the fluid, D is the diameter or the pipe, μ is viscosity, ρ is density and ε is a geometric scale for roughness. The dimensions of the variables in the M, L, T basis are

$$
\begin{array}{c|ccccccc}
 & J & g & V & D & \mu & \rho & \varepsilon \\
\hline
\Delta & 1 & LT^{-2} & LT^{-1} & L & ML^{-1}T^{-1} & ML^{-3} & L
\end{array},
\tag{1.165}
$$

where Δ stands for "is dimensionally equal".

We first wish to eliminate the equation of length, e.g., selecting D as the mandate variable. The typical equation can be written as

$$
J = f\left(\frac{g}{D}, \frac{V}{D}, D, \mu D, \rho D^3, \frac{\varepsilon}{D}\right),
\tag{1.166}
$$

and the dimensions of the new variables are

$$
\begin{array}{c|ccccccc}
 & J & \dfrac{g}{D} & \dfrac{V}{D} & D & \mu D & \rho D^3 & \dfrac{\varepsilon}{D} \\
\hline
\Delta & 1 & T^{-2} & T^{-1} & L & MT^{-1} & M & 1
\end{array}.
\tag{1.167}
$$

This is a general equation, but we observe that D cannot remain since there is no other variable containing a length that would balance it. Therefore, the new typical equation becomes

$$
J = f\left(\frac{g}{D}, \frac{V}{D}, \mu D, \rho D^3, \frac{\varepsilon}{D}\right).
\tag{1.168}
$$

Gibbings (2011) gives more details on the reason why D must be removed. We can say that this is the same reason why dimensional variables cannot remain as arguments in the presence of dimensionless groups. In a similar fashion, wishing to eliminate the equation of mass dimension, we recombine the variables as

$$
J = f\left(\frac{g}{D}, \frac{V}{D}, \frac{\mu}{\rho D^2}, \rho D^3, \frac{\varepsilon}{D}\right),
\tag{1.169}
$$

and the dimensions of the new variables are

$$
\begin{array}{c|cccccc}
 & J & \dfrac{g}{D} & \dfrac{V}{D} & \dfrac{\mu}{\rho D^2} & \rho D^3 & \dfrac{\varepsilon}{D} \\
\hline
\Delta & 1 & T^{-2} & T^{-1} & T^{-1} & M & 1
\end{array},
\tag{1.170}
$$

where ρD^3 must be removed since is the only variable containing the mass dimension. The new typical equation becomes

$$
J = f\left(\frac{g}{D}, \frac{V}{D}, \frac{\mu}{\rho D^2}, \frac{\varepsilon}{D}\right).
\tag{1.171}
$$

As a last step, the elimination of the time dimension yields

$$J = f\left(\frac{gD}{V^2}, \frac{V}{D}, \frac{\mu}{\rho D V}, \frac{\varepsilon}{D}\right), \tag{1.172}$$

and the dimensions of the new variables are

$$\begin{array}{c|ccccc} & J & \dfrac{gD}{V^2} & \dfrac{V}{D} & \dfrac{\mu}{\rho D V} & \dfrac{\varepsilon}{D} \\ \hline \Delta & 1 & 1 & T^{-1} & 1 & 1 \end{array}, \tag{1.173}$$

where the variable V/D must be removed, being the only one still containing the time dimension. The final expression of the typical equation is

$$J = f\left(\frac{gD}{V^2}, \frac{\mu}{\rho D V}, \frac{\varepsilon}{D}\right), \tag{1.174}$$

which is usually reformulated as

$$J = \frac{V^2}{2gD} f\left(\frac{\rho D V}{\mu}, \frac{\varepsilon}{D}\right). \tag{1.175}$$

The procedure requires a number of steps equal to the number of independent dimensions. As remarked by Gibbings, this should be considered the simplest proof of Buckingham's Theorem since no hypothesis is formulated about the differentiability or continuity of the function representing the typical equation.

1.4.6 The Criterion of Linear Proportionality

A useful indication of the selection of the most appropriate dimensionless groups can be obtained by applying a criterion proposed by Barr (1969). It is not a new method compared to those already described but rather a variant, in particular, of Buckingham's Theorem, in which instead of carrying out a transformation of the quantities to reorganise them into a smaller number of dimensionless groups, a transformation is performed to reorganise these quantities into a smaller number of power functions having the dimensions of a length (a generalisation of the procedure is described in Sect. 2.3.1).

From the reorganisation of the variables in terms of lengths, we obtain the definition of *linear* proportionality. The procedure can be retrieved within Ipsen's procedure as explained in Sect. 1.4.5.

The proportionality was initially developed by Barr in terms of speed, but subsequent applications revealed an excessive complication of the analysis; hence, length was preferentially selected. However, the choice of length as the dimension of homogenisation is in principle equivalent to the choice of any other dimension.

A term obtained by combining 2 or more quantities and having the dimension of a length is defined as a linear *proportionality*. For example, U^2/g is a linear proportionality. When the term contains only one quantity with a nonzero mass dimension, it is necessary to include the density in the proportionality. In fact, the presence of 2 or more quantities containing the mass allows them to be combined to obtain a power function in which the mass does not appear.

Since the initial functional relationship is transformed into homogeneous power functions having the dimension of a length, all the initial quantities, which already have the dimension of a length, are simply added to the formulation, as we do for the terms that are already dimensionless in the application of the Buckingham' Theorem.

For example, let us consider a fluid flowing with velocity U and assume that the typical equation of the process is

$$U = f(g, \, v, \, l), \tag{1.176}$$

where g is the acceleration of gravity, v is the kinematic viscosity and l is a geometric scale. A linear proportionality can be identified for each term (except l, already a length) by combining U with g, U with v and g with v:

$$\Phi \left(\underbrace{\frac{U^2}{g}, \, \frac{v}{U}, \, \frac{v^{2/3}}{g^{1/3}}}_{\substack{\text{combination of 2} \\ \text{of the 3 terms}}}, \, l \right) = 0. \tag{1.177}$$

The dimensional matrix is ranked 2 (the problem is purely kinematic), and by virtue of Buckingham's Theorem, we expect no more than 2 dimensionless groups, which are obtained by dividing the combination of power functions by the length l. Therefore, we have to choose only 2 out of the 3 linear proportionalities. There are 3 possible combinations, namely,

$$\Phi_1 \left(\frac{U^2}{g}, \, \frac{v}{U}, \, l \right) = 0,$$

$$\Phi_2 \left(\frac{U^2}{g}, \, \frac{v^{2/3}}{g^{1/3}}, \, l \right) = 0, \tag{1.178}$$

$$\Phi_3 \left(\frac{v}{U}, \, \frac{v^{2/3}}{g^{1/3}}, \, l \right) = 0,$$

providing 3 possible combinations of dimensionless groups:

$$\Phi_1'\left(\frac{U^2}{gl}, \frac{v}{Ul}\right) = 0,$$

$$\Phi_2'\left(\frac{U^2}{gl}, \frac{v^{2/3}}{g^{1/3}l}\right) = 0, \qquad\qquad (1.179)$$

$$\Phi_3'\left(\frac{v}{Ul}, \frac{v^{2/3}}{g^{1/3}l}\right) = 0.$$

The combinations of dimensionless groups increase by increasing the quantities involved in the physical process. In general, if m quantities are involved (excluding density and all characteristic lengths), it is possible to generate $(m-1) + (m-2) + \ldots + 1$ linear proportionalities, from which $(m-1)$ must be chosen for each combination. To this aim it is necessary:

- that each of the involved quantities (other than lengths) appears in at least one of the dimensionless groups;
- that the number of dimensionless terms is equal to the number of dimensional terms (lengths) minus one;
- that all dimensionless terms should be correlated through one or more dimensional terms (lengths).

Each of the possible dimensionless groups is the relationship between a geometric characteristic of the physical process and a length scale controlled by at least one of the involved quantities. Of course, we can use a criterion of *temporal* proportionality, choosing to homogenise the terms with respect to the time variable, or a criterion of proportionality with respect to any other quantity.

Despite the apparent complications, there are some advantages offered by such an approach.

The choice of a single fundamental quantity (for instance, the length) determines the number of dimensional groups having the dimension of a length equal to $(n - k + 1)$, greater than the number of dimensionless groups foreseen by Buckingham's Theorem. The degrees of freedom in excess allow us to select the linear terms to be involved in the functional relationship. This facilitates a solution in which the dependent variable appears in as few dimensionless groups as possible (ideally, it should appear in only one dimensionless group).

1.4.7 A Corollary of Buckingham's Theorem

Let us assume that in a given physical process, some variables assume a constant value. We consider whether this allows a reduction in the number of dimensionless groups.

It can be demonstrated that:

Given a functional relationship between n quantities, k of which are independent, if n_f quantities assume a constant value (by hypothesis or because they are constant *in the analysed data set), with k_f being the number of independent constant quantities (equal to the rank of their dimensional matrix), it is possible to express the process as a function of $(n - k) - (n_f - k_f)$ dimensionless groups (Sonin 2004).*

We consider a physical process expressed by the following typical equation:

$$f(a_1,, a_2, \ldots, a_n) = 0, \qquad (1.180)$$

and we assume $n_f < n$ are the variables, indicated with $(b_1, b_2, \ldots, b_{n_f})$, that have a constant value even in a single realisation of the process. To emphasise these variables, we re-write the typical equation as:

$$f(a_1,, a_2, \ldots, a_{n-n_f}, b_1, b_2, \ldots, b_{n_f}) = 0. \qquad (1.181)$$

If k_f of the n_f variables are independent, then we can express $(n_f - k_f)$ dimensional variables in dimensionless form:

$$f\left(a_1,, a_2, \ldots, a_{n-n_f}, \underbrace{b_1, b_2, \ldots, b_{k_f}}_{k_f}, \underbrace{\tilde{b}_{k_f+1}, \tilde{b}_{k_f+2}, \ldots, \tilde{b}_{n_f}}_{n_f-k_f} \right) = 0,$$
$$\qquad (1.182)$$

where the symbol $\sim$ indicates the dimensionless value. Since $(\tilde{b}_{k_f+1}, \tilde{b}_{k_f+2}, \ldots, \tilde{b}_{n_f})$ are dimensionless constants by hypothesis, the process can be re-written as a function of only $(n - n_f + k_f)$ variables:

$$f_1(a_1,, a_2, \ldots, a_{n-n_f}, b_1, b_2, \ldots, b_{k_f}) = 0. \qquad (1.183)$$

The rank of the dimensional matrix is still k, and by applying Buckingham's Theorem, it is possible to reduce the set of variables to $(n - k) - (n_f - k_f)$ dimensionless groups. This concludes the demonstration.

The minimum number of constant variables that can give an expectation of a reduction in the number of dimensionless groups is equal to 2, and it is improper to eliminate a dimensionless group only because it contains a constant quantity. However, the reduction of the number of dimensionless groups sometimes requires the reformulation of the groups, and this is an important limitation since the choice of dimensionless groups is made in a way that gives them a physical meaning, which could be lost in the new dimensionless groups.

The following examples help the comprehension of the Corollary of Buckingham's Theorem.

Example 1.10 We consider a cable of length l and diameter d dragged by a moving motorboat with speed U, as shown in Fig. 1.10. The drag force F depends on l, d and U, on the density of the water ρ and on the dynamic viscosity μ, and the typical equation of the process is

$$F = f(\rho, \mu, U, l, d). \qquad (1.184)$$

Fig. 1.10 Schematic for the analysis of the drag of a cable pulled in water by a motorboat

The dimensional matrix of the 6 variables

$$
\begin{array}{c|cccccc}
 & F & \rho & \mu & U & l & d \\
\hline
M & 1 & 1 & 1 & 0 & 0 & 0 \\
L & 1 & -3 & -1 & 1 & 1 & 1 \\
T & -2 & 0 & -1 & -1 & 0 & 0
\end{array}
\tag{1.185}
$$

has rank 3. Buckingham's Theorem guarantees that the physical process can be expressed as a function of $(6 - 3) = 3$ dimensionless groups, for example, the following groups:

$$
\frac{F}{\rho U^2 l^2} = \tilde{f}\left(\mathrm{Re},\ \frac{l}{d}\right).
\tag{1.186}
$$

Suppose that the density and the dynamic viscosity are constant and equal to the values for seawater. The 2 variables ρ and μ are linearly independent since their dimensional matrix

$$
\begin{array}{c|cc}
 & \rho & \mu \\
\hline
M & 1 & 1 \\
L & -3 & -1 \\
T & 0 & -1
\end{array}
\tag{1.187}
$$

has rank 2 and, therefore, $n_f = k_f = 2$. This means that the maximum number of dimensionless groups describing the physical process, equal to $(6 - 3) - (2 - 2) = 3$, does not change even if ρ and μ are constant.

In fact, it is not possible to write any dimensionless group that contains only ρ and μ and that assumes a constant value such that it can be excluded from the typical equation.

Example 1.11 We analyse the process of heat transfer from a sphere of radius R to an infinitely extended fluid medium at uniform pressure and temperature in a gravity field. We assume the following typical equation:

$$Q = f\left(R,\ \Delta\theta,\ g,\ \rho,\ \nu,\ c_p,\ \alpha,\ \beta\right), \qquad (1.188)$$

where Q is the heat flux, $\Delta\theta$ is the temperature difference between the sphere and the ambient fluid, g is the acceleration of gravity, ρ is the density of the fluid, ν is the kinematic viscosity of the fluid, c_p is the specific heat at constant pressure, α is the thermal diffusivity, and β is the coefficient of thermal expansion. The last 5 variables are fluid properties. Of the 9 variables, 4 are linearly independent, and the physical process can be described as a function of $(9-4) = 5$ dimensionless groups, for example:

$$\frac{Q}{\rho\, c_p\, \alpha\, \Delta\theta\, R} = \tilde{f}\left(\frac{\beta\, \Delta\theta\, g\, R^3}{\nu^2},\ \beta\, \Delta\theta,\ \frac{\nu}{\alpha},\ \frac{c_p\, \Delta\theta}{g\, R}\right). \qquad (1.189)$$

If we are interested in estimating the heat flux as a function of the sphere radius and the temperature difference, for the same fluid and for the same acceleration of gravity, $n_f = 6$ quantities have a constant value; these quantities are the acceleration of gravity and the 5 properties of the fluid. To evaluate the minimum number of dimensionless groups, we calculate k_f, the rank of the matrix of the n_f constant quantities:

	g	ρ	ν	c_p	α	β
M	0	1	0	0	0	0
L	1	-3	2	2	2	0
T	-2	0	-1	-2	-1	0
Θ	0	0	0	-1	0	-1

$$(1.190)$$

which is equal to 4. Therefore, we can describe the physical process as a function of $(9-4) - (6-4) = 3$ dimensionless groups.

In fact, if we choose the k_f quantities as fundamental, the physical process can be described by 5 dimensionless groups:

$$\frac{Q}{\rho\, c_p\, \alpha\, \Delta\theta\, R} = \tilde{f}\left(\frac{g\, R^3}{\nu^2},\ \beta\, \Delta\theta,\ \frac{\nu}{\alpha},\ \frac{c_p}{\beta\, (\nu\, g)^{2/3}}\right), \qquad (1.191)$$

and the last 2 groups are constant. These last 2 groups can be excluded from the typical equation, which is simplified as follows:

$$\frac{Q}{\rho\, c_p\, \alpha\, \Delta\theta\, R} = \tilde{f}\left(\frac{g\, R^3}{\nu^2},\ \beta\, \Delta\theta\right). \qquad \therefore \ (1.192)$$

It is always possible to compose $(n_f - k_f)$ dimensionless groups in which only those quantities that assume a constant value appear and that, as a consequence, can be eliminated from the typical equation.

Summarising Concepts

- The logical and introductory path to dimensional analysis requires the classification of physical quantities, defined by some authors as 'concepts' that do not require a measurement procedure. However, to be applicable, physical quantities must be accompanied by the definition of a 'system of units', which establishes the conventions necessary for the consistency of further operations on the variables and their measurement.
- Unit systems have many similarities with coordinate systems: they require a basis that must be complete and made up of mutually independent elements. The number of elements in the basis is largely conventional, and it is not convenient to use one-dimensional and omnidimensional systems: to progress, it is necessary to use multidimensional systems.
- The need to preserve the 'intrinsic' value of quantities leads to dimensional equations in power function form: every derived quantity must be expressed as a power function of the fundamental quantities; if not, the intrinsic value of the quantity would vary when passing from one system of units to another system of the same 'class' (which shares the fundamental quantities but not the units of the same). This is a form of the general covariance principle in physics.
- A fundamental element of the analysis is the principle of dimensional homogeneity, allowing immediate verification of the correctness of the equations describing a physical process, which has important applicative consequences.
- Each physical process can be symbolically expressed through a typical equation involving the relevant variables. Rayleigh's method, Buckingham's Theorem and corollaries allow us to reduce the number of variables involved for the benefit of theoretical analysis and experimental verification. At the base of these reductions is always the principle of covariance of physics, applied directly in the demonstration of Buckingham's Theorem through group theory.
- The variables in the typical equation can be reduced by combining them into a smaller number of both dimensional and dimensionless groups. Typically, the goal is to obtain dimensionless groups, which are in a minimum number sufficient to describe the space of the physical process variables.
- It is possible to know a priori the number of dimensionless groups, but the groups are not uniquely defined, and their identification is based on physical intuition and experiments. In lucky cases, the dimensionless groups are identified through calculus. Experiments may reveal that not all dimensionless groups are necessary.

References

Barenblatt, G. I. (2003). *Scaling*. Cambridge University Press.
Barr, D. I. H. (1969). Method of synthesis–basic procedures for the new approach to similitude. *Water Power, 21*(4), 148–153.
Benbow, J. J. (1960). Cone cracks in fused silica. *Proceedings of the Physical Society, 75*(5), 697.
Birkhoff, G. (1950). *Hydrodynamics: A Study in Logic, Fact and Similitude*. Dover Publications.
Bluman, G. W., & Kumei, S. (1989). *Symmetries and Differential Equations*. Springer.
Bridgman, P. W. (1922). *Dimensional Analysis*. Yale University Press.
Buckingham, E. (1914). On physically similar systems; illustrations of the use of dimensional equations. *Physical Review, 4*(4), 345–376.
Duncan, W. J. (1953). *Physical Similarity and Dimensional Analysis*. Edward Arnold & Co.
Fourier, J. B. (1822). *Theorie analytique de la chaleur, par* M. Fourier: Chez Firmin Didot, père et fils, reprinted by Cambridge University Press, 2009.
Gibbings, J. C. (2011). *Dimensional analysis*. Springer Science & Business Media.
Ipsen, D. C. (1960). *Units, Dimensions, and Dimensionless Numbers*. McGraw-Hill Book Co.
Langhaar, H. L. (1951). *Dimensional Analysis and Theory of Models*. Wiley.
Maxwell, C. J. (1869). Remarks on the mathematical classification of physical quantities. *Proceedings of the London Mathematical Society, 1*(1), 224–233.
Sedov, L. I. (1959). *Similarity and Dimensional Methods in Mechanics*. Academic Press.
Sonin, A. A. (2004). A generalization of the Π-theorem and dimensional analysis. *Proceedings of the National Academy of Sciences, 101*(23), 8525–8526.
Strutt, J. W. (Lord Rayleigh) (1915). The principle of similitude. *Nature, 95*, 66.
Taylor, G. I. (1950). The formation of a blast wave by a very intense explosion I. Theoretical discussion. *Proceedings of the Royal Society London A 201*(1065), 159–174.
Thomson, W. (Lord Kelvin) (1883). Electrical Units of Measurement. *Popular Lectures and Addresses 1*(73).
Van Driest, E. R. (1946). On dimensional analysis and the presentation of data in fluid-flow problems. *Journal of Applied Mechanics-Transactions of the ASME, 13*(1), A34–A40.
Vaschy, A. (1892). Sur les lois de similitude en physique. *Annales Télégraphiques, 19*, 25–28.
Warner, F. W. (2013). *Foundations of Differentiable Manifolds and Lie Groups* (Vol. 94). Springer Science & Business Media.
Woisin, G. (1992). On J. J. Sharp et al. *Application of Matrix Manipulation in Dimensional Analysis Involving Large Numbers of Variables, 5*(4), 333–348; *Marine Structures 5*(4), 349–356.

Chapter 2
Handling Dimensionless Groups in Dimensional Analysis

> *An experiment is a question which science poses to Nature and a measurement is the recording of Nature's answer.*
>
> *Max Planck*

Buckingham's Theorem indicates only the number of dimensionless groups; it is necessary to go into some concepts that allow a reduction of these groups to make the results useful. In the case of a large number of variables, it is necessary to adopt formal tools that allow these variables to be easily processed and then focus only on the interpretation of the data and not on the purely computational aspects.

Matrix methods are particularly suitable for quick and formally elegant calculations, especially in cases where the number of quantities involved is particularly high. Matrix notation may be advisable, for example, to reduce the possibility of errors and allow automatic implementation (Sharp et al. 1992). For this purpose, it is useful to go into more detail here, also in consideration of the fact that some commercial software offers toolboxes for dimensional analysis. However, we should bear in mind that the matrix methods applied to dimensional analysis are only calculation tools, which in no way increase our level of insight into the problem.

2.1 The Dimensional and Physical Relevance of Variables

The identification of the variables involved in a physical process is based on intuition, experience, and experimental results. Some of the selected variables can be *dimensionally* or *physically irrelevant*.

S. G. Longo, *Principles and Applications of Dimensional Analysis and Similarity*, Mathematical Engineering, https://doi.org/10.1007/978-3-030-79217-6_2

2.1.1   Dimensionally Irrelevant Variables

A variable is *dimensionally irrelevant* if, by virtue of its dimensions alone, it cannot be part of any relation between the other variables.

We demonstrate that:

A sufficient condition for a variable to be dimensionally irrelevant is that its dimension contains a fundamental quantity (i.e., one of the set of quantities chosen as fundamental) that is not contained in any other variable.

Let us assume that the physical process depends on 4 variables with 3 fundamental quantities d_1, d_2 and d_3. The typical equation is

$$f(a_1,\ a_2,\ a_3,\ a_4) = 0. \tag{2.1}$$

By virtue of Buckingham's Theorem, we can group all the variables in a single dimensionless group with the unknowns obtained by solving the following dimensional equation:

$$[\,a_1\,]^{\alpha_1}\,[\,a_2\,]^{\alpha_2}\,[\,a_3\,]^{\alpha_3}\,[\,a_4\,]^{\alpha_4} = d_1^0\,d_2^0\,d_3^0, \tag{2.2}$$

where the parentheses for the fundamental dimensions have been omitted, as by convention. If a_4 is the only variable in which the fundamental quantity d_3 appears, Eq. (2.2) becomes

$$[\,a_1\,]^{\alpha_1}\,[\,a_2\,]^{\alpha_2}\,[\,a_3\,]^{\alpha_3}\,\left(d_1^{\beta_1}\,d_2^{\beta_2}\,d_3^{\beta_3}\right)^{\alpha_4} = d_1^0\,d_2^0\,d_3^0, \tag{2.3}$$

which is satisfied if

$$\beta_3\,\alpha_4 = 0, \tag{2.4}$$

since a_1, a_2 and a_3 do not contain d_3. By hypothesis $\beta_3 \neq 0$, we must necessarily have $\alpha_4 = 0$, and the variable a_4 is irrelevant and can be deleted. As a result, the quantity d_3 is no longer required and can be eliminated, with a consequent reduction in the rank of the dimensional matrix.

Example 2.1 Let us consider a mass-spring system in the presence of gravity, assuming that the period of oscillation t is a function of the mass m, the stiffness of the spring k, and the acceleration of gravity g:

$$t = f(m,\ k,\ g). \tag{2.5}$$

The dimensional matrix

	t	m	k	g
M	0	1	1	0
L	0	0	0	1
T	1	0	-2	-2

$$\tag{2.6}$$

has rank 3. All the minors of rank 3 contain g, which is the only variable including the fundamental L dimension. We can neglect g and L since they do not appear in any of the other variables, and a single dimensionless group remains, for example,

$$\Pi_1 = t\sqrt{\frac{k}{m}}, \tag{2.7}$$

that must be constant:

$$t = C_1\sqrt{\frac{m}{k}}, \qquad\qquad \therefore \tag{2.8}$$

where C_1 is a dimensionless coefficient. Therefore, the gravity acceleration is dimensionally irrelevant. We observe that the elimination of gravity acceleration reduces the rank of the dimensional matrix, and the row corresponding to the L dimension can be eliminated.

The condition is also necessary, not only sufficient:

A variable is dimensionally irrelevant if its elimination results in a reduction in the rank of the dimensional matrix.

In fact, the elimination of a variable would bring us to $(n - 1 - k)$ dimensionless groups, but the reduction in the rank reduces k by one unit; therefore, the number of dimensionless groups is still equal to $(n - k)$, even without the deleted variable. Therefore, the deleted variable is irrelevant.

Once the dimensionally irrelevant variable has been identified, it is eliminated from the typical equation and from the dimensional matrix; the dimension that reduces the rank of the new dimensional matrix by a single unit is also eliminated.

If there are several dimensionless groups, a variable is dimensionally irrelevant if it does not appear in any of the selected groups.

It can be demonstrated that if a variable is dimensionally irrelevant, it remains irrelevant regardless of its expression. In fact, if it does not appear in any dimensionless group, it means that the other variables are sufficient to describe the problem and that the dimension of the irrelevant variable (and of any other variable that should be added) is not essential.

In brief, the dimensional irrelevance of a variable is a consequence of the dimensional sufficiency of the other variables to describe the phenomenon.

2.1.1.1 The Cascade Effect in Dimensionally Irrelevant Variables

It may happen that the procedure of eliminating a dimensionally irrelevant variable leads to a problem in which another variable becomes dimensionally irrelevant.

Let us assume that a physical process is described by 5 variables with 4 fundamental quantities, with the following dimensional matrix:

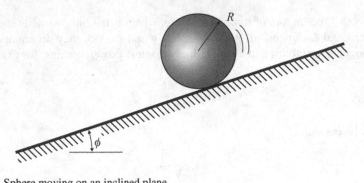

Fig. 2.1 Sphere moving on an inclined plane

$$
\begin{array}{c|ccccc}
 & a_1 & a_2 & a_3 & a_4 & a_5 \\
\hline
d_1 & 2 & 2 & 1 & 2 & 3 \\
d_2 & 0 & 1 & 1 & 1 & 0. \\
d_3 & 0 & 0 & -2 & 3 & 0 \\
d_4 & 0 & 0 & 1 & 0 & 0
\end{array}
\tag{2.9}
$$

The matrix has rank 4, but a_3 is dimensionally irrelevant since by deleting this variable, the rank is reduced to 3 and d_4 can be deleted. If we delete a_3 and d_4, the variable a_4 is also dimensionally irrelevant and can be deleted together with d_3. Finally, a_2 is also irrelevant, and the problem is reduced to a function of a_1 and a_5, with d_1 being the fundamental quantity:

$$
f(a_1, a_5) = 0 \rightarrow \Pi_1 = \frac{a_1}{a_5^{2/3}} = \text{const.}
\tag{2.10}
$$

Therefore, the dimensional irrelevance of a variable entails, in cascade, the dimensional irrelevance of other variables.

This situation can be advantageously used to reduce the complexity of the problem or to detect errors and omissions in the formulation. The following example may be illustrative.

Example 2.2 Let us consider a sphere that moves on an inclined plane; see Fig. 2.1 (from Szirtes 2007). If friction is less than a critical value, the sphere slides; otherwise, it rolls. We can assume that the critical friction angle μ_c depends on the sphere radius R, the plane inclination ϕ, the gravity acceleration g, and the density ρ:

$$
\mu_c = f(\phi, R, g, \rho).
\tag{2.11}
$$

The dimensional matrix is

$$
\begin{array}{c|ccccc}
 & \mu_c & \phi & R & g & \rho \\
\hline
M & 0 & 0 & 0 & 0 & 1 \\
L & 0 & 0 & 1 & 1 & -3 \\
T & 0 & 0 & 0 & -2 & 0
\end{array}
\tag{2.12}
$$

and it can be shown that, in cascade, g, ρ and R are dimensionally irrelevant. It follows that the physical process can be expressed as

$$\mu_c = \tilde{f}(\phi). \qquad \qquad \therefore \quad (2.13)$$

For a sphere, we find that $\mu_c = 7/2 \tan \phi$. We observe that excluding the dependence on the inclination angle of the plane ϕ, we reach an absurd conclusion, even though the formulation of the problem seems correct.

2.1.2 Physically Irrelevant Variables

The definition of physical irrelevance of a variable is less precise than the definition of dimensional irrelevance.

A governing variable is *physically irrelevant* if its influence on the governed variable is below a threshold. We remark on a *governing variable,* since it does not make sense to deal with the case of a physically irrelevant governed variable.

Such a definition shifts the problem to the choice of the threshold value. Common sense helpfully suggests, for example, that the thickness of paint on the outside of a pipeline where the fluid flows is physically irrelevant. Luckily, there are some criteria that do not require a threshold.

The identification of a physically irrelevant variable is based:

- on the existence of dimensional irrelevance;
- on heuristic reasoning;
- on experiments combined with data interpretation.

2.1.2.1 Physical Irrelevance as a Result of Dimensional Irrelevance

A sufficient condition for a variable to be physically irrelevant is that it is *dimensionally irrelevant*. In fact, if a variable is physically relevant, it must be in the dimensional matrix without the possibility of elimination, which is instead allowed for a dimensionally irrelevant variable.

If a variable is dimensionally relevant, its effect on the process may be so modest to justify its classification as a physically irrelevant variable.

Example 2.3 Let us consider the phenomenon of aquaplaning.

If the road surface is wet and the speed of the vehicle is below a critical speed, the tire features are efficient in removing water, and there is a sufficient grip (see Fig. 2.2). At critical speed, a film of water forms between the tire and the road surface, and the tire loses grip.

The variables of interest should include the weight of the vehicle supported by the tire and the footprint area of the tire. However, experimental evidence suggests that only the ratio between the two quantities, almost coincident with the inflation

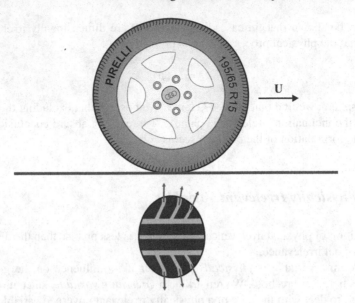

Fig. 2.2 The footprint of a tire on a wet surface

pressure of the tire, is relevant. This pressure also coincides with the hydrostatic pressure of the water film that grows under incipient aquaplaning conditions. We can assume that the critical speed U_c is a function of the tire pressure p, the water density ρ, and the gravity acceleration g:

$$U_c = f(p, \rho, g). \tag{2.14}$$

The dimensional matrix

$$
\begin{array}{c|cccc}
 & U_c & p & \rho & g \\
\hline
M & 0 & 1 & 1 & 0 \\
L & 1 & -1 & -3 & 1 \\
T & -1 & -2 & 0 & -2 \\
\end{array}
\tag{2.15}
$$

has rank 3, but if we eliminate g, the rank becomes 2; hence, g is dimensionally irrelevant, and the reduced dimensional matrix involves only U_c, p and ρ:

$$
\begin{array}{c|ccc}
 & U_c & p & \rho \\
\hline
M & 0 & 1 & 1 \\
L & 1 & -1 & -3 \\
T & -1 & -2 & 0 \\
\end{array}
\cdot
\tag{2.16}
$$

If we merge the variables p and ρ as p/ρ, M is not essential:

$$
\begin{array}{c|cc}
 & U_c & p/\rho \\
\hline
M & 0 & 0 \\
L & 1 & 2 \\
T & -1 & -2
\end{array}
\tag{2.17}
$$

With this last expression, the matrix has rank 1, and we can calculate a single dimensionless group, for instance

$$
\Pi_1 = U_c \sqrt{\frac{\rho}{p}},
\tag{2.18}
$$

which has a constant value. Thus,

$$
\Pi_1 = \text{const} \rightarrow U_c \propto \sqrt{\frac{p}{\rho}}.
\qquad \therefore \quad (2.19)
$$

This relationship shows that, in wet conditions, it is advisable to increase tire pressure.

For very high speeds, airplaning may also occur, a physical process that should be analysed by including the exponent of the polytropic transformation of air.

2.1.2.2 Physical Irrelevance Following Heuristic Reasoning

To demonstrate the physical irrelevance of a variable, sometimes it may be sufficient to perform thought experiments, often of remarkable elegance and effectiveness. In this regard, we report a famous reasoning by Galilei (1687) which is configured as a thought experiment able to demonstrate that the speed of a falling body does not depend on its mass if the drag force is negligible.

In this example, the demonstration tool is a *reductio ad absurdum*.

Example 2.4 Let us assume that two bodies of mass m and M, with $m < M$, fall with speed V_1 and V_2, respectively, and with $V_1 < V_2$; see Fig. 2.3. If we tether the two bodies with a wire, the m body mass would be accelerated by the pull of the wire and would increase the speed over its own V_1 speed, while the M body mass would be slowed down with a final speed below its own V_2 speed. The common speed V_3 of tethered bodies would then be between V_1 and V_2, i.e., $V_1 < V_3 < V_2$. However, if the speed is proportional to the mass, the body equivalent to the two tied bodies of mass $m + M$ (neglecting the mass of the rope) should fall with a speed V_4 even greater than V_2, that is, $V_4 > V_2 > V_3 > V_1$ and V_4 cannot be equal to V_3. The conflicting results can be reconciled only by assuming that the falling speed is independent of the body mass, i.e., $V_1 = V_2 \equiv V_3 \equiv V_4$.

As a consequence, the mass is physically irrelevant. (The concept of mass was still absent in Galilei's conception, and in the original manuscript, the mass was confused with the volume.)

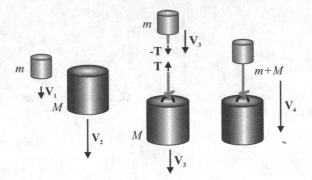

Fig. 2.3 Thought experiment by Galilei to prove that the falling speed of a body is not related to its mass

2.1.2.3 Physical Irrelevance Following Experiments Combined with Data Interpretation

Sometimes, neither of the two previous methods allows us to identify a physically irrelevant quantity; therefore, experimental tests are required.

We demonstrate that if in a relationship between dimensionless groups $\Pi_1 = f(\Pi_2)$, a variation in Π_2 leaves Π_1 unchanged, then all the quantities present in Π_2 and absent in Π_1 are physically irrelevant.

Let us suppose that the dimensionless groups are:

$$\Pi_1 = a_1^{\alpha_1} \, a_2^{\alpha_2} \, a_3^{\alpha_3}, \qquad \Pi_2 = a_2^{\beta_2} \, a_3^{\beta_3} \, a_4^{\beta_4}. \tag{2.20}$$

If $\Pi_1 = C_1 = $ const for any value assumed by Π_2, selecting a_2 and a_3 (common to the two dimensionless groups), we can write

$$a_1 = \left(\frac{C_1}{a_2^{\alpha_2} \, a_3^{\alpha_3}} \right)^{1/\alpha_1}. \tag{2.21}$$

Once a value of a_1 has been set, the values a_2 and a_3 cannot take on arbitrary values but must satisfy Eq. (2.21). In contrast, the variable a_4 can take any value to generate, with a_2 and a_3, the sequence of values of Π_2 without changing the value of Π_1. This coincides with the definition of physical irrelevance of a_4: a change in the numerical value of a_4 changes the numerical value of Π_2 but not the numerical value of $f(\Pi_2)$ and leaves the numerical value of Π_1 unchanged.

However, it does not hold that the variable a_1, present in Π_1 and not in Π_2, is physically irrelevant.

Fig. 2.4 Three-arm rotor
sprinkler

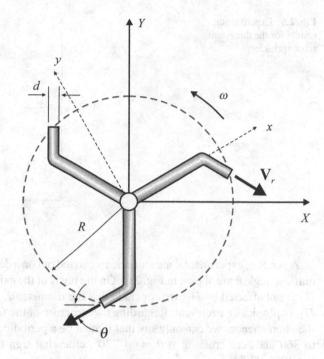

Example 2.5 Let us consider the spinning wheel sprinkler shown in Fig. 2.4. We
wish to calculate the rotational rate ω at full speed, depending on the relative velocity
of the water V_r, the diameter of the pipes d, the radius R, and the outflow angle θ,
by assuming the following typical equation:

$$\omega = f(d, \ V_r, \ R, \ \theta). \tag{2.22}$$

The dimensional matrix

$$
\begin{array}{c|ccccc}
 & \omega & d & V_r & R & \theta \\
\hline
L & 0 & 1 & 1 & 1 & 0 \\
T & -1 & 0 & -1 & 0 & 0
\end{array}
\tag{2.23}
$$

has rank 2. The outflow angle is dimensionless, and the typical equation can be
reduced to a function of 2 dimensionless groups and θ:

$$\Pi_1 = \tilde{f}(\Pi_2, \ \theta). \tag{2.24}$$

Two possible dimensionless groups are $\Pi_1 = \omega R/V_r$ and $\Pi_2 = d/R$. Assuming
a power function for Eq. (2.24), we can write

$$\Pi_1 = \Pi_2^{\alpha} \, \Phi(\theta) \rightarrow \omega = \frac{V_r}{R} \left(\frac{d}{R}\right)^{\alpha} \Phi(\theta). \tag{2.25}$$

Fig. 2.5 Experimental
results for the three-arm
rotor sprinkler

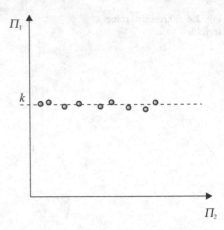

A series of experimental measurements carried out on a device for a given constant outflow angle θ are shown in Fig. 2.5. On the basis of the experimental evidence that Π_1 is not affected by Π_2, we conclude that the diameter d, present in Π_2 but not in Π_1, is physically irrelevant. Regarding the structure of the function $\Phi(\theta)$, analysing the phenomenon, we can only say that it must be a periodic function of period equal to 360°and zero crossing at $\theta = 90$–270°, changing sign from $\theta \in [-90°, 90°]$ to $\theta \in [90°, 270°]$.

The analytical result is obtained by applying the balance of the angular momentum, with

$$\omega = \frac{V_r \cos \theta}{R}, \qquad \qquad \therefore \ (2.26)$$

where the diameter d does not appear.

We observe that the invariance of a dimensionless group based on experimental data, as concluded by observing Fig. 2.5, is verified only for an assigned confidence level that quantifies in probabilistic terms the uncertainties in the measurements. This means that, in principle, more accurate measurements could reveal a weak but still present relevance of d.

2.2   Reducing the Number of Dimensionless Groups

The functional relations between more than 3 dimensionless groups are almost useless, since without indications deriving from principles of symmetry, of conservation, or of different nature (see Sect. 3), that permit the identification of the function (possibly combining certain groups), the number of experiments required to identify the structure of the function is excessive. For this reason, it may be convenient to reduce the number of groups by increasing whenever possible the number of fundamental quantities without changing the number of variables or by grouping the relevant variables involved.

To this end, we can:

- differentiate the fundamental dimension length, according to the direction, a process named *vectorisation*;
- attribute different meanings to the same fundamental quantity, according to its role in the variables, a process named *discrimination*;
- select the fundamental variables in a suitable manner, in order to merge some variables, a process named *rationalisation*.

In the following, there are some applications of the three processes, bearing in mind that the first two have been debated for a long time and are not unanimously accepted.

2.2.1   Vectorisation of Quantities

A criticism against classical dimensional analysis is its scalar nature that does not consider, for example, that velocity has three components and that some quantities involved in physical processes may be anisotropic, in fact or in a conceptual scheme of the physical process. Making dimensional analysis vectorial requires the introduction of fundamental quantities of lengths different in the three directions. Therefore, a system of the class $M\,L\,T$ in Cartesian coordinates becomes a system of the class $M\,L_x\,L_y\,L_z\,T$ with three distinct fundamental quantities for the length; in cylindrical coordinates, it becomes a system of the class $M\,L_r\,L_\theta\,L_z\,T$. The density, for example, in an orthogonal Cartesian coordinate system is $[\,\rho\,] = M\,L_x^{-1}\,L_y^{-1}\,L_z^{-1}$, and so on in other coordinate systems. For all other variables involving lengths only in one or two directions, the variables themselves become directional, and we need to pay attention to the selection of the direction of action. For example, the kinematic viscosity for a shear in the plane of normal x along z is $[\,v_z\,] = L_z^2\,T^{-1}$, while that for a shear in the same plane along y is $[\,v_y\,] = L_y^2\,T^{-1}$.

A similar approach is adopted in the study of differential problems of multiscale processes, with the introduction of different geometric scales in different directions. A classic example is the study of the boundary layer in fluid mechanics.

The identification of the correct dimension of variables, in the presence of an extended set of fundamental quantities (the extension can be for any quantity), requires a deeper knowledge of the phenomenon and is not error-free. Vectorial dimensional analysis can be an instrument of refinement of investigation, but there are criticisms on the attribution of a directionality to length (and to derived quantities containing a length); many researchers consider it an instrument that makes analysis more complex and artificial and that aims only to justify already known results (Massey 1971).

Fig. 2.6 The capillary rising in a tube

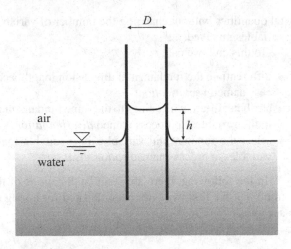

For example, let us consider the process of capillary rising of a liquid in a vertical tube (from Szirtes 2007), and apply classic dimensional analysis first. The schematic is shown in Fig. 2.6. The climbing height h can be expressed as

$$h = f(\rho, \; D, \; g, \; \sigma), \tag{2.27}$$

where ρ is the density of the liquid, D is the inner diameter of the vertical tube, g is the acceleration of gravity, and σ is the surface tension of the liquid with respect to the air (the real process also involves the air–glass and liquid–glass interface tensions if the capillary is glass). The dimensional matrix

$$
\begin{array}{c|ccccc}
 & h & \rho & D & g & \sigma \\
\hline
M & 0 & 1 & 0 & 0 & 1 \\
L & 1 & -3 & 1 & 1 & 0 \\
T & 0 & 0 & 0 & -2 & -2
\end{array}
\tag{2.28}
$$

has rank 3, and we can re-write the typical equation with only two dimensionless arguments; for example,

$$\Pi_1 = \frac{h}{D}, \quad \Pi_2 = \frac{\rho \, D^2 \, g}{\sigma}. \tag{2.29}$$

Equation (2.27) can be written as

$$\Pi_1 = \tilde{f}(\Pi_2), \tag{2.30}$$

or

$$h = D \, \tilde{f}\left(\frac{\rho \, D^2 \, g}{\sigma}\right). \tag{2.31}$$

We have no indication of the shape of the function, although on a logical basis, we can hypothesise that, for very large diameters and for $\sigma \to 0$, the capillary rise is negligible. Therefore, we could propose a power function structure of the function in the inverse of Π_2, namely,

$$\Pi_1 = \frac{const}{\Pi_2} \to h = C_1\, D\, \frac{\sigma}{\rho\, D^2\, g} \equiv C_1\, \frac{\sigma}{\rho\, D\, g}, \tag{2.32}$$

which indeed corresponds to the theoretical relation with a value of the dimensionless coefficient C_1 equal to $4\cos\psi$, where ψ is the contact angle between the liquid and the inner wall of the tube.

We now apply vector dimensional analysis in cylindrical coordinates, with the length component in the vertical direction indicated with L_z and in the radial direction indicated with L_r.

The linear dimensions of h, D and g are intuitive. The linear dimensions of ρ derive from the evidence that ρ coupled with gravity balances the surface tension force, with a rise of a volume of fluid having a scale equal to the product of the scale base L_r^2 and of the scale height L_z, hence $[\rho] = M\, L_r^{-2}\, L_z^{-1}$.

Finally, the surface tension exerts a force along the z-axis, with dimensions $[F_z] = M\, L_z\, T^{-2}$, acting on a circumference of geometric scale L_r; hence, $[\sigma] = [F_z]\, L_r^{-1} \equiv M\, L_r^{-1}\, L_z\, T^{-2}$.

The dimensional matrix becomes

$$
\begin{array}{c|ccccc}
 & h & \rho & D & g & \sigma \\
\hline
M & 0 & 1 & 0 & 0 & 1 \\
L_r & 0 & -2 & 1 & 0 & -1 \\
L_z & 1 & -1 & 0 & 1 & 1 \\
T & 0 & 0 & 0 & -2 & -2
\end{array}
\tag{2.33}
$$

and has rank 4. Applying Buckingham's Theorem, we can reduce the process to a function of a single dimensionless and constant group (see Sect. 2.4.1):

$$\Pi_1 = \frac{h\,\rho\,D\,g}{\sigma} \to \tilde{f}(\Pi_1) = 0 \to \Pi_1 = const. \tag{2.34}$$

This result coincides with the theoretical relationship. We need a limited number of hypotheses, and the result is more immediate than that of the classic dimensional analysis.

Example 2.6 We wish to calculate the rotation angle ϕ of a prismatic bar with circular cross-section and length l due to an applied torque M_t; see Fig. 2.7 (from Szirtes 2007). We can write the typical equation as

$$\phi = f\left(l,\ M_t,\ I_p,\ G\right), \tag{2.35}$$

Fig. 2.7 Prismatic bar with circular cross-section loaded by a torque

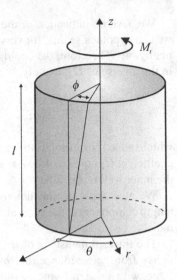

where I_p is the moment of inertia of the cross-section of the bar and G is the shear modulus. The dimensional matrix

$$
\begin{array}{c|ccccc}
 & \phi & l & M_t & I_p & G \\
\hline
M & 0 & 0 & 1 & 0 & 1 \\
L & 0 & 1 & 2 & 4 & -1 \\
T & 0 & 0 & -2 & 0 & -2
\end{array}
\tag{2.36}
$$

has rank 2, and applying Buckingham's Theorem, we can write the functional relationship with only $(5 - 2) = 3$ dimensionless groups as arguments, for example, the following groups:

$$
\Pi_1 = \phi, \quad \Pi_2 = \frac{l}{I_p^{1/4}}, \quad \Pi_3 = \frac{M_t}{G\, I_p^{3/4}}.
\tag{2.37}
$$

Hence, results

$$
\phi = \tilde{f}\left(\frac{l}{I_p^{1/4}}, \frac{M_t}{G\, I_p^{3/4}}\right).
\tag{2.38}
$$

This relationship is of limited practical application and would require a vast number of experiments to identify the structure of the function.

If we apply the criteria of the vector dimensional analysis, we must first specify to which axis to refer all the lengths that pertain to the variables, having the dimensions L_r, L_θ and L_z with obvious meaning of the subscripts. The angle of rotation ϕ can be expressed as the ratio between the tangential and radial lengths, $[\phi] = L_\theta L_r^{-1}$. The dimension of l is $[l] = L_z$. The dimension of the moment of inertia is calculated from the definition of I_p:

$$I_p = \int 2\pi r^3 \, dr \rightarrow [I_p] = L_r^4. \tag{2.39}$$

The dimension of the torque is that of a force applied in the tangential direction multiplied by the distance in the radial direction, $[M_t] = M \, L_\theta \, L_r \, T^{-2}$; the dimension of the shear modulus G is calculated on the basis of its definition:

$$G \equiv \frac{\text{shear stress}}{\text{shear strain}}. \tag{2.40}$$

The shear stress is the ratio of the force in the θ direction and the area of the surface that scales according to L_r^2:

$$[\tau] = \frac{M \, L_\theta \, T^{-2}}{L_r^2}. \tag{2.41}$$

The shear strain is on cylindrical surfaces coaxial to the bar and has dimensions

$$[\psi] = \frac{L_\theta}{L_z}. \tag{2.42}$$

Ultimately, it appears that

$$[G] = M \, L_r^{-2} \, L_z \, T^{-2}. \tag{2.43}$$

The dimensional matrix becomes

$$
\begin{array}{c|ccccc}
 & \phi & l & M_t & I_p & G \\
\hline
M & 0 & 0 & 1 & 0 & 1 \\
L_r & -1 & 0 & 1 & 4 & -2 \\
L_\theta & 1 & 0 & 1 & 0 & 0 \\
L_z & 0 & 1 & 0 & 0 & 1 \\
T & 0 & 0 & -2 & 0 & -2 \\
\end{array}
\tag{2.44}
$$

and has rank 4. We can identify a single dimensionless group; for example, the following:

$$\Pi_1 = \frac{\phi \, I_p \, G}{l \, M_t}; \tag{2.45}$$

hence,

$$\tilde{f}(\Pi_1) = 0 \rightarrow \Pi_1 = \text{const} \rightarrow \phi = C_1 \frac{l \, M_t}{I_p \, G}, \tag{2.46}$$

where C_1 is a dimensionless coefficient.

This result is much more immediate than that obtained by applying standard dimensional analysis. However, a modest variation in the selection of directional quantities leads to incorrect results, and there is fierce criticism against directionality; see, e.g., Massey (1978), Massey (1986), Gibbings (1981). For example, if we calculate the dimension of the moment of inertia as $[I_p] = L_r^3 L_\theta$, the dimensional matrix becomes

$$
\begin{array}{c|ccccc}
 & \phi & l & M_t & I_p & G \\
\hline
M & 0 & 0 & 1 & 0 & 1 \\
L_r & -1 & 0 & 1 & 3 & -2 \\
L_\theta & 1 & 0 & 1 & 1 & 0 \\
L_z & 0 & 1 & 0 & 0 & -1 \\
T & 0 & 0 & -2 & 0 & -2
\end{array}
\tag{2.47}
$$

which still has rank 4, but the variable ϕ appears to be dimensionally irrelevant (see Sect. 2.1) since its elimination results in a reduction in the rank to 3.

2.2.2 The Discrimination of Fundamental Quantities

The increase in the number of fundamental quantities can also be obtained by attributing a different physical meaning to the same quantity, i.e., by performing a *discrimination*. For example, the mass dimensionally involved in the dynamic viscosity or in the pressure can be considered an inertial mass (on the basis of the physical meaning of the viscosity and of the pressure), while the mass involved in the density can be considered a quantity of matter (again on the basis of the physical meaning of the density). In the presence of acceleration of gravity, it is also possible to identify a gravitational mass, which is relevant by virtue of the action of gravity.

Example 2.7 Let us consider the flow of a liquid in a cylindrical pipeline of circular cross-section. The mass flow rate Q_m can be expressed as

$$
Q_m = f(\Delta p / l, \ \rho, \ \mu, \ D),
\tag{2.48}
$$

where $\Delta p / l$ is the pressure variation per unit length, ρ is the density, μ is the dynamic viscosity, and D is the pipeline inner diameter. The dimensional matrix

$$
\begin{array}{c|ccccc}
 & Q_m & \Delta p/l & \rho & \mu & D \\
\hline
M & 1 & 1 & 1 & 1 & 0 \\
L & 0 & -2 & -3 & -1 & 1 \\
T & -1 & -2 & 0 & -1 & 0
\end{array}
\tag{2.49}
$$

has rank 3 and, selecting ρ, μ and D as the fundamental quantities, we have $(5 - 3) = 2$ dimensionless groups, for instance, the following:

$$
\Pi_1 = \frac{Q_m}{\mu D}, \quad \Pi_2 = \frac{\Delta p \, \rho \, D^3}{l \, \mu^2}.
\tag{2.50}
$$

Therefore, we can write

$$\Pi_1 = \tilde{f}(\Pi_2) \rightarrow Q_m = \mu D \tilde{f}\left(\frac{\Delta p\, \rho\, D^3}{l\, \mu^2}\right). \tag{2.51}$$

If we distinguish between inertial mass M_i and mass considered as the quantity of matter M_q, and if we assume that the former is involved in the variables $\Delta p/l$ and μ, the latter in the variables Q_m and ρ, the dimensional matrix becomes

$$
\begin{array}{c|ccccc}
 & Q_m & \Delta p/l & \rho & \mu & D \\
\hline
M_i & 0 & 1 & 0 & 1 & 0 \\
M_q & 1 & 0 & 1 & 0 & 0 \\
L & 0 & -2 & -3 & -1 & 1 \\
T & -1 & -2 & 0 & -1 & 0
\end{array} \tag{2.52}
$$

and has rank 4. A single dimensionless group, for example the following, is sufficient:

$$\Pi_1 = \frac{Q_m\, \mu\, l}{\Delta p\, \rho\, D^4}. \tag{2.53}$$

Hence,

$$\tilde{f}(\Pi_1) = 0 \rightarrow \Pi_1 = \text{const} \rightarrow Q_m = C_1 \frac{\Delta p\, \rho\, D^4}{\mu\, l}, \qquad \therefore \tag{2.54}$$

which represents the law of Poiseuille (valid in laminar motion), with a value of the coefficient C_1 equal to $\pi/128$.

2.2.3 The Process of Rationalisation with the Change in the Fundamental Quantities and the Grouping of the Variables

In other processes, a change in the fundamental quantities allows the grouping of some variables and a consequent reduction in their number.

Let us consider a sphere subject to gravity immersed in a fluid that has reached the terminal speed V. This speed can be expressed as

$$V = f(D,\ \rho,\ \rho_s,\ \mu,\ g), \tag{2.55}$$

where D is the diameter of the sphere, ρ is the density of the fluid, ρ_s is the density of the material of the sphere, μ is the dynamic viscosity and g is the acceleration of gravity. The dimensional matrix of the 6 variables is

$$
\begin{array}{c|cccccc}
 & V & D & \rho & \rho_s & \mu & g \\
\hline
M & 0 & 0 & 1 & 1 & 1 & 0 \\
L & 1 & 1 & -3 & -3 & -1 & 1 \\
T & -1 & 0 & 0 & 0 & -1 & -2
\end{array}
\tag{2.56}
$$

and has rank 3. The functional relationship can be expressed using only $(6 - 3) = 3$ dimensionless groups. Selecting D, ρ and μ as a triad of fundamental quantities, we calculate, for example, the groups:

$$
\Pi_1 = \frac{V D \rho}{\mu}, \quad \Pi_2 = \frac{\rho_s}{\rho}, \quad \Pi_3 = \frac{g D}{V^2}.
\tag{2.57}
$$

In general, we can write

$$
\Pi_1 = \tilde{f}(\Pi_2, \Pi_3),
\tag{2.58}
$$

although it is preferable to combine dimensionless groups such as

$$
\Pi_3' \equiv \frac{1}{\Pi_1 \Pi_3} = \tilde{f}_1(\Pi_1, \Pi_2) \to \frac{V \mu}{\rho g D^2} = \tilde{f}_1\left(\frac{V D \rho}{\mu}, \frac{\rho_s}{\rho}\right).
\tag{2.59}
$$

If we introduce the force F instead of mass, we can eliminate the acceleration of gravity by replacing the specific gravity γ with the density. The new dimensional matrix becomes

$$
\begin{array}{c|ccccc}
 & V & D & \gamma & \gamma_s & \mu \\
\hline
F & 0 & 0 & 1 & 1 & 1 \\
L & 1 & 1 & -3 & -3 & -2 \\
T & -1 & 0 & 0 & 0 & 1
\end{array}
\tag{2.60}
$$

and has rank 3. The functional relationship involves only $(5 - 3) = 2$ dimensionless groups and can be reduced to the equation

$$
V = \frac{D^2 \gamma}{\mu} f\left(\frac{\gamma}{\gamma_s}\right),
\tag{2.61}
$$

which is more compact than Eq. (2.59), with only two dimensionless groups.

The rationalisation method is free of critics, but its application is an exception.

2.3   Formalisation of Matrix Methods

Let us assume that the physical process of interest is expressed by the functional relationship between n variables:

$$
f(x_1, x_2, \ldots, x_n) = 0.
\tag{2.62}
$$

Defining $(y_1, y_2, \ldots, y_k)$ as the k fundamental quantities, we can build a matrix in which a_{ij} is the dimension of x_j, $(j = 1, 2, \ldots, n)$ relative to y_i, $(i = 1, 2, \ldots, k)$. For example, if the k selected fundamental quantities are M, L and T, we have:

$$\mathbf{M} = \begin{array}{c|ccccccc} & x_1 & x_2 & x_3 & x_4 & \ldots & x_n \\ \hline M & a_{11} & a_{12} & a_{13} & a_{14} & \ldots & a_{1n} \\ L & a_{21} & a_{22} & a_{23} & a_{24} & \ldots & a_{2n} \\ T & a_{31} & a_{32} & a_{33} & a_{34} & \ldots & a_{3n} \end{array}. \tag{2.63}$$

Let us assume that the rank of the matrix is equal to the number of fundamental quantities (3 in the present case). The dimensional matrix $\mathbf{M}$ can be split into two matrices $\mathbf{A}$ and $\mathbf{B}$, where $\mathbf{A}$ is a minor of order 3 and $\mathbf{B}$ is the remaining part:

$$\mathbf{A} = \begin{bmatrix} a_{1(n-2)} & a_{1(n-1)} & a_{1n} \\ a_{2(n-2)} & a_{2(n-1)} & a_{2n} \\ a_{3(n-2)} & a_{3(n-1)} & a_{3n} \end{bmatrix}, \tag{2.64}$$

$$\mathbf{B} = \begin{bmatrix} a_{11} & \ldots & a_{1(n-3)} \\ a_{21} & \ldots & a_{2(n-3)} \\ a_{31} & \ldots & a_{3(n-3)} \end{bmatrix}, \tag{2.65}$$

where $\mathbf{M} = [\mathbf{B} \ \ \mathbf{A}]$. Then, we obtain the matrix

$$\mathbf{C} = \mathbf{A}^{-1} \cdot \mathbf{B}, \tag{2.66}$$

with $\mathbf{A}$ invertible. The matrix $\mathbf{C}$ contains the exponents of the quantities $(x_1, x_2, \ldots, x_{n-3})$ with respect to the fundamental quantities (x_{n-2}, x_{n-1}, x_n):

$$\mathbf{C} = \begin{array}{c|ccc} & x_1 & \ldots & x_{n-3} \\ \hline x_{n-2} & c_{11} & \ldots & c_{1(n-3)} \\ x_{n-1} & c_{21} & \ldots & c_{2(n-3)} \\ x_n & c_{31} & \ldots & c_{3(n-3)} \end{array}, \tag{2.67}$$

and the i-th dimensionless group is

$$\Pi_i = \frac{x_i}{x_{n-2}^{c_{1i}} x_{n-1}^{c_{2i}} x_n^{c_{3i}}}, \qquad (i = 1, 2, \ldots, n - 3). \tag{2.68}$$

The matrix technique, like any other possible technique in dimensional analysis, does not automatically identify the most suitable dimensionless groups but only a set of possible dimensionless groups. Every combination or power is still a dimensionless group. Moreover, if there is more than one minor with a nonzero determinant, the $\mathbf{C}$ matrix is not unique: the choice of one minor instead of another is equivalent to selecting a set of fundamental quantities (among the quantities involved in the process) instead of another possible set.

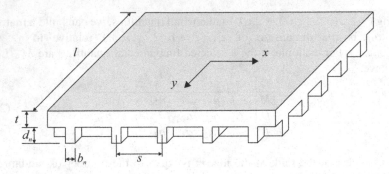

Fig. 2.8 Geometry of a reinforced orthotropic plate

The choice of the order of writing the quantities in the dimensional matrix is not obvious, although it is generally advantageous to write the dependent quantities first.

Example 2.8 We consider a square orthotropic plate reinforced with ribs (see Fig. 2.8), and let us assume that the ribs are identical and equally spaced in both directions. The maximum deflection of the plate η is a function of the geometric and mechanical characteristics of the plate and the ribs, and the typical equation is:

$$\eta = f(l,\ t,\ b_n,\ d_n,\ s,\ E,\ E_n,\ v,\ v_n,\ \text{l.c.},\ \text{b.g.c.},\ F), \qquad (2.69)$$

where E is Young's modulus of the plate material and v is Poisson's ratio. The n subscript refers to the same variables as the material of the ribs. The symbol l.c. indicates the load conditions (concentrated, evenly distributed), b.g.c. indicates the boundary geometric conditions (frictionless support, free rotation), and F is the applied load.

For given load conditions and geometric boundary conditions, the typical equation simplifies as

$$\eta = f(l,\ t,\ b_n,\ d_n,\ s,\ E,\ E_n,\ v,\ v_n,\ F). \qquad (2.70)$$

The dimensional matrix, as a function of mass, length and time,

$$
\begin{array}{c|ccccccccccc}
  & t & b_n & d_n & s & E_n & v & v_n & \eta & F & l & E \\
\hline
M & 0 & 0 & 0 & 0 & 1 & 0 & 0 & 0 & 1 & 0 & 1 \\
L & 1 & 1 & 1 & 1 & 1 & 0 & 0 & 1 & 1 & 1 & -1 \\
T & 0 & 0 & 0 & 0 & -2 & 0 & 0 & 0 & -2 & 0 & -2 \\
\end{array}
\qquad (2.71)
$$

has rank 2 since the T row and the M row are linearly dependent. Then, we can express the typical equation as a function of $(11 - 2) = 9$ dimensionless groups. Given the large number of variables involved, it is convenient to use matrix notation. Therefore, after reducing the matrix (2.71) by merging M and T into a single line

$M T^{-2}$ and finding a matrix extracted with a nonzero determinant, for example, the one with columns l and E,

$$
\mathbf{M} = \begin{array}{c} \\ L \\ MT^{-2} \end{array}
\begin{array}{|ccccccccc|cc|}
t & b_n & d_n & s & E_n & v & v_n & \eta & F & l & E \\
\hline
1 & 1 & 1 & 1 & 1 & 0 & 0 & 1 & 1 & 1 & -1 \\
0 & 0 & 0 & 0 & 1 & 0 & 0 & 0 & 1 & 0 & 1
\end{array} \equiv [\mathbf{B} \ \ \mathbf{A}], \quad (2.72)
$$

the submatrices are

$$
\mathbf{A} = \begin{bmatrix} 1 & -1 \\ 0 & 1 \end{bmatrix} \tag{2.73}
$$

and

$$
\mathbf{B} = \begin{bmatrix} 1 & 1 & 1 & 1 & 1 & 0 & 0 & 1 & 1 \\ 0 & 0 & 0 & 0 & 1 & 0 & 0 & 0 & 1 \end{bmatrix}. \tag{2.74}
$$

Then, we calculate the matrix

$$
\mathbf{C} = \mathbf{A}^{-1} \cdot \mathbf{B} \equiv \begin{bmatrix} 1 & 1 & 1 & 1 & 0 & 0 & 0 & 1 & 2 \\ 0 & 0 & 0 & 0 & 1 & 0 & 0 & 0 & 1 \end{bmatrix}. \tag{2.75}
$$

We observe that the matrix $\mathbf{C}$ is the dimensional matrix of the remaining variables in the new basis l, E, i.e.,

$$
\begin{array}{c|ccccccccc}
 & t & b_n & d_n & s & E_n & v & v_n & \eta & F \\
\hline
l & 1 & 1 & 1 & 1 & 0 & 0 & 0 & 1 & 2 \\
E & 0 & 0 & 0 & 0 & 1 & 0 & 0 & 0 & 1
\end{array}. \tag{2.76}
$$

The i-th dimensionless group is

$$
\Pi_i = \frac{x_i}{l^{c_{1i}} E^{c_{2i}}}, \quad (i = 1, 2, \ldots, 9). \tag{2.77}
$$

Hence,

$$
\begin{aligned}
\Pi_1 &= \frac{t}{l^1 E^0} \equiv \frac{t}{l}, \quad \Pi_2 = \frac{b_n}{l^1 E^0} \equiv \frac{b_n}{l}, \quad \Pi_3 = \frac{d_n}{l^1 E^0} \equiv \frac{d_n}{l}, \\
\Pi_4 &= \frac{s}{l^1 E^0} \equiv \frac{s}{l}, \quad \Pi_5 = \frac{E_n}{l^0 E^1} \equiv \frac{E_n}{E}, \quad \Pi_6 = \frac{v}{l^0 E^0} \equiv v, \\
\Pi_7 &= \frac{v_n}{l^0 E^0} \equiv v_n, \quad \Pi_8 = \frac{\eta}{l^1 E^0} \equiv \frac{\eta}{l}, \quad \Pi_9 = \frac{F}{l^2 E}.
\end{aligned} \tag{2.78}
$$

2.3.1   A Further Generalisation of the Matrix Technique for the Calculation of Nonzero-dimension Power Functions

We have already described the method of linear proportionality, which reduces power functions to lengths (see Sect. 1.4.6). Now we can generalise the approach. Let us suppose we wish to calculate the exponents of a power function expression between quantities so that the result has an assigned dimension, for example, a length or a velocity (as a particular case, we could treat a null dimension, and the power function would become a dimensionless group). If $(x_1, x_2, \ldots, x_n)$ are the n quantities and (y_1, y_2, y_3) are the 3 generic selected dimensions, the exponents of the desired power function must satisfy the equation

$$x_1^{\alpha_1} x_2^{\alpha_2} \cdots x_n^{\alpha_n} = y_1^{\beta_1} y_2^{\beta_2} y_3^{\beta_3}, \tag{2.79}$$

where $y_1^{\beta_1} y_2^{\beta_2} y_3^{\beta_3}$ is the *target* power function. We know the 3 dimensions y_i, the dimensions a_{ij} of the n quantities x_j with respect to the y_i, i.e., the exponents of the dimensional equation

$$[x_j] = y_1^{a_{1j}} y_2^{a_{2j}} y_3^{a_{3j}}, \tag{2.80}$$

and we have set the three nonzero exponents β_1, β_2, β_3. We wish to calculate the n exponents α_1, α_2, α_3, $\ldots$, α_n. We must solve a system of equations that, in compact form, reads

$$\begin{bmatrix} a_{11} & a_{12} & \ldots & a_{1n} \\ a_{21} & a_{22} & \ldots & a_{2n} \\ a_{31} & a_{32} & \ldots & a_{3n} \end{bmatrix} \cdot \begin{Bmatrix} \alpha_1 \\ \alpha_2 \\ \alpha_3 \\ \vdots \\ \alpha_n \end{Bmatrix} = \begin{Bmatrix} \beta_1 \\ \beta_2 \\ \beta_3 \end{Bmatrix}, \tag{2.81}$$

or

$$\underbrace{\mathbf{M}}_{3 \times n} \cdot \underbrace{\boldsymbol{\alpha}}_{n \times 1} = \underbrace{\boldsymbol{\beta}}_{3 \times 1}, \tag{2.82}$$

where $\boldsymbol{\beta}$ is the vector of the exponents of the target power function.

The system of equations allows a solution if, and only if, the rank of the matrix $\mathbf{M}$ is equal to the rank of the matrix $[\mathbf{M} \ \boldsymbol{\beta}]$; if not, it means that the target power function is not a combination of the $(x_1, x_2, \ldots, x_n)$ quantities, and therefore it is not possible to satisfy the dimensional homogeneity of Eq. (2.79).

The $\mathbf{M}$ matrix can be split into two submatrixes $\mathbf{A}$ and $\mathbf{B}$, $\mathbf{M} \equiv [\mathbf{B} \ \mathbf{A}]$ with $\mathbf{A}$ square (3×3) and not singular and $\mathbf{B}$ with dimensions of $3 \times (n-3)$. In the general case, the dimension of $\mathbf{A}$ is $(k \times k)$, and the dimension of $\mathbf{B}$ is $k \times (n-k)$. Then, we can compose the following system of equations equivalent to the system (2.81):

$$
\begin{bmatrix} \mathbf{I} & \mathbf{0} \\ \mathbf{B} & \mathbf{A} \end{bmatrix} \cdot \begin{Bmatrix} \alpha_1 \\ \alpha_2 \\ \alpha_3 \\ \vdots \\ \alpha_n \end{Bmatrix} = \begin{Bmatrix} \alpha_1 \\ \vdots \\ \beta_1 \\ \beta_2 \\ \beta_3 \end{Bmatrix} \equiv \mathbf{Z}, \tag{2.83}
$$

where $\mathbf{I}$ is the identity matrix $(n-3) \times (n-3)$ and $\mathbf{0}$ is the zero-matrix of dimension $(n-3) \times 3$. In the general case, $\mathbf{I}$ is $(n-k) \times (n-k)$, and $\mathbf{0}$ is $(n-k) \times k$. The vector $\mathbf{Z}$ contains, in its tail, the exponents of the dimensions that must have all the power functions of grouping variables. By inverting the matrix, we obtain:

$$
\begin{Bmatrix} \alpha_1 \\ \alpha_2 \\ \alpha_3 \\ \vdots \\ \alpha_n \end{Bmatrix} = \begin{bmatrix} \mathbf{I} & \mathbf{0} \\ -\mathbf{A}^{-1} \cdot \mathbf{B} & \mathbf{A}^{-1} \end{bmatrix} \cdot \mathbf{Z} \equiv \mathbf{E} \cdot \mathbf{Z}. \tag{2.84}
$$

The $\mathbf{E}$ matrix is an $(n \times n)$ exponent matrix that, multiplied by the vector $\mathbf{Z}$, allows us to calculate the exponents of a power function of the desired dimensions.

If $\mathbf{A}$ is singular, it may happen that:

- the rank of $\mathbf{M}$ is equal to the number of dimensions (rows). In this case, it is possible to swap the columns of $\mathbf{M}$ until we obtain a nonunique extracted matrix $\mathbf{A}$;
- the rank of $\mathbf{M}$ is less than the number of dimensions (rows), and it is not possible to find a permutation that allows us to satisfy the previous condition. This means that the dimensional space is smaller than that assumed and that it is necessary to merge the rows in a linear combination until the desired condition is met.

Example 2.9 We consider a heat diffusion process from a source that involves the following 7 variables:

$$
c_p, \ k, \ Q, \ \theta, \ x, \ t, \ \rho, \tag{2.85}
$$

where c_p is the specific heat, k is the thermal conductivity, Q is the source intensity, θ is the temperature, x is the space coordinate, t is time, and ρ is the density.

The dimensional matrix

	c_p	k	Q	θ	x	t	ρ
M	0	1	1	0	0	0	1
L	2	1	-1	0	1	0	-3
T	-2	-3	-2	0	0	1	0
Θ	-1	-3	0	1	0	0	0

$$\tag{2.86}$$

has rank 4. The minor extract of the last 4 columns is nonvanishing. Then, we define
the extracted matrices **A** and **B**:

$$
\underbrace{\mathbf{A}}_{k \times k} =
\begin{bmatrix}
0 & 0 & 0 & 1 \\
0 & 1 & 0 & -3 \\
0 & 0 & 1 & 0 \\
1 & 0 & 0 & 0
\end{bmatrix},
\tag{2.87}
$$

$$
\underbrace{\mathbf{B}}_{k \times (n-k)} =
\begin{bmatrix}
0 & 1 & 1 \\
2 & 1 & -1 \\
-2 & -3 & -2 \\
-1 & -3 & 0
\end{bmatrix}.
\tag{2.88}
$$

The matrix of the exponents is

$$
\underbrace{\mathbf{E}}_{n \times n} =
\left[
\begin{array}{ccc|cccc}
1 & 0 & 0 & 0 & 0 & 0 & 0 \\
0 & 1 & 0 & 0 & 0 & 0 & 0 \\
0 & 0 & 1 & 0 & 0 & 0 & 0 \\
\hline
1 & 3 & 0 & 0 & 0 & 0 & 1 \\
-2 & -4 & -2 & 3 & 1 & 0 & 0 \\
2 & 3 & 2 & 0 & 0 & 1 & 0 \\
0 & -1 & -1 & 1 & 0 & 0 & 0
\end{array}
\right].
\tag{2.89}
$$

If we wish to calculate the exponents of the power function of 7 variables that have
the dimension of a force,

$$
[F] = M\,L\,T^{-2}\,\Theta^{0} \rightarrow \beta_1 = 1,\ \beta_2 = 1,\ \beta_3 = -2,\ \beta_4 = 0,
\tag{2.90}
$$

we simply multiply the matrix **E** and the vector **Z**, generated, for example, by placing
the first terms equal to zero

$$
\mathbf{Z} \equiv
\begin{Bmatrix}
\alpha_1 \\
\alpha_2 \\
\alpha_3 \\
\beta_1 \\
\beta_2 \\
\beta_3 \\
\beta_4
\end{Bmatrix}
\equiv
\begin{Bmatrix}
0 \\
0 \\
0 \\
1 \\
1 \\
-2 \\
0
\end{Bmatrix},
\tag{2.91}
$$

with the following result:

$$
\begin{Bmatrix} \alpha_1 \\ \alpha_2 \\ \alpha_3 \\ \alpha_4 \\ \alpha_5 \\ \alpha_6 \\ \alpha_7 \end{Bmatrix} =
\begin{bmatrix}
1 & 0 & 0 & 0 & 0 & 0 & 0 \\
0 & 1 & 0 & 0 & 0 & 0 & 0 \\
0 & 0 & 1 & 0 & 0 & 0 & 0 \\
1 & 3 & 0 & 0 & 0 & 0 & 1 \\
-2 & -4 & -2 & 3 & 1 & 0 & 0 \\
2 & 3 & 2 & 0 & 0 & 1 & 0 \\
0 & -1 & -1 & 1 & 0 & 0 & 0
\end{bmatrix}
\cdot
\begin{Bmatrix} 0 \\ 0 \\ 0 \\ 1 \\ 1 \\ -2 \\ 0 \end{Bmatrix}
\rightarrow
$$

$$
\begin{Bmatrix} \alpha_1 \\ \alpha_2 \\ \alpha_3 \\ \alpha_4 \\ \alpha_5 \\ \alpha_6 \\ \alpha_7 \end{Bmatrix} =
\begin{Bmatrix} 0 \\ 0 \\ 0 \\ 0 \\ 4 \\ -2 \\ 1 \end{Bmatrix} . \qquad (2.92)
$$

Therefore,

$$
[\, c_p^0 \, k^0 \, Q^0 \, \theta^0 \, x^4 \, t^{-2} \, \rho \,] \equiv [\, x^4 t^{-2} \rho \,] \equiv [\, F \,]. \qquad (2.93)
$$

The vector $\mathbf{Z}$ can be created by choosing all possible combinations of the first $(n - k)$ values. For example, if the vector $\mathbf{Z}$ is equal to

$$
\mathbf{Z} \equiv
\begin{Bmatrix} \alpha_1 \\ \alpha_2 \\ \alpha_3 \\ \beta_1 \\ \beta_2 \\ \beta_3 \\ \beta_4 \end{Bmatrix} \equiv
\begin{Bmatrix} 2 \\ 0 \\ 0 \\ 1 \\ 1 \\ -2 \\ 0 \end{Bmatrix} , \qquad (2.94)
$$

we compute

$$
\begin{Bmatrix} \alpha_1 \\ \alpha_2 \\ \alpha_3 \\ \alpha_4 \\ \alpha_5 \\ \alpha_6 \\ \alpha_7 \end{Bmatrix} = \mathbf{E} \cdot \mathbf{Z} \rightarrow
\begin{Bmatrix} \alpha_1 \\ \alpha_2 \\ \alpha_3 \\ \alpha_4 \\ \alpha_5 \\ \alpha_6 \\ \alpha_7 \end{Bmatrix} =
\begin{Bmatrix} 2 \\ 0 \\ 0 \\ 2 \\ 0 \\ 2 \\ 1 \end{Bmatrix} \qquad (2.95)
$$

resulting in

$$
[\, c_p^2 \, k^0 \, Q^0 \, \theta^2 \, x^0 \, t^2 \, \rho \,] \equiv [\, c_p^2 \theta^2 t^2 \,] \equiv [\, F \,]. \qquad \therefore \quad (2.96)
$$

This power function is different and independent from that shown in Eq. (2.93).

The matrix method is certainly an excellent aid when the number of variables involved is particularly high, but it is often considered with sufficiency because it favours, according to critics, the purely mathematical aspect over the physical aspect. In fact, dimensionless groups calculated by matrix methods do not always have an immediate physical meaning, but this problem occurs regardless of the method chosen for the application of Buckingham's Theorem.

2.3.2 The Number of Independent Solutions

We wonder how many independent solutions admit the problem of **Example** 2.9.

We distinguish two cases.

If all the exponents are zero (i.e., if we wish to find the number of solutions of the problem that render dimensionless the power function combination of all the variables involved), then the system of equations (2.81) is homogeneous and admits ∞^{n-k} solutions, where n is the number of variables and k is the rank of the dimensional matrix. Of these infinities, only $(n - k)$ are independent, as stated by Buckingham's Theorem.

If at least one exponent is nonzero, the system is nonhomogeneous and can be transformed as:

$$\mathbf{M} \cdot \boldsymbol{\alpha} = \boldsymbol{\beta} \rightarrow [\mathbf{M}\ \boldsymbol{\beta}] \cdot \left\{ \begin{array}{c} \boldsymbol{\alpha} \\ -\alpha_{n+1} \end{array} \right\} = \mathbf{0}. \tag{2.97}$$

The original system has been written in the form of an equivalent homogeneous system by moving to the left the vector of the exponents of the target power function and can be treated in the usual way. Assuming that the target power function $\boldsymbol{\beta}$ is dimensionally homogeneous with a power function of the quantities, the rank of the matrix $[\mathbf{M}\ \boldsymbol{\beta}]$ is equal to that of the matrix $\mathbf{M}$, while the number of unknowns has increased by one unit. Therefore, the number of independent solutions is $(n - k + 1)$, and the solutions are calculated as follows.

We consider the previous example, with 7 variables, of which 4 are fundamental. We expect $(7 - 4 + 1) = 4$ combinations of exponents to generate 4 independent power functions that have the dimension of a force. The independence of power functions implies that they cannot be derived from each other through product operations (including power elevations and ratios; see Sect. 1.4.2.2). It is therefore possible to formally write down the 4 solutions:

$$
\begin{Bmatrix} \alpha_{11} \\ \alpha_{21} \\ \alpha_{31} \\ \alpha_{41} \\ \alpha_{51} \\ \alpha_{61} \\ \alpha_{71} \end{Bmatrix} = \mathbf{E} \cdot \begin{Bmatrix} \alpha_{11} \\ \alpha_{21} \\ \alpha_{31} \\ \beta_1 \\ \beta_2 \\ \beta_3 \\ \beta_4 \end{Bmatrix}, \quad
\begin{Bmatrix} \alpha_{12} \\ \alpha_{22} \\ \alpha_{32} \\ \alpha_{42} \\ \alpha_{52} \\ \alpha_{62} \\ \alpha_{72} \end{Bmatrix} = \mathbf{E} \cdot \begin{Bmatrix} \alpha_{12} \\ \alpha_{22} \\ \alpha_{32} \\ \beta_1 \\ \beta_2 \\ \beta_3 \\ \beta_4 \end{Bmatrix},
$$

$$
\begin{Bmatrix} \alpha_{13} \\ \alpha_{23} \\ \alpha_{33} \\ \alpha_{43} \\ \alpha_{53} \\ \alpha_{63} \\ \alpha_{73} \end{Bmatrix} = \mathbf{E} \cdot \begin{Bmatrix} \alpha_{13} \\ \alpha_{23} \\ \alpha_{33} \\ \beta_1 \\ \beta_2 \\ \beta_3 \\ \beta_4 \end{Bmatrix}, \quad
\begin{Bmatrix} \alpha_{14} \\ \alpha_{24} \\ \alpha_{34} \\ \alpha_{44} \\ \alpha_{54} \\ \alpha_{64} \\ \alpha_{74} \end{Bmatrix} = \mathbf{E} \cdot \begin{Bmatrix} \alpha_{14} \\ \alpha_{24} \\ \alpha_{34} \\ \beta_1 \\ \beta_2 \\ \beta_3 \\ \beta_4 \end{Bmatrix},
\tag{2.98}
$$

obtained by multiplying the matrix of the exponents $\mathbf{E}$ by 4 column vectors, in which the first 3 elements are the 4 combinations that provide the result and the last 4 elements are the dimensional exponents of the variable to which we wish to refer to the power function variables of the process.

In compact form:

$$
\mathbf{P} = \mathbf{E} \cdot \mathbf{H} \quad \text{with} \quad \mathbf{H} = \begin{bmatrix} \alpha_{11} & \alpha_{12} & \alpha_{13} & \alpha_{14} \\ \alpha_{21} & \alpha_{22} & \alpha_{23} & \alpha_{24} \\ \alpha_{31} & \alpha_{32} & \alpha_{33} & \alpha_{34} \\ \beta_1 & \beta_1 & \beta_1 & \beta_1 \\ \beta_2 & \beta_2 & \beta_2 & \beta_2 \\ \beta_3 & \beta_3 & \beta_3 & \beta_3 \\ \beta_4 & \beta_4 & \beta_4 & \beta_4 \end{bmatrix}.
\tag{2.99}
$$

The two matrices $\mathbf{P}$ and $\mathbf{H}$ have the dimension $n \times (n - k + 1)$ and must have the same rank since the matrix $\mathbf{E}$ is not singular. The first $(n - k)$ rows of the $\mathbf{H}$ matrix are fixed arbitrarily, with the only condition that the square matrix obtained by adding a nonzero row to the other k rows is not singular. This is equivalent to imposing that the rank of $\mathbf{H}$ be equal to $(n - k + 1)$, since to obtain $(n - k + 1) = 4$ independent solutions, the rank of $\mathbf{P}$ must be equal to $(n - k + 1) = 4$. Since the values are arbitrary, for simplicity, we choose them as 0 or 1, for example:

$$
\begin{bmatrix} \alpha_{11} & \alpha_{12} & \alpha_{13} & \alpha_{14} \\ \alpha_{21} & \alpha_{22} & \alpha_{23} & \alpha_{24} \\ \alpha_{31} & \alpha_{32} & \alpha_{33} & \alpha_{34} \end{bmatrix} \equiv \begin{bmatrix} 1 & 0 & 0 & 0 \\ 0 & 1 & 0 & 0 \\ 0 & 0 & 1 & 0 \end{bmatrix}.
\tag{2.100}
$$

Adding one of the nonzero rows, for example, [1 1 1 1], the rank is 4, as required. Therefore, the compound matrix

$$
\mathbf{H} =
\begin{bmatrix}
1 & 0 & 0 & 0 \\
0 & 1 & 0 & 0 \\
0 & 0 & 1 & 0 \\
1 & 1 & 1 & 1 \\
1 & 1 & 1 & 1 \\
-2 & -2 & -2 & -2 \\
0 & 0 & 0 & 0
\end{bmatrix}
\tag{2.101}
$$

has rank 4. Resolving, we obtain

$$
\mathbf{P} = \mathbf{E} \cdot \mathbf{H} \rightarrow \mathbf{P} =
\begin{bmatrix}
1 & 0 & 0 & 0 \\
0 & 1 & 0 & 0 \\
0 & 0 & 1 & 0 \\
1 & 3 & 0 & 0 \\
2 & 0 & 2 & 4 \\
0 & 1 & 0 & -2 \\
1 & 0 & 0 & 1
\end{bmatrix},
\tag{2.102}
$$

where $\mathbf{P}$ is the 7-row, 4-column matrix containing the unknown exponents of the 7 variables of the power function. Each column contains exponents of independent power functions with the dimension of a force:

$$
c^1 K^0 Q^0 \theta^1 x^2 t^0 \rho^1, \quad c^0 K^1 Q^0 \theta^3 x^0 t^1 \rho^0,
$$
$$
c^0 K^0 Q^1 \theta^0 x^2 t^0 \rho^0, \quad c^0 K^0 Q^0 \theta^0 x^4 t^{-2} \rho^1, \tag{2.103}
$$

or

$$
c\,\theta\, x^2 \rho, \quad K\,\theta^3 t, \quad Q\,x^2, \quad x^4 t^{-2} \rho. \tag{2.104}
$$

We can verify that all 4 power functions have the dimension $[\,F\,] = M\,L\,T^{-2}$ and are independent.

In a general description with 3 fundamental quantities (for simplicity), the matrix $\mathbf{P}$ can be transposed and composed of the matrix $[\,\mathbf{B} \quad \mathbf{A}\,]$ as follows:

$$
\begin{bmatrix}
\mathbf{B} & \mathbf{A} \\
\mathbf{P}^T
\end{bmatrix}
\equiv
\begin{bmatrix}
\mathbf{B} & \mathbf{A} \\
\mathbf{D} & \mathbf{K}
\end{bmatrix}.
\tag{2.105}
$$

The resulting matrix, in explicit form, has the expression

	x_1	x_2	x_3	$\ldots$	x_{n-2}	x_{n-1}	x_n	
y_1	a_{11}	a_{12}	a_{13}	$\ldots$	$a_{1(n-2)}$	$a_{1(n-1)}$	a_{1n}	
y_2	a_{21}	a_{32}	a_{23}	$\ldots$	$a_{2(n-2)}$	$a_{2(n-1)}$	a_{2n}	
y_3	a_{31}	a_{32}	a_{33}	$\ldots$	$a_{3(n-2)}$	$a_{3(n-1)}$	a_{3n}	
Π_1	d_{11}	d_{12}	d_{13}	$\ldots$	k_{11}	k_{12}	k_{13}	(2.106)
Π_2	d_{21}	d_{22}	d_{23}	$\ldots$	k_{21}	k_{22}	k_{23}	
$\ldots$	$\ldots$	$\ldots$	$\ldots$		$\ldots$	$\ldots$	$\ldots$	
Π_{n-3}	$d_{(n-3)1}$	$d_{(n-3)2}$	$d_{(n-3)3}$	$\ldots$	$k_{(n-3)1}$	$k_{(n-3)2}$	$k_{(n-3)3}$	
Π_{n-2}	$d_{(n-2)1}$	$d_{(n-2)2}$	$d_{(n-2)3}$	$\ldots$	$k_{(n-2)1}$	$k_{(n-2)2}$	$k_{(n-2)3}$	

Hence, the matrix $[\mathbf{D} \quad \mathbf{K}]$ contains the exponents of the variables in the $(n - k + 1)$ groups Π_1, Π_2, $\ldots$, Π_{n-2}, all with the imposed dimension $y_1^{\beta_1} y_2^{\beta_2} y_3^{\beta_3}$. The matrix $\mathbf{D}$ is $(n - k + 1) \times (n - k)$, while the matrix $\mathbf{K}$ is $(n - k + 1) \times k$.

In the example with 3 fundamental quantities, the groups Π_1, Π_2, $\ldots$, Π_{n-2} are $(n - 2)$, $\mathbf{D}$ is a matrix with dimensions $(n - 2) \times (n - 3)$ and $\mathbf{K}$ is a matrix with dimensions $(n - 2) \times 3$.

If the dimension of the power functions is set to zero (i.e., if $\beta_1 = \beta_2 = \beta_3 = 0$), the power functions are dimensionless and number $(n - k)$, according to Buckingham's Theorem. Then, the matrix $\mathbf{D}$ becomes square $(n - k) \times (n - k)$, and the matrix $\mathbf{K}$ becomes $(n - k) \times k$.

In the example with 3 fundamental quantities, if the dimension of the power functions is set to zero, the power functions are dimensionless and number $(n - 3)$, the matrix $\mathbf{D}$ becomes square $(n - 3) \times (n - 3)$ and the matrix $\mathbf{K}$ becomes $(n - 3) \times 3$.

The set of dimensional (or dimensionless) power functions calculated with this procedure is, therefore, a complete set and sufficient to describe the functional space of the process in lieu of the initial quantities.

The advantage in the adoption of this set is the reduction of the dimension, which decreases from n to $(n - k + 1)$. An even more evident advantage, with the further reduction of variables to $(n - k)$, is the choice of a set of zero-dimensional power functions (Buckingham's Theorem).

Let us assume that we wish to identify the exponents of a power function of the quantities involved in the physical process, such that the power function can be expressed as a combination of two power functions of different dimensions, with exponents different from zero. For simplicity, we assume that the number of fundamental quantities is $k = 3$.

If $(x_1, x_2, \ldots, x_n)$ are the n quantities and (y_1, y_2, y_3) are the 3 dimensions, the two power functions are $y_1^{\beta_1} y_2^{\beta_2} y_3^{\beta_3}$ and $y_1^{\beta_4} y_2^{\beta_5} y_3^{\beta_6}$, where β_1, β_2, β_3 are the known exponents of the first power function and β_4, β_5, β_6 are the known exponents of the second power function. The following equation should be satisfied:

$$x_1^{\alpha_1} x_2^{\alpha_2} \cdots x_n^{\alpha_n} = \left(y_1^{\beta_1} y_2^{\beta_2} y_3^{\beta_3} \right)^{\delta_1} \left(y_1^{\beta_4} y_2^{\beta_5} y_3^{\beta_6} \right)^{\delta_2}, \qquad (2.107)$$

where δ_1, δ_2 are the unknown exponents. The target power function is a power combination of the power functions $\left(y_1^{\beta_1} y_2^{\beta_2} y_3^{\beta_3} \right)$ and $\left(y_1^{\beta_4} y_2^{\beta_5} y_3^{\beta_6} \right)$. We know the

3 dimensions y_i, the dimensions a_{ij} of the n quantities x_j with respect to the y_i, i.e., the exponents of the dimensional equation

$$[x_j] = y_1^{a_{1j}} \, y_2^{a_{2j}} \, y_3^{a_{3j}}. \tag{2.108}$$

Since the rank of the matrix $\mathbf{M}$ does not vary by adding the two columns containing the dimensions of the two power functions (if this were not the case, it would not be possible to satisfy the dimensional homogeneity of Eq. (2.107)), we aim to calculate how many independent solutions are available for the n exponents $\alpha_1, \ \alpha_2, \ \alpha_3, \ \ldots, \ \alpha_n$. The procedure is identical to the one adopted in Sect. 2.3.1. By imposing dimensional homogeneity, the following system of equations is obtained:

$$\mathbf{M} \cdot \boldsymbol{\alpha} = \delta_1 \, \boldsymbol{\beta}_{(1)} + \delta_2 \, \boldsymbol{\beta}_{(2)}, \quad \boldsymbol{\beta}_{(1)} = \begin{Bmatrix} \beta_1 \\ \beta_2 \\ \beta_3 \end{Bmatrix}, \quad \boldsymbol{\beta}_{(2)} = \begin{Bmatrix} \beta_4 \\ \beta_5 \\ \beta_6 \end{Bmatrix}. \tag{2.109}$$

This nonhomogeneous system in the two unknowns δ_1 and δ_2 can be rewritten in the form of an equivalent homogeneous system:

$$\mathbf{M} \cdot \boldsymbol{\alpha} = \delta_1 \, \boldsymbol{\beta}_{(1)} + \delta_2 \, \boldsymbol{\beta}_{(2)} \rightarrow \begin{bmatrix} \mathbf{M} & \boldsymbol{\beta}_{(1)} & \boldsymbol{\beta}_{(2)} \end{bmatrix} \cdot \begin{Bmatrix} \boldsymbol{\alpha} \\ -\delta_1 \\ -\delta_2 \end{Bmatrix} = \mathbf{0}. \tag{2.110}$$

For the hypothesis of dimensional homogeneity, the rank of the new matrix $\begin{bmatrix} \mathbf{M} & \boldsymbol{\beta}_{(1)} & \boldsymbol{\beta}_{(2)} \end{bmatrix}$ is the same as that of the matrix $\mathbf{M}$, but the number of unknowns has increased to $(n + 2)$. The system admits ∞^{n-k+2} solutions, of which $(n - k + 2)$ are independent. The extension to the case of r variables is immediate, with $r \leq k$.

In summary, the reduction of the n variables to a set of power functions, having a number $r \leq k$ of different and independent nonzero dimensions, leads to $(n - k + r)$ groups. For $r = 0$, we calculate $(n - k)$ groups; for $r = 1$, we calculate $(n - k + 1)$ groups; and for $r = k$, we calculate n groups, equal to the number of variables.

2.3.2.1 Selectable or Constrained Exponents of Dimensional Power Functions

In Sect. 2.2, we already learned how to operate if the submatrix $\mathbf{A}$ selected from the $\mathbf{M}$ matrix is unique, noting that if with a permutation of columns (i.e., with a different selection of the quantities to be assumed as fundamental), the matrix $\mathbf{M}$ cannot be made nonsingular, it means that the rank of the matrix is lower than the number of rows and, therefore, that it is necessary to delete one or more rows; more precisely, it is necessary to incorporate in a row the combination of the two rows linearly dependent. This means that the number of fundamental quantities is smaller than assumed. A typical case is that of physical processes involving forces, i.e., mass, length and time, in steady state, without excitation: the process is 2-dimensional even if the dimensional matrix has 3 rows.

The rows to be recombined must be selected in the minimum number necessary to obtain the dimensional matrix with a maximum number of rows equal to its rank (which is equivalent to a full-rank matrix). Therefore, it is not possible to delete an arbitrary row or recombine any pair of rows.

Example 2.10 Let us consider the deflection of a shelf with a rectangular cross-section and loaded at the free end. The physical process can be described with the typical equation

$$\eta = f(F, h, b, E, l), \tag{2.111}$$

where η is the vertical displacement of a section, F is the load, h is the height, b is the width, E is Young's modulus, and l is the length. The dimensional matrix

$$
\begin{array}{c|cccccc}
 & \eta & F & h & b & E & l \\
\hline
M & 0 & 1 & 0 & 0 & 1 & 0 \\
L & 1 & 1 & 1 & 1 & -1 & 1 \\
T & 0 & -2 & 0 & 0 & -2 & 0
\end{array}
\tag{2.112}
$$

has rank 2; therefore, it is not possible to extract a minor of order 3 regardless of the permutation of the columns. It is necessary to delete a row, for example, the first or the third one, to obtain a dimensional matrix reduced to two rows, still of rank 2. It is not possible to delete the second row because the two remaining rows are in linear combination.

If we remove the third row, the (wrong) reduced dimensional matrix is

$$
\begin{array}{c|cccccc}
 & \eta & F & h & b & E & l \\
\hline
M & 0 & 1 & 0 & 0 & 1 & 0 \\
L & 1 & 1 & 1 & 1 & -1 & 1
\end{array}
\tag{2.113}
$$

and the two submatrices are

$$
\mathbf{B} = \begin{bmatrix} 0 & 1 & 0 & 0 \\ 1 & 1 & 1 & 1 \end{bmatrix}, \quad \mathbf{A} = \begin{bmatrix} 1 & 0 \\ -1 & 1 \end{bmatrix}. \tag{2.114}
$$

The matrix of the exponents is

$$
\mathbf{E} = \begin{bmatrix}
1 & 0 & 0 & 0 & 0 & 0 \\
0 & 1 & 0 & 0 & 0 & 0 \\
0 & 0 & 1 & 0 & 0 & 0 \\
0 & 0 & 0 & 1 & 0 & 0 \\
0 & -1 & 0 & 0 & 1 & 0 \\
-1 & -2 & -1 & -1 & 1 & 1
\end{bmatrix}. \tag{2.115}
$$

Suppose that we wish to calculate power functions with the dimension $M^{-1} L^3$. The matrix of the known terms is equal to

$$\mathbf{H} = \begin{bmatrix} 1 & 0 & 0 & 0 & 0 \\ 0 & 1 & 0 & 0 & 0 \\ 0 & 0 & 1 & 0 & 0 \\ 0 & 0 & 0 & 1 & 0 \\ -1 & -1 & -1 & -1 & -1 \\ 3 & 3 & 3 & 3 & 3 \end{bmatrix}, \qquad (2.116)$$

and the matrix of the exponents of the 5 groups having the dimensions $M^{-1} L^3$ is equal to

$$\mathbf{P} = \mathbf{E} \cdot \mathbf{H} \equiv \begin{bmatrix} 1 & 0 & 0 & 0 & 0 \\ 0 & 1 & 0 & 0 & 0 \\ 0 & 0 & 1 & 0 & 0 \\ 0 & 0 & 0 & 1 & 0 \\ -1 & -2 & -1 & -1 & -1 \\ 1 & 0 & 1 & 1 & 2 \end{bmatrix}. \qquad (2.117)$$

The dimensionless groups are:

$$\Pi_1 = \frac{\eta\, l}{E}, \quad \Pi_2 = \frac{F}{E^2}, \quad \Pi_3 = \frac{h\, l}{E}, \quad \Pi_4 = \frac{b\, l}{E}, \quad \Pi_5 = \frac{l^2}{E}. \quad \therefore \quad (2.118)$$

We can demonstrate that all groups have dimension $M^{-1} L^3 T^2$ instead of the imposed dimension $M^{-1} L^3$. This fact indicates that the exponent of the dimension T is not selectable, but derives from the value attributed to the other exponents. On a different basis, it can be shown that by adding to the dimensional matrix (2.117), having rank 2, a column with the dimensions of the target power function, a new matrix is obtained that has rank 3; therefore, the target power function with dimensions $M^{-1} L^3$ does not satisfy the dimensional homogeneity with a power function of the other 6 variables. In contrast, when adding a column corresponding to the dimension $M^{-1} L^3 T^2$, the rank is unvaried.

In general, the dimension associated with the deleted row (that of the time, in the previous example) can no longer be fixed arbitrarily, and the number of constrained exponents (which also cannot be fixed arbitrarily) is equal to the difference between the number of dimensions involved and the rank of the dimensional matrix. The exponents of the deleted rows are still constrained. In fact, the matrix (2.113) is not a correct dimensional matrix since, for example, the force appears with the dimension $[F] = M L$, which is wrong. Instead of deleting the T row, it is necessary to combine it with the L row to obtain a new $L T^{-2}$ row. It can be verified that the exponent of T is not selectable at our pleasure, but always assumes a value equal to -2 times the value of the exponent of L.

2.4   A Recipe for Dimensionless Groups

To perform a correct dimensional analysis and to identify the dimensionless groups that could be the most representative, it is advisable to proceed as follows:

- list the variables that appear intuitively relevant or are included in an equation that describes the physical process;
- check the absence of dimensionally irrelevant variables;
- write the dimensional matrix with reference to a set of fundamental quantities that are certainly independent, for example, the fundamental quantities adopted in the International System;
- calculate the rank of the dimensional matrix and the number of dimensionless groups on the basis of Buckingham's Theorem;
- calculate a set of possible dimensionless groups by applying Rayleigh's method, Buckingham's method, or linear proportionality;
- if possible, combine the calculated dimensionless groups to obtain another set of dimensionless groups that appear to be more representative of the physical process, having a physical meaning also compared to the dimensionless groups reported in the literature;
- verify that the new set of dimensionless groups is complete.

2.4.1   Some Properties of Dimensional and Dimensionless Power Functions

On the basis of the previous relationships, the minimum number of independent *dimensionless* groups is equal to 1, and the minimum number of independent *dimensional* groups (having the same dimension) is equal to 2.

If a functional relationship has a single dimensionless argument, then the argument must be constant.

In fact, if we can select a single dimensionless group, then we can write either $f(\Pi_1) = 0$ or $f(\Pi_1) = $ const, and given the arbitrariness of the f function, it must be $\Pi_1 = $ const.

If the dimensional matrix is square, then the matrix must be singular; otherwise, it is not possible to identify any relation between the quantities.

The demonstration is immediate: the number of dimensionless groups is $(n - k)$ and is null if $n = k$. Similarly, if the number of rows of the dimensional matrix is greater than the number of columns (i.e., if the dimensions are greater than the quantities), then to obtain at least 1 dimensionless group, the rank of the matrix must be equal to the number of quantities minus 1.

Example 2.11 Let us consider the three variables k, h and k_B, representing the gravitational constant, the Planck constant and the Boltzmann constant, respectively. The dimensional matrix is

$$
\begin{array}{c|ccc}
 & k & h & k_B \\
\hline
M & -1 & 1 & 1 \\
L & 3 & 2 & 2 \\
T & -2 & -1 & -2 \\
\Theta & 0 & 0 & -1
\end{array}
\tag{2.119}
$$

with a number of rows (the dimensions) greater than the number of columns (the variables). The rank is 3, and it is not possible to find any relation among the 3 quantities.

If we are looking for a physical process that also involves the speed of light c and temperature θ, the typical equation becomes:

$$
\theta = f(c,\ k,\ h,\ k_B),
\tag{2.120}
$$

and the dimensional matrix is

$$
\begin{array}{c|ccccc}
 & \theta & c & k & h & k_B \\
\hline
M & 0 & 0 & -1 & 1 & 1 \\
L & 0 & 1 & 3 & 2 & 2 \\
T & 0 & -1 & -2 & -1 & -2 \\
\Theta & 1 & 0 & 0 & 0 & -1
\end{array}
\tag{2.121}
$$

which has rank 4. The dimensionless group of practical interest is

$$
\frac{\theta^2\, k\, k_B^2}{h\, c^5},
\tag{2.122}
$$

which must be constant. Thus, we calculate

$$
\theta = C_1 \sqrt{\frac{h\, c^5}{k\, k_B^2}},
\tag{2.123}
$$

where C_1 is a dimensionless coefficient. For $C_1 = 1/\sqrt{2\,\pi}$, the temperature θ is called the *Planck temperature* and is the maximum thermodynamic temperature admissible within the limits of the validity of the gravitational theory and quantum theory. The Planck temperature is equal to

$$
\theta_P = \frac{1}{\sqrt{2\pi}} \sqrt{\frac{h\, c^5}{k\, k_B^2}} = 1.416\,785\,(\pm 0.000\,071) \times 10^{32}\ \text{K}. \qquad \therefore \tag{2.124}
$$

Summarising Concepts

- The criteria for dimensional analysis require the identification of the variables involved in the process as a first step. The list of these variables is drawn on the basis of the researcher's knowledge, indications from the literature, and solutions from similar processes. Screening is necessary to validate the eligibility of these variables on a dimensional basis and on a physical basis. In retrospect, experimental evidence may indicate the irrelevance of some variables.
- The reduction in the number of dimensional or dimensionless groups is of fundamental importance for a viable interpretation of the process and for planning the experimental activity. This reduction comes through (i) vectorisation of the fundamental quantities, (ii) discrimination of the fundamental quantities, and (iii) rationalisation of the variables by combining them. Vectorisation and discrimination are widely debated and contested processes. Rationalisation, mainly supported by experimental results, is the preferred method, although it can seldom be applied.
- Matrix methods are part of dimensional calculus and add nothing of conceptual relevance. They are particularly useful when there are a large number of variables in the physical process and a preparatory step to subsequent interpretation is necessary.
- The rank of dimensional matrices indicates the dimension of the space of the variables involved in the physical process. This results in a treatment along the lines of matrix algebra, which allows generalisation by reducing the number of variables through the reduction in the number of dimensions involved in the new variables. The most frequent case is that the new variables are dimensionless, which guarantees the maximum reduction of variables.
- In the case where we choose the maximum reduction of variables, transforming the original variables into dimensionless groups, it turns out that a typical equation that is a function of a single dimensionless group necessarily requires that the dimensionless group be a constant. This is a lucky case where dimensional analysis also brings the structure of the typical equation.

References

Galilei, G. (1687). *De Motu Antiquiora* (1589–1592).
Gibbings, J. C. (1981). Directional attributes of length in dimensional analysis. *International Journal of Mechanical Engineering Education, 8*(3), 263–272.
Massey, B. S. (1971). *Units, dimensional analysis and physical similarity*. Van Nostrand Reinhold Company.
Massey, B. S. (1978). Directional analysis? *International Journal of Mechanical Engineering Education, 6*(1), 33–36.

Massey, B. S. (1986). *Measures in science and engineering: Their expression, relation and inter-pretation*. Ellis Horwood Ltd.

Sharp, J. J., Deb, A., & Deb, M. K. (1992). Applications of matrix manipulation in dimensional analysis involving large numbers of variables. *Marine Structures, 5*(4), 333–348.

Szirtes, T. (2007). *Applied dimensional analysis and modeling*. Butterworth-Heinemann.

Chapter 3
The Structure of the Functions of the Dimensionless Groups, Symmetry and Affine Transformations

No great discovery was ever made without a bold guess.

Sir Isaac Newton

As we have often mentioned, the structure of the typical equation between dimensionless groups is generally unknown. A wide class of typical equations is represented by power functions. There are cases in which some information can be inferred from possible symmetries of the physical process or from some general principles, such as the second principle of thermodynamics.

Symmetry has become a powerful and effective tool for constraining problems and finding solutions. A kind of internal symmetry, with the presence of transformations that leave the differential problem unchanged, allows the identification of self-similar solutions, sometimes with analytical expression.

3.1 The Structure of the Functions of the Dimensionless Groups

Classification of the typical equation, already reduced to a function of dimensionless groups, is either as a *power function* or as a *non-power function* structure.

A functional relation is a *power function* if it is expressed as a product of powers of the variables.

A functional relation is a *non-power function* if it contains symbols of sum or subtraction (algebraic sum) or if it contains transcendental functions (trigonometric, exponential, logarithmic, hyperbolic functions). This latter inclusion derives from the former, since a transcendental function can be expanded in a Taylor series as a linear combination of power functions.

S. G. Longo, *Principles and Applications of Dimensional Analysis and Similarity*, Mathematical Engineering, https://doi.org/10.1007/978-3-030-79217-6_3

A power function relation between dimensionless groups contains the fewest number of constants, considered both as exponents and as multiplicative constants. In fact, if the dimensionless groups are $(n - k)$ and are in a power function relationship with each other, for example,

$$\Pi_1^{\alpha_1} \Pi_2^{\alpha_2} \cdots \Pi_{n-k}^{\alpha_{n-k}} = C_1, \qquad (3.1)$$

there is only one dimensionless constant C_1 and $(n - k - 1)$ independent exponents, for a total of $(n - k)$ unknowns, equal to the number of dimensionless groups.

If the dimensionless groups are in a non-power function relation, the number of constants is at least equal to 2, and the number of exponents is still equal to $(n - k - 1)$. The sum of the number of constants and the number of independent exponents is at least $(n - k + 1)$, which is at least one more than the number of dimensionless groups. For example, this happens for the following relationship with two terms:

$$\Pi_1^{\alpha_1} = C_1 \, \Pi_2^{\alpha_2} + C_2 \left(\Pi_3^{\alpha_3} \cdots \Pi_{n-k}^{\alpha_{n-k}} \right). \qquad (3.2)$$

In addition to reducing the number of unknowns, there are other advantages in identifying a power function between dimensionless groups. In fact, if the relationship is a power function between $(n - k)$ dimensionless groups, it is sufficient to perform $(n - k)$ experiments to estimate the numerical value of the constants and exponents. Instead, if the relation is a non-power function, the minimum number of experiments is indefinite, just as the form of the function itself is indefinite.

For these reasons, the search for a power function between dimensionless groups is the first attempt to represent the function.

In the following paragraphs, we analyse some cases that can help the identification of the structure of the functional relationship, possibly a power function.

3.1.1   The Structure of the Function of Dimensionless Groups is Necessarily a Power Function

There are some cases in which the power function structure of the function can be deduced directly from the analysis of the physical process.

Let us consider the outflow process from a broad-crested Bélanger weir; see Fig. 3.1, which can be described with the typical equation

$$Q = f(b, \ h, \ g), \qquad (3.3)$$

where Q is the volumetric flow rate, b is the width, h is the water level with respect to the crest, and g is the acceleration of gravity. The dimensional matrix

Fig. 3.1 Broad-crested
Bélanger weir

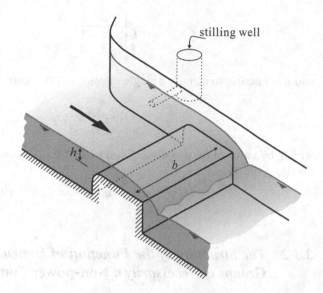

has rank 2 and, with h and g selected as fundamental quantities, the following two dimensionless groups are calculated:

$$\Pi_1 = \frac{Q}{h^{5/2}\, g^{1/2}}, \quad \Pi_2 = \frac{b}{h}. \tag{3.5}$$

The typical equation involving the two dimensionless groups

$$\Phi\left(\frac{Q}{h^{5/2}\, g^{1/2}}, \frac{b}{h}\right) = 0 \tag{3.6}$$

must be a power function. We know, in fact, that doubling the width b doubles the flow rate Q. The linear dependence between Q and b (on a logical basis) forces a power function as

$$\frac{Q}{h^{5/2}\, g^{1/2}} = C_1 \frac{b}{h} \;\rightarrow\; Q = C_1\, b h^{3/2}\, g^{1/2}, \tag{3.7}$$

where C_1 is a dimensionless coefficient.

The same result can be achieved more directly by selecting the volumetric flow rate per unit width as the governed variable $q = Q/b$. In this case, the number of variables involved is reduced to 3 in a 2-dimension space

$$\begin{array}{c|ccc} & q & h & g \\ \hline L & 2 & 1 & 1 \\ T & -1 & 0 & -2 \end{array},$$

(3.8)

and it is possible to identify a single dimensionless group

$$\Pi_1 = \frac{q}{h^{3/2}\,g^{1/2}}$$

(3.9)

which, by virtue of the properties listed in Sect. 2.4.1, must necessarily be constant, that is,

$$\Pi_1 = \text{const} \rightarrow q \equiv \frac{Q}{b} = C_1\,h^{3/2}\,g^{1/2}.$$

(3.10)

3.1.2   The Structure of the Function of Dimensionless Groups is Necessarily a Non-power Function

In other cases, the structure of the functional relationship between dimensionless groups cannot be a power function on a logical or physical basis.

Let us consider the process:

probability that a person in Roma Termini station can see a train directed to Florence.

The probability p depends on the time interval T_t between two subsequent trains both convenient to reach Florence, the time the train stays in the station (including the time it takes to cross the station) Δt_t, and the waiting time of the person Δt_p:

$$p = f\left(\Delta t_t,\ \Delta t_p,\ T_t\right).$$

(3.11)

The dimensional matrix of the physical process is

$$\begin{array}{c|cccc} & p & \Delta t_t & \Delta t_p & T_t \\ \hline T & 0 & 1 & 1 & 1 \end{array},$$

(3.12)

where p is already dimensionless. The two dimensionless groups in addition to p are

$$\Pi_1 = \frac{\Delta t_t}{T_t},\quad \Pi_2 = \frac{\Delta t_p}{T_t}$$

(3.13)

and the typical equation can be written as

$$p = \Phi(\Pi_1,\ \Pi_2).$$

(3.14)

Let us suppose Φ is a power function:

$$p = C_1 \, \Pi_1^\alpha \, \Pi_2^\beta \equiv C_1 \left(\frac{\Delta t_t}{T_t} \right)^\alpha \left(\frac{\Delta t_p}{T_t} \right)^\beta . \tag{3.15}$$

The process is symmetrical since it can be expressed as *probability that while the train to Florence is at a stop or is crossing Roma Termini station, the person is waiting in the station.* As a consequence, the exponent of the two groups must be the same, and it results

$$p = C_1 \left(\frac{\Delta t_t \, \Delta t_p}{T_t^2} \right)^\gamma . \tag{3.16}$$

The exponent γ can only be positive since the probability p increases monotonically with increasing Δt_t or Δt_p. However, we observe that even if the time the person is waiting is zero, the probability of seeing the train cannot be null: upon arrival at the station, followed by immediate departure, the person sees the train. If the waiting time of the train is zero (the train is only in transit), again, the probability p cannot be null: the train crosses the station when the person is waiting. For this reason, the relationship cannot have the structure of Eq. (3.16) or any other power function structure. In fact, the calculus of probability reveals the following expression for p:

$$p = \frac{\Delta t_t}{T_t} + \frac{\Delta t_p}{T_t} - \frac{1}{2} \left[\left(\frac{\Delta t_t}{T_t} \right)^2 + \left(\frac{\Delta t_p}{T_t} \right)^2 \right] . \tag{3.17}$$

3.1.3 The Structure of the Function of Dimensionless Groups is Possibly a Power Function

Finally, there are physical processes that are described with a functional relationship having a structure that, on the basis of physical intuition and imposed constraints, could also be a power function, while then it is *a non-power function*.

Let us consider the following scenario (Szirtes 2007). A pedestrian walks on the footpath of a building of height h, when a vase falls, accidentally knocked over by the apartment owner on the top floor. The owner who has jostled the vase shouts a warning to allow the pedestrian to escape. To be effective, the warning must arrive before the vase, assuming that the reaction time is zero for both persons. The physical process can be expressed as

$$\Delta t = f(h, \ c, \ g), \tag{3.18}$$

where Δt is the difference between the flight time of the vase and the transit time of the warning, c is the speed of sound and g is the acceleration of gravity. The safety of the pedestrian requires $\Delta t > 0$. The dimensional matrix

$$
\begin{array}{c|cccc}
 & \Delta t & h & c & g \\
\hline
L & 0 & 1 & 1 & 1 \\
T & 1 & 0 & -1 & -2
\end{array}
\tag{3.19}
$$

has rank 2. If we select c and g as the fundamental quantities, we calculate the two dimensionless groups:

$$
\Pi_1 = \frac{\Delta t\, g}{c}, \quad \Pi_2 = \frac{h\,g}{c^2};
\tag{3.20}
$$

hence,

$$
\Pi_1 = \Phi(\Pi_2).
\tag{3.21}
$$

A power function structure for Φ has the expression

$$
\Pi_1 = C_1\, \Pi_2^{\alpha} \rightarrow \Delta t = C_1\, \frac{c}{g}\left(\frac{h\,g}{c^2}\right)^{\alpha},
\tag{3.22}
$$

where C_1 is a dimensionless coefficient. If $h = 0$, then $\Delta t = 0$ and then $\alpha > 0$. Additionally, if the speed c of sound increases, Δt must also increase, and therefore, $(1 - 2\alpha) > 0 \rightarrow \alpha < 1/2$. If g increases, then Δt should decrease and, therefore, $\alpha < 1$.

The final result is that the power function structure is compatible with the physical process, as long as the exponent α is positive and less than $1/2$.

However, neglecting flow resistance in vase motion, the problem has an analytical solution:

$$
\Delta t = \sqrt{\frac{2h}{g}} - \frac{h}{c} \rightarrow \Pi_1 = \sqrt{2\,\Pi_2} - \Pi_2,
\tag{3.23}
$$

which is not a power function.

3.2   The Use of Symmetry to Specify the Expression of the Function

In general, the dimensional approach has the disadvantage of not providing hints on the selection of dimensionless groups and on the structure of the functional relationship. In some cases, however, it is possible to go into more detail. This is the case of a physical process with symmetries.

Symmetry is one of the cornerstones of physics and is closely linked to the conservation of certain quantities. For example, invariance with respect to a translation in time (stationarity) yields, as a consequence, conservation of energy. The opposite is also true: energy conservation requires invariance to a shift of time. The invariances with respect to a translation in space (homogeneity) and to a rotation (isotropy) yield, as a consequence, conservation of momentum and angular momentum, respectively.

Symmetry is mathematically defined in a rigorous way, but for simplicity, we mean by symmetry any transformation that leaves unchanged one or more properties of the system.

In the following, there are some applicative examples where the symmetry properties are used to specify the structure of the typical equation of dimensionless groups.

Example 3.1 Let us consider a piston that moves in a cylinder full of gas, generating a wave that propagates with finite speed. We neglect the friction of the gas on the wall. In front of the piston, there is a region of moving gas separated by an undisturbed region downstream. The wave front position x depends on the time t, the speed of the piston U, the density of the gas ρ, the pressure p and the internal specific energy e in the undisturbed region, and has the following typical equation:

$$x = f(t, \ U, \ \rho, \ p, \ e). \tag{3.24}$$

The dimensional matrix

$$\begin{array}{c|cccccc} & x & t & U & \rho & p & e \\ \hline M & 0 & 0 & 0 & 1 & 1 & 0 \\ L & 1 & 0 & 1 & -3 & -1 & 2 \\ T & 0 & 1 & -1 & 0 & -2 & -2 \end{array} \tag{3.25}$$

has rank 3. Selecting x, t and ρ as a triad of fundamental quantities, the following 3 dimensionless groups are identified:

$$\Pi_1 = \frac{Ut}{x}, \quad \Pi_2 = \frac{p\,t^2}{x^2\,\rho}, \quad \Pi_3 = \frac{e\,t^2}{x^2}. \tag{3.26}$$

We observe that x and t always appear as a ratio; therefore, we can reduce the number of variables by defining $c = x/t$, the forward speed of the wave front.

The new dimensional matrix becomes

$$\begin{array}{c|ccccc} & c & U & \rho & p & e \\ \hline M & 0 & 0 & 1 & 1 & 0 \\ L & 1 & 1 & -3 & -1 & 2 \\ T & -1 & -1 & 0 & -2 & -2 \end{array} \tag{3.27}$$

and has rank 2. Again, we observe that the mass appears only in ρ and p, which, therefore, can collapse in the ratio p/ρ.

Based on these observations, we can group p and ρ and delete the M row. The dimensional matrix becomes

$$\begin{array}{c|cccc} & c & U & p/\rho & e \\ \hline L & 1 & 1 & 2 & 2 \\ T & -1 & -1 & -2 & -2 \end{array} \tag{3.28}$$

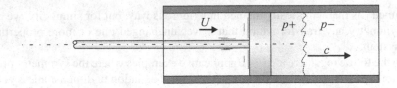

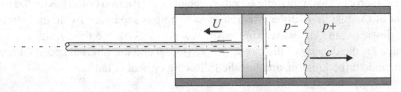

Fig. 3.2 Compression and rarefaction waves for a slow piston speed

and has rank 1. Choosing p/ρ as the fundamental quantity, we obtain the functional relationship

$$\frac{c}{\sqrt{p/\rho}} = \Phi\left(\frac{U}{\sqrt{p/\rho}}, \frac{p}{\rho e}\right). \tag{3.29}$$

If the piston moves to the right ($U > 0$), a compression wave is generated that propagates to the right, with $c > 0$; see Fig. 3.2.

If the piston moves to the left ($U < 0$), the wave is a rarefaction but continues to propagate to the right, i.e., $c > 0$, so the function Φ is even with respect to $U/\sqrt{p/\rho}$. For low piston speeds, the waves propagate with the speed of sound, i.e.,

$$\lim_{(U/\sqrt{p/\rho}) \to 0} \frac{c}{\sqrt{p/\rho}} = \tilde{\Phi}_1\left(\frac{p}{\rho e}\right). \tag{3.30}$$

For very fast speed to the right, the shock wave has a speed proportional to the speed of the piston:

$$\lim_{(U/\sqrt{p/\rho}) \to \infty} \frac{c}{\sqrt{p/\rho}} = \frac{U}{\sqrt{p/\rho}} \tilde{\Phi}_2\left(\frac{p}{\rho e}\right) \to \frac{c}{U} = \tilde{\Phi}_2\left(\frac{p}{\rho e}\right). \qquad \therefore \quad (3.31)$$

Gas dynamics analysis brings us to $\tilde{\Phi}_1(p/\rho e) = \sqrt{\gamma}$ and $\tilde{\Phi}_2(p/\rho e) = (\gamma + 1)/2$, with γ being the ratio of specific heat at constant pressure to that at constant volume.

Example 3.2 Let us consider a spherical particle of radius r and density equal to that of the fluid in a flow field between two parallel planes; see Fig. 3.3 (modified from Hornung 2006). The bottom wall is at rest, and the ceiling is moving to the

Fig. 3.3 A sphere in a flow field between two parallel planes

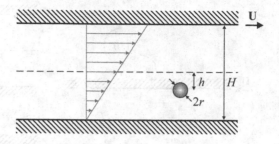

right at a constant speed. The flow field has a uniform shear rate equal to $\dot{\gamma} = U/H$. The transverse force exerted by the fluid on the particle can be expressed as

$$F = f(h,\ H,\ r,\ \rho,\ \mu,\ \dot{\gamma}),\tag{3.32}$$

where ρ is the common density of fluid and particle and μ is the dynamic viscosity. The dimensional matrix

$$
\begin{array}{c|ccccccc}
 & F & h & H & r & \rho & \mu & \dot{\gamma} \\
\hline
M & 1 & 0 & 0 & 0 & 1 & 1 & 0 \\
L & 1 & 1 & 1 & 1 & -3 & -1 & 0 \\
T & -2 & 0 & 0 & 0 & 0 & -1 & -1
\end{array}
\tag{3.33}
$$

has rank 3, and it is possible to calculate $(7-3) = 4$ dimensionless groups. Ultimately, we can write

$$\frac{F}{\mu\,\dot{\gamma}\,r^2} = \Phi\left(\frac{\rho\,\dot{\gamma}\,r^2}{\mu},\ \frac{r}{H},\ \frac{h}{H}\right),\tag{3.34}$$

where $(\rho\,\dot{\gamma}\,r^2/\mu)$ is the Reynolds number.

When reversing the motion ($\dot{\gamma}$ changes sign), the sign of the force F must not change; hence, F is an odd function of the Reynolds number, which, in the present analysis, has its own sign due to $\dot{\gamma}$.

In addition, for Re $\to 0$, the force must be zero, and Eq. (3.34) can be re-written as

$$\frac{F}{\mu\,\dot{\gamma}\,r^2} = \mathrm{Re}^n\,\frac{|\dot{\gamma}|}{\dot{\gamma}}\,\Phi_1\left(\frac{r}{H},\ \frac{h}{H}\right),\tag{3.35}$$

where n is any exponent other than zero; otherwise, ρ becomes dimensionally irrelevant.

More information about the structure of Φ_1 comes from symmetry.

Rotating the system of 180° around an axis orthogonal to the page, we obtain a flow field in which the particle, initially below the mid-plane, is located above it and receives a thrust toward it (see Fig. 3.4a). If we perform a Galilean transformation, with a new coordinate system that moves with the speed of the ceiling, see Fig. 3.4b,

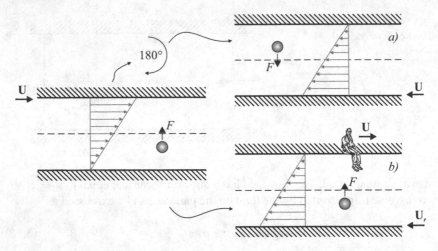

Fig. 3.4 Transformation as a result of *a)* a 180° rotation and *b)* a Galilean transformation

we observe a flow field identical to that obtained by rotation of 180°, but with the particle still below the mid-plane, toward which it receives a thrust. It follows that the direction of the force is always directed toward the mid-plane, and F is an odd function of h/H and is null for $h = 0$; hence,

$$\frac{F}{\mu \dot{\gamma} r^2} = \mathrm{Re}^n \frac{|\dot{\gamma}|}{\dot{\gamma}} \left(\frac{h}{H}\right)^{(2m-1)} \Phi_2 \left(\frac{r}{H}\right), \qquad\qquad \therefore \quad (3.36)$$

with *m* a positive integer.

Example 3.3 Let us consider the fluctuations of the interface between a liquid and a gas that occur in the form of waves. Their speed c can be expressed as

$$c = f(l, \ h, \ H, \ g, \ \sigma, \ \rho), \qquad\qquad (3.37)$$

where l is the wavelength, h is the liquid depth (assumed constant), H is the wave height, g is the acceleration of gravity, σ is the surface tension and ρ is the density of the liquid. We are neglecting the gas effects.

The dimensional matrix

$$
\begin{array}{c|ccccccc}
 & c & l & h & H & g & \sigma & \rho \\
\hline
M & 0 & 0 & 0 & 0 & 0 & 1 & 1 \\
L & 1 & 1 & 1 & 1 & 1 & 0 & -3 \\
T & -1 & 0 & 0 & 0 & -2 & -2 & 0
\end{array}
\tag{3.38}
$$

has rank 3, and it is therefore possible to express Eq. (3.37) as a function of $(7 - 3) =$ 4 dimensionless groups, for example

$$
\frac{c}{\sqrt{l g}} = \Phi \left(\frac{\sigma}{\rho l^2 g}, \frac{h}{l}, \frac{H}{l} \right),
\tag{3.39}
$$

where H/l is named the *steepness* of the wave. For small steepness, H/l has negligible effects, and Eq. (3.39) is simplified:

$$
\frac{c}{\sqrt{l g}} = \Phi_1 \left(\frac{\sigma}{\rho l^2 g}, \frac{h}{l} \right).
\tag{3.40}
$$

If gravity is less important than surface tension, the former can be eliminated from the dimensional matrix, which still has rank 3. The effect of gravity is null only if the function Φ_1 has the structure

$$
\Phi_1 \left(\frac{\sigma}{\rho l^2 g}, \frac{h}{l} \right) \equiv \sqrt{\frac{\sigma}{\rho l^2 g}} \, \Phi_2 \left(\frac{h}{l} \right).
\tag{3.41}
$$

In this way, the result is

$$
c = \Phi_2 \left(\frac{h}{l} \right) \sqrt{\frac{\sigma}{\rho l}}.
\tag{3.42}
$$

If h/l is large enough, it is physically irrelevant, and the function Φ_2 becomes a constant. These waves are called *capillary waves*.

If surface tension is less important than gravity, to eliminate the dependence on σ, it is necessary that the function Φ_1 is a function of h/l only. The result is the following expression of speed:

$$
c = \sqrt{g l} \, \Phi_2 \left(\frac{h}{l} \right),
\tag{3.43}
$$

and these waves are called *gravity waves*.

The result of the reduction in the number of dimensionless groups is consistent with the fact that the elimination of σ leads to the dimensional irrelevance of ρ. Therefore, the mass M does not intervene, the dimensional matrix has rank 2, and the process is purely kinematic. The 4 residual variables can be organised into 2 dimensionless groups, one of which is the steepness of the wave that does not intervene if it is small enough.

Dimensional analysis no longer helps to identify the structure of Φ_1 when gravity and surface tension are equally important. However, we may try to find an equivalence between the action of gravity and that of surface tension.

The stabilising force of gravity on a crest of equation

$$z = a \sin\left(\frac{2\pi}{l}x\right), \quad 0 < x < l/2, \tag{3.44}$$

where a is the amplitude of the sinusoidal wave, is equal to its weight per unit front,

$$\rho g \frac{a l}{\pi}. \tag{3.45}$$

The stabilising force per unit front due to surface tension is equal to

$$\int_0^{l/2} \frac{\sigma \frac{d^2 z}{dx^2}}{\left[1 + \left(\frac{dy}{dx}\right)^2\right]^{3/2}} dx \approx \int_0^{l/2} \sigma \frac{d^2 z}{dx^2} dx = \frac{4 a \pi \sigma}{l}. \tag{3.46}$$

The contribution of surface tension can be considered a gravity correction with g' given by

$$\frac{\rho a l}{\pi} g' = \frac{\rho a l}{\pi} \left(g + \frac{4\pi^2 \sigma}{\rho l^2}\right); \tag{3.47}$$

inserting g' into Eq. (3.43) yields

$$c = \sqrt{g l + \frac{4\pi^2 \sigma}{\rho l}} \; \Phi_2\left(\frac{h}{l}\right). \qquad \therefore \quad (3.48)$$

If we consider a trough instead of a crest of the sinusoidal wave, the results are identical, with stabilising forces of gravity and tension surface acting in the same direction.

Example 3.4 (from Hornung 2006) Let us consider a viscoelastic fluid in a cylindrical container with a spinning rod. If the diameter of the rod is small enough, the properties of the fluid produce a rise in correspondence of the axis (see Fig. 3.5). This is the *Weissenberg effect*, and it is exactly the opposite of what happens in a Newtonian fluid, for which the centrifugal component generates a depression of the free surface in the same region.

With reference to the schematic shown in Fig. 3.6, the elevation h is a function of several variables characterising the geometry, the flow field and the fluid:

Fig. 3.5 Weissenberg effect, viscoelastic fluid rising in correspondence with the cylindrical circular bar in rotation (from http://web.mit.edu/nnf/, courtesy of Gareth McKinley)

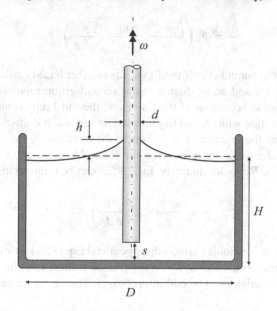

Fig. 3.6 Weissenberg effect, schematic for the analysis of the physical process

$$h = f(\omega, \ \mu, \ \nu_1, \ \rho, \ g, \ d, \ s, \ D, \ H), \tag{3.49}$$

where ω is the angular velocity, μ is the dynamic viscosity, ρ is the density, g is the acceleration of gravity, and ν_1 is the parameter that characterises the deviation from the hydrostatic distribution of one of the components of the normal stress, with $\sigma_{22} = \nu_1 \dot{\gamma}^2 - p$, $[\nu_1] = M \, L^{-1}$; the deviation of the other components is neglected.

The dimensional matrix of the 10 variables

	h	ω	μ	v_1	ρ	g	d	s	D	H
M	0	0	1	1	1	0	0	0	0	0
L	1	0	-1	-1	-3	1	1	1	1	1
T	0	-1	-1	0	0	-2	0	0	0	0

$$(3.50)$$

has rank 3, and it is possible to express the relationship using only $(10 - 3) = 7$ dimensionless groups, for example

$$\frac{h}{d} = \Phi \left(\frac{\rho \omega d^2}{\mu}, \frac{v_1}{\rho d^2}, \frac{g}{\omega^2 d}, \frac{s}{d}, \frac{D}{d}, \frac{H}{d} \right). \tag{3.51}$$

The experiments indicate the independence on D/d and H/d if $D \gg d$ and $H \gg d$; in addition, we wish to restrict our analysis to a fixed value of s/d, and the relationship is simplified as follows:

$$\frac{h}{d} = \Phi_1 \left(\frac{\rho \omega d^2}{\mu}, \frac{v_1}{\rho d^2}, \frac{g}{\omega^2 d} \right). \tag{3.52}$$

The first group in parentheses is the Reynolds number based on the lateral speed of the cylindrical bar and on its diameter; the second group represents the effects of the non-Newtonian behaviour of the fluid; and the third represents the effect of centrifugal acceleration with respect to gravity. The variable h is always positive and does not depend on the orientation of ω; hence, the function Φ_1 must be even in ω. Additionally, h has a weak dependence on the Reynolds number for Re $\to$ 0. As a consequence, at low Reynolds numbers, Eq. (3.52) can be reduced to

$$\frac{h}{d} = \frac{\omega^2 d}{g} \Phi_2 \left(\frac{v_1}{\rho d^2} \right). \tag{3.53}$$

For $v_1 \to 0$, the non-Newtonian effects disappear and Eq. (3.53) should include only the centrifugal component. The sign of h is reversed if there is a change in the sign of v_1; hence, the function must be odd in $v_1/(\rho d^2)$. Therefore, we can write

$$\frac{h}{d} = -C_1 \frac{\omega^2 d}{g} \left[1 - C_2 \left(\frac{v_1}{\rho d^2} \right)^{2m-1} \right], \tag{3.54}$$

where C_1 and C_2 are positive dimensionless coefficients, m is a positive integer and the group $g/(\omega^2 d)$ appears with its reciprocal because $h \to 0$ if $\omega \to 0$; the negative sign before C_1 occurs to ensure that $h < 0$ if $v_1 = 0$. On the basis of Eq. (3.54), the Weissenberg effect occurs only if

$$\frac{\nu_1}{\rho\, d^2} > \frac{1}{C_2^{1/(2m-1)}}, \qquad \therefore \quad (3.55)$$

otherwise, depression due to the centrifugal component is dominant.

The theory indicates $m = 1$, $C_2 = 8$; hence, $d < \sqrt{8\,\nu_1/\rho}$.

We observe that the condition $s/d = $ const does not erase the s/d group or reduce the number of variables but simply allows the analysis of the process in a subdomain.

Example 3.5 Let us consider a Coriolis mass flow meter, schematically represented in Fig. 3.7 by a straight tube of length l clamped at the extremes through which fluid flows with velocity U and density ρ_f. We first mathematically analyse the behaviour of the pipeline-fluid system in the presence of forced oscillations due to, for example, an external electromagnetic pick-up.

The equation that describes the transverse vibrations $\delta_t(x, t)$ of the pipeline, schematically represented as a straight beam of Euler, derives from the principle of Lagrange's stationary action of the function (Raszillier and Durst 1991):

$$\mathscr{L}_t(\delta_t) = \frac{1}{2}\left[m_t \left(\frac{\partial \delta_t}{\partial t}\right)^2 - T_t \left(\frac{\partial \delta_t}{\partial x}\right)^2 - E\,I \left(\frac{\partial^2 \delta_t}{\partial x^2}\right)^2 \right], \qquad (3.56)$$

where m_t is the mass of the pipe per unit length, T_t is the axial force, E is Young's modulus and I is the moment of inertia of the cross-section in the plane of oscillation. Similarly, the equation describing the transverse vibrations $\delta_f(x, t)$ of the fluid, represented as a string, i.e., a continuous whose only interaction with the inner walls of the pipe can be reduced to pressure forces, derives from the Lagrange principle of stationary action of the function:

$$\mathscr{L}_f(\delta_f) = \frac{1}{2}\left[m_f \left(\frac{\partial \delta_f}{\partial t} + U \frac{\partial \delta_f}{\partial x}\right)^2 - T_f \left(\frac{\partial \delta_f}{\partial x}\right)^2 \right], \qquad (3.57)$$

where m_f is the mass of fluid per unit length and T_f is the axial force that stresses the fluid string (compression only). The pipe and the fluid oscillate with equal amplitude

$$\delta_t(x, t) = \delta_f(x, t). \qquad (3.58)$$

The Lagrange function of the coupled system becomes

$$\mathscr{L}(\delta_t, \delta_f, \lambda) = \mathscr{L}_t + \mathscr{L}_f + \lambda\,(\delta_t - \delta_f), \qquad (3.59)$$

where λ is the Lagrange multiplier. Minimising the Lagrange multiplier brings us to the equation

$$(m_t + m_f)\,\frac{\partial^2 \delta}{\partial t^2} + 2\,m_f\,U\,\frac{\partial^2 \delta}{\partial t\,\partial x} + m_f\,U^2\,\frac{\partial^2 \delta}{\partial x^2} + E\,I\,\frac{\partial^4 \delta}{\partial x^4} = 0, \qquad (3.60)$$

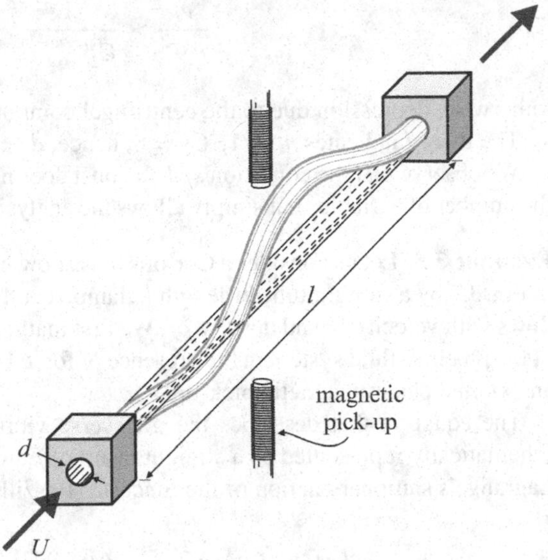

Fig. 3.7 Coriolis mass flow meter. The dashed lines refer to the undisturbed configuration, with the pick-ups off

where $\delta \equiv \delta_t \equiv \delta_f$ represents the amplitude of oscillation common to the fluid and the pipe, while the axial force $T = T_t + T_f$ has been eliminated since it does not play a relevant role. In this equation, the second and third terms are non-inertial contributions; in particular, the second term is the component of Coriolis, and the third term is the centrifugal component.

The boundary conditions (small pipe clamped at the extremities) are:

$$\begin{cases} \delta(0,\ t) = \delta(l,\ t) = 0, \\ \left.\dfrac{\partial \delta}{\partial x}\right|_{(0,\ t)} = \left.\dfrac{\partial \delta}{\partial x}\right|_{(l,\ t)} = 0. \end{cases} \tag{3.61}$$

The analytical solution of the problem is not elementary and is calculated considering the effect of the fluid flow as a perturbation of the oscillating modes of the pipe only when the fluid is at rest.

With fluid at rest, Eq. (3.60) simplifies as

$$\left(m_t + m_f\right) \frac{\partial^2 \delta}{\partial t^2} + E I \frac{\partial^4 \delta}{\partial x^4} = 0. \tag{3.62}$$

The solution is obtained by separating the variables and calculating the eigenvalues that meet the boundary conditions. Two families of self-oscillations are identified: the first contains symmetrical modes, and the second contains antisymmetrical modes. The first two symmetrical and antisymmetrical modes are shown in Fig. 3.8.

Introducing the effects of the fluid in motion, we calculate the first mode deformation of the pipe shown in Fig. 3.9 at different times.

Fig. 3.8 The first two symmetrical (s_1 and s_2) and antisymmetrical (a_1 and a_2) modes. The amplitude is arbitrarily normalised to the same value for all modes

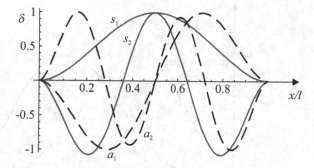

Fig. 3.9 Symmetric deformation of the first mode with fluid in motion at different times. The amplitude is normalised

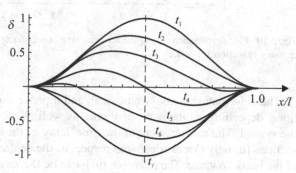

The solution of the first symmetrical oscillating mode deformation is (except for a coefficient)

$$\delta(x,\ t) = \delta_0(x)\ \sin \omega_1 t - \frac{U}{l\, \omega_1}\ \frac{m_f}{m_f + m_t}\ \delta_1(x)\ \cos \omega_1 t, \qquad (3.63)$$

with

$$
\begin{cases}
\delta_0(x) = \cosh \gamma_1 \cos\left[\gamma_1 \left(\frac{2x}{l} - 1\right)\right] - \cos \gamma_1 \cosh\left[\gamma_1 \left(\frac{2x}{l} - 1\right)\right], \\[2mm]
\delta_1(x) = \frac{l}{4}\left[(2x - l)\,\frac{d^2\delta_0}{dx^2} + C_1\,\frac{d\delta_0}{dx} + l^2 C_2\,\frac{d^3\delta_0}{dx^3}\right], \\[2mm]
\omega_1 = \left(\frac{2\gamma_1}{l}\right)^2 \sqrt{\frac{EI}{m_f + m_t}}, \\[2mm]
C_1 = -2\,(1 + \gamma_1 \tanh \gamma_1), \\[2mm]
C_2 = -\frac{1}{2}\,\frac{1}{\gamma_1}\,\coth \gamma_1,
\end{cases}
\qquad (3.64)
$$

and $\gamma_1 = 2.356\ldots$ is the smallest positive solution of the transcendent equation $\tan \gamma = -\tanh \gamma$.

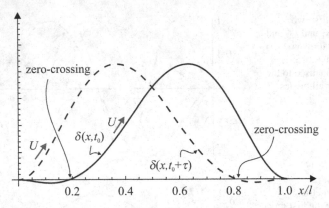

Fig. 3.10 Pipe deformation at t_0 of the zero crossing at section $x/l = 0.2$ and at $t_0 + \tau$ of the zero crossing at section $x/l = 0.8$

The deformation is not symmetrical, and a phase shift occurs between two sections, depending on the mass flow rate, as well as all the other characteristics of the system. The measurement of the time delay of the deformed axis between two sections (usually symmetrical with respect to the centre line) allows the estimation of the mass flow rate. The reference time can be the zero-crossing time. In addition, if the fluid is at rest, the frequency of oscillation depends on the density of the fluid, and the device can be used as a densimeter.

In Fig. 3.10, the axis of the pipe crossing the zero at section $x/l = 0.2$ at time t_0 and crossing the zero at section $x/l = 0.8$ at time $(t_0 + \tau)$ is shown. For each oscillation cycle of the tube, two zero crossings occur for each section, and it is possible to take two readings of τ. The time delay between two symmetrical sections x and $(l - x)$ is

$$\tau(x) = \frac{Q_m}{16\,\gamma_1{}^4} \frac{l^3}{8EI} \frac{\delta_1(x)}{\delta_0(x)}. \tag{3.65}$$

The measurement of τ is averaged over numerous readings. The frequency of the oscillations is generally very high: for a tube of length $l = 0.25$ m, inner diameter $d_i = 20$ mm, thickness $s = 1$ mm made of aluminium ($E_{Al} = 70$ GPa, $\rho_{Al} = 2700$ kg m^{-3}), containing flowing water at $20\,°C$ ($\rho_f = 998.21$ kg m^{-3}), the frequency of the first oscillating mode is 1300 Hz.

In summary, this is a problem that we can solve analytically but with a relatively high level of complexity.

We again analyse the same problem using the criteria of dimensional analysis (see Raszillier and Raszillier 1991, for details).

The equations that describe the physical process can be expressed as

$$\tau = f\big(l,\ U,\ m_f,\ m_g,\ EI,\ x,\ n\big), \tag{3.66}$$

where n is the oscillation mode and the mass flow rate $Q_m = m_f U$ is implicitly included. Young's modulus and moment of inertia appear coupled, as happens in many similar problems. It is also advantageous to sum up the linear mass of the tube and the linear mass of the fluid $m_g = (m_f + m_t)$.

The dimensional matrix

$$
\begin{array}{c|ccccccccc}
 & \tau & l & U & m_f & m_g & EI & x & n \\
\hline
M & 0 & 0 & 0 & 1 & 1 & 1 & 0 & 0 \\
L & 0 & 1 & 1 & -1 & -1 & 3 & 1 & 0 \\
T & 1 & 0 & -1 & 0 & 0 & -2 & 0 & 0
\end{array}
\tag{3.67}
$$

has rank 3. The minor corresponding to the columns l, m_g and (EI) is nonzero and can be selected as a new basis:

$$
\begin{array}{c|ccccc}
 & \tau & U & m_f & x & n \\
\hline
l & 2 & -1 & 0 & 1 & 0 \\
m_g & 1/2 & -1/2 & 1 & 0 & 0 \\
EI & -1/2 & 1/2 & 0 & 0 & 0
\end{array}'
\tag{3.68}
$$

computing the following dimensionless groups:

$$
\Pi_1 = \frac{\tau}{l^2}\sqrt{\frac{EI}{m_g}}, \quad \Pi_2 = U l \sqrt{\frac{m_g}{EI}},
$$

$$
\Pi_3 = \frac{m_f}{m_g}, \quad \Pi_4 = \frac{x}{l},
\tag{3.69}
$$

where n is already dimensionless.

The functional relationship (3.66) can be rewritten as

$$
\Pi_1 = \tilde{f}(\Pi_2, \Pi_3, \Pi_4, n).
\tag{3.70}
$$

Some useful information on the structure of the function $\tilde{f}$ comes from the symmetry properties of the physical process.

The starting equations (3.60) and the boundary conditions (3.61) are invariant for a reflection of the space and of fluid velocity:

$$
\begin{cases}
t \to t, \\
x \to l - x, \\
U \to -U,
\end{cases}
\tag{3.71}
$$

and the solution satisfies the condition

$$
\delta(x, U) = \delta(l - x, -U).
\tag{3.72}
$$

The time delay τ is usually measured between two symmetrical sections with respect to the centreline. If the origin of the time t_0 is the zero-crossing in $x_s < l/2$, then by definition $\delta[x_s, t_0(x_s, U), U] = 0$ and, for symmetry, we also have $\delta[l - x_s, t_0(l - x_s, -U), -U] = 0$.

The time delay is equal to

$$\tau(x_s, U) = t_0(x_s, U) - t_0(l - x_s, U). \tag{3.73}$$

If the flow is reversed,

$$\tau(x_s, -U) = t_0(l - x_s, -U) - t_0(x_s, -U). \tag{3.74}$$

This is because the delay has the same absolute value for direct or reverse flow (i.e., for U and for $-U$), but while with the direct flow ($U > 0$), the zero-crossing in $(l - x_s)$ is delayed with respect to x_s, with the reversed flow ($U < 0$), the opposite is true, and the time delay changes sign. Therefore,

$$\tau(x_s, U) = -\tau(l - x_s, -U). \tag{3.75}$$

To meet these conditions, the group Π_1 must already appear combined with an odd power of Π_2 (to make the power function antimetric with respect to U), and the function must depend on Π_2^{2s-2} with s a positive integer. Additionally, since for $U = 0$ the delay τ is zero, we expect τ to be a monotonically increasing function of U. Thus,

$$\Pi_1 \Pi_2^{(1-2r)} = \Phi\left(\Pi_2^{2s-2}, \Pi_3, \Pi_4, n\right), \tag{3.76}$$

where r is a positive integer. For the first mode ($n = 1$), with the minimum values $r = 1$ and $s = 1$,

$$\tau = U \, l^3 \frac{m_f + m_t}{EI} \Phi_1\left(\frac{m_f}{m_f + m_t}, \frac{x}{l}\right), \tag{3.77}$$

where x indicates the measurement section of the zero crossing and $\tau(x) = t_0(l - x) - t_0(x)$.

The delay τ must be symmetric for $x \to (l - x)$ and odd; this requires that it is an odd function of the argument $(2x/l - 1)$.

The comparison with theory in Eq. (3.65) indicates that, in fact,

$$\tau = U \, l^3 \frac{m_f + m_t}{EI} \frac{m_f}{m_f + m_t} \Phi_1\left(\frac{2x - l}{l}\right) \equiv U \, l^3 \frac{m_f}{EI} \Phi_1\left(\frac{2x - l}{l}\right). \tag{3.78}$$

A similar analysis for the pulsation of the self-oscillation, ω, brings us to the typical equation

$$\omega = f\left(l, U, m_f, m_f + m_t, EI, n\right). \tag{3.79}$$

Here, ω is a global property of the system and does not depend on a specific abscissa section x.

Applying the criteria of dimensional analysis, the following dimensionless groups are calculated:

$$\Pi_1 = \omega\, l^2 \sqrt{\frac{m_f + m_t}{E I}}, \quad \Pi_2 = U\, l \sqrt{\frac{m_f + m_t}{E I}},$$

$$\Pi_3 = \frac{m_f}{m_f + m_t}, \quad \Pi_4 = n. \tag{3.80}$$

Since ω must be an even function of U, the functional relationship must involve square terms in U of the form Π_2^{2s-2} with s a positive integer:

$$\omega = \frac{1}{l^2} \sqrt{\frac{E I}{m_f + m_t}}\, \Phi_1 \left(U^2\, l^2\, \frac{m_f + m_t}{E I}, \ \frac{m_f + m_t}{m_f} \right), \quad \therefore \tag{3.81}$$

where Φ_1 is the Φ function calculated for $n = 1$. A comparison with theory (3.64) indicates that Φ_1 is constant and equal to $4\gamma_1^2$.

Coriolis mass flow meters are calibrated, but at the end of the present analysis, it is evident that the calibration procedure can be designed on the basis of dimensional analysis only, without knowledge of the exact solution of the problem.

3.3   Group Theory and Affine Transformations for Self-similar Solutions

The procedure to reduce the number of arguments of a functional relationship that describes a physical process has a much more general mathematical interpretation, framed within group theory and, in particular, groups of Lie and affine transformations (Bluman and Kumei 1989). We have seen in Sect. 1.4.4 that Buckingham's Theorem is a consequence of the general covariance principle in physics, and we have also defined the transformation that brings us to a functional relationship where only dimensionless groups are allowed as a consequence of the invariance principle with respect to the units of measurements in systems belonging to the same class.

If we can find a transformation leaving invariant a differential or algebraic problem, we can expect a reduction in the variables depending on the number of parameters of the transformation, allowing similarity solutions to be obtained, possibly also self-similar. Hence, invariance with respect to units of measurements is the first step, and looking for invariance with respect to a supplementary group of transformations is a further step toward the simplification and possibly the solution of the problems.

Before providing details on the procedure to recover self-similar solutions, we briefly define an *affine transformation* between two Euclid spaces, symbolically

$$f : \mathbb{R}^n \to \mathbb{R}^m, \tag{3.82}$$

as a linear transformation of the type

$$\mathbf{x} \mapsto \mathbf{A} \cdot \mathbf{x} + \mathbf{b}, \tag{3.83}$$

where $\mathbf{A}$ is a matrix $(m \times n)$ and $\mathbf{b}$ is a vector of $\mathbb{R}^m$; the transformation allows linear mapping of a vector $\mathbf{x}$ of $\mathbb{R}^n$ to a new vector of $\mathbb{R}^m$ through the matrix $\mathbf{A}$, plus a translation (eventually null) represented by the vector $\mathbf{b}$. An affine transformation preserves lines and parallelism, but not necessarily distances and angles. Examples of affine transformations are translation, scaling, homothety, similarity, reflection, rotation, and shear mapping. Any composition of affine transformations is still an affine transformation.

In the following example, we detail the procedure for detecting self-similarity in a classical diffusion problem.

Example 3.6 Let us consider the process of heat diffusion in a linear bar with a localised (point) source in the origin, described by the following differential problem:

$$\begin{cases} \rho \, c_p \dfrac{\partial u}{\partial t} - k \dfrac{\partial^2 u}{\partial x^2} = 0, & -\infty < x < \infty, \ t > 0, \\[2mm] u(x, 0) = \dfrac{Q}{\rho \, c_p \, S} \delta(x), \\[2mm] \lim\limits_{x \to \pm\infty} u(x, \ t) = 0, \end{cases} \tag{3.84}$$

where ρ is the density, c_p is the specific heat, S is the cross section of the bar, u is the temperature, k is the thermal conductivity, Q is the intensity of the source, and $\delta(x)$ is the Dirac function, having a dimension equal to the inverse of the dimension of its argument. We assume that $u = \Theta(x, t)$ is a solution of this problem. We consider the following transformation:

$$\begin{cases} x^* = \alpha \, x, \\ t^* = \beta \, t, \\ u^* = \gamma \, u, \end{cases} \tag{3.85}$$

where α, β and γ are positive constants. The transformation leaves the differential problem invariant for any solution $u = \Theta(x, t)$ if $v = \gamma \, \Theta(x, t)$ is a solution of

$$\begin{cases} \rho \, c_p \dfrac{\partial v}{\partial t^*} - k \dfrac{\partial^2 v}{\partial x^{*2}} = 0, & -\infty < x^* < \infty, \ t > 0, \\[2mm] v(x^*, 0) = \dfrac{Q}{\rho \, c_p \, S} \delta(x^*), \\[2mm] \lim\limits_{x^* \to \pm\infty} v(x^*, \ t^*) = 0. \end{cases} \tag{3.86}$$

If the scaling leaves the differential problem (3.84) invariant and $u = \Theta(x, t)$ is one of its solutions, then $u = \gamma \, \Theta(x/\alpha, t/\beta)$ is also a solution.

To evaluate the coefficients, we consider the invariance of each of the three equations describing the problem. Here, $u = \Theta(x, t)$ solves the first equation in (3.84) only if $v = \gamma \Theta(x, t)$ solves the first equation in (3.86); hence, $\gamma \beta^{-1} = \gamma \alpha^{-2}$. In a similar way, $\gamma = \alpha^{-1}$ (these results derive from the definition of the Dirac function, with $\delta(\alpha x) = \delta(x)/|\alpha|$.

The solution is $\beta = \alpha^2$, $\gamma = \alpha^{-1}$, and the required affine transformation is

$$\begin{cases} x^* = \alpha x, \\ t^* = \alpha^2 t, \\ u^* = \dfrac{1}{\alpha} u, \end{cases} \qquad (3.87)$$

which is a one-parameter Lie group of scalings.

The two systems (3.84)–(3.86) share the same solutions; hence,

$$\Theta(x^*, t^*) = \frac{1}{\alpha}\Theta(x, t) \rightarrow \Theta(\alpha x, \alpha^2 t) = \frac{1}{\alpha}\Theta(x, t), \qquad (3.88)$$

which is defined as a similarity or invariant solution. In particular, since the scalings belong to a one-parameter Lie group, the solution is called self-similar. A self-similar solution is sought as

$$\Theta(x, t) = t^r U(\xi), \quad \xi = x t^s. \qquad (3.89)$$

Based on the results of the affine transformations, the self-similar solution requires that:

$$\Theta(x, t) \equiv \alpha \Theta(\alpha x, \alpha^2 t) \rightarrow t^r U(x t^s) \equiv \alpha(\alpha^2 t)^r U(\alpha x(\alpha^2 t)^s) \rightarrow$$
$$t^r U(x t^s) \equiv \alpha^{2r+1} t^r U(\alpha^{2s+1} x t^s), \qquad (3.90)$$

which is satisfied for any value of α if $s = r = -1/2$.

Thus, the self-similar solution has the expression

$$\Theta(x, t) = \frac{U(x/\sqrt{t})}{\sqrt{t}} \equiv \frac{U(\xi)}{\sqrt{t}}, \quad \xi = \frac{x}{\sqrt{t}}. \qquad (3.91)$$

More straightforwardly, it turns out that since α is arbitrary and positive, setting $\alpha = 1/\sqrt{t}$, we obtain

$$\Theta(x, t) \equiv \alpha \Theta(\alpha x, \alpha^2 t) \rightarrow$$
$$\Theta(x, t) = \frac{\Theta(x/\sqrt{t}, 1)}{\sqrt{t}} \rightarrow \Theta(x, t) = \frac{U(x/\sqrt{t})}{\sqrt{t}}. \qquad (3.92)$$

Inserting (3.91) into the system (3.84) yields

$$\begin{cases} 2\kappa\, U'' + \xi\, U' + U = 0, \\[2mm] U(\xi) = \dfrac{Q}{\rho\, c_p\, S}\,\delta(\xi), \\[2mm] \lim_{\xi \to \pm\infty} U(\xi) = 0, \end{cases} \qquad (3.93)$$

(where $\kappa = k/(\rho\, c_p)$ is the thermal diffusivity), which is an ordinary differential equation (ODE) problem in the dependent variable $U(\xi)$ and in the single independent variable $\xi = x/\sqrt{t}$. The prime and the double prime indicate the first and the second derivatives, respectively, with respect to the argument ξ. To check the dimensional homogeneity, we recall that $[U] = \Theta\, T^{1/2}$, $[\xi] = L\, T^{-1/2}$, $[\delta(\xi)] \equiv [\xi]^{-1} = L^{-1}\, T^{1/2}$.

The differential problem (3.93) admits the following analytical solution:

$$\Theta(x,\, t) = \frac{Q}{\rho\, c_p\, S\, \sqrt{4\pi\,\kappa\, t}}\,\exp\left(-\frac{x^2}{4\,\kappa\, t}\right). \qquad \therefore \quad (3.94)$$

In summary, the affine transformation has allowed a reduction of the number of variables that explicitly appear in the physical process, grouping them in a smaller number of variables; in particular, it has grouped the independent variables, allowing us to transform a partial derivative equation into a total derivative equation.

3.3.1 The Non-dimensionalisation of Algebraic Equations and Differential Problems

Expressing equations and differential problems in dimensionless form can be convenient to identify the order of magnitude of terms, allowing us to ignore some of these terms in certain conditions. The criteria of the dimensional analysis can also be used for this operation.

We analyse the procedure applied to the example of heat diffusion (see Example 3.6). We are dealing with a physical process where the following variables appear:

$$\Theta = f\left(x,\, t,\, \rho,\, c_p,\, k,\, Q'\right), \qquad (3.95)$$

where $Q' = Q/(S\, l_0)$ are grouped since the problem involves a bar with constant cross section S with a reference volume $S\, l_0$, and we expect these terms to appear always grouped. The dimensional matrix

	θ	x	t	ρ	c_p	k	Q'
M	0	0	0	1	0	1	1
L	0	1	0	-3	2	1	-1
T	0	0	1	0	-2	-3	-2
Θ	1	0	0	0	-1	-1	0

$$(3.96)$$

has rank 4. We can find $(7 - 4) = 3$ dimensionless groups sufficient to fully describe the differential problem.

By selecting the fundamental quantities ρ, c_p, k and Q' (they are independent), the following dimensionless groups are calculated:

$$\Pi_1 = \frac{\rho\, c_p}{Q'}\,\theta, \quad \Pi_2 = \frac{c_p\,\sqrt{\rho\, Q'}}{k}\,x, \quad \Pi_3 = \frac{Q'\, c_p}{k}\,t, \qquad (3.97)$$

which can be more conveniently redefined as the *dimensionless temperature, dimensionless abscissa* and *dimensionless time*:

$$\tilde{\theta} = \frac{\rho\, c_p}{Q'}\,\theta, \quad \tilde{x} = \frac{c_p\,\sqrt{\rho\, Q'}}{k}\,x, \quad \tilde{t} = \frac{Q'\, c_p}{k}\,t. \qquad (3.98)$$

The differential problem can be rewritten as a function of the new dimensionless variables:

$$\begin{cases} \dfrac{\partial \tilde{\theta}}{\partial \tilde{t}} - \dfrac{\partial^2 \tilde{\theta}}{\partial \tilde{x}^2} = 0, & -\infty < \tilde{x} < \infty,\ \tilde{t} > 0, \\[2mm] \tilde{\theta}(\tilde{x},\,0) = \tilde{\delta}(\tilde{x}), \\[2mm] \lim_{\tilde{x} \to \pm\infty} \tilde{\theta}(\tilde{x}) = 0. \end{cases} \qquad (3.99)$$

By performing the same analysis as for the differential problem with dimensional variables, a self-similar solution with the following expression can be found:

$$\tilde{\theta}\left(\tilde{x},\,\tilde{t}\right) = \frac{\tilde{U}\left(\tilde{x}/\sqrt{\tilde{t}}\right)}{\sqrt{\tilde{t}}}. \qquad (3.100)$$

As an alternative, we can avoid fixing the scales, and we can write Eq. (3.84) as

$$\begin{cases} \dfrac{\rho\, c_p\, l_0^2}{k\, t_0}\,\dfrac{\partial \tilde{\theta}}{\partial \tilde{t}} - \dfrac{\partial^2 \tilde{\theta}}{\partial \tilde{x}^2} = 0, & -\infty < \tilde{x} < \infty,\ \tilde{t} > 0, \\[2mm] \tilde{\theta}(\tilde{x},0) = \dfrac{Q}{\rho\, c_p\, S\, \theta_0\, l_0}\,\tilde{\delta}(\tilde{x}), \\[2mm] \lim_{\tilde{x} \to \pm\infty} \tilde{\theta}(\tilde{x},\,\tilde{t}) = 0, \end{cases} \qquad (3.101)$$

where the two dimensionless groups $\rho\, c_p\, l_0^2 / (k\, t_0)$ and $Q/(\rho\, c_p\, S\, \theta_0\, l_0)$ are defined in terms of the three scales l_0, t_0 and θ_0 for length, time, and temperature, respectively. In the presence of scales different from those listed in (3.98), the dimensionless groups are not unitary, which means a different ratio between the different contributions in the differential problem. In fact, in some cases, the scales naturally arise; in other cases, there are several possible combinations amongst scales possibly with different relevance of the dimensionless groups and with different approximate solutions.

3.3.2 Methods of Identification of Self-similar Variables for Complete (First-Kind) Similarity

We recall the statement that

"Self-similar solutions of the first type, whose existence is related only to the conservation laws and to the invariance of the degenerate problem with respect to the group of similarity transformations of quantities with independent dimensions are well known" (Barenblatt and Zel'Dovich 1972).

More generally, self-similar solutions are typical of degenerate problems, where degeneracy indicates that all parameters in the initial and boundary conditions and with the dimensions of the independent variables vanish or become infinite.

The search for the structure of the self-similar variables for a first-kind similarity problem must be appropriately preceded by an analysis of the differential problem to determine a priori whether we can expect a self-similar solution.

Example 3.7 Let us consider a gravity current of a non-Newtonian fluid propagating in a fracture of width L_y, see Fig. 3.11, according to the scheme reported in Di Federico et al. (2017). The fluid rheology is described by a Herschel–Bulkley model that is also suitable for describing a power-law fluid, including a Newtonian fluid:

$$\begin{cases} \tau = \left(\mu_0 |\dot{\gamma}|^{n-1} + \tau_p |\dot{\gamma}|^{-1}\right) \dot{\gamma}, & \tau \geq \tau_p, \\ \dot{\gamma} = 0, & \tau < \tau_p, \end{cases} \tag{3.102}$$

given in terms of the shear stress τ and of the shear rate $\dot{\gamma}$. We adopt the one-dimensional formulation, neglecting the tensorial formulation. The parameter μ_0, the consistency index, represents a viscosity-like parameter, whilst τ_p is the yield stress of the fluid, and n, the fluid behaviour index, controls the extent of shear-thinning ($n < 1$) or shear-thickening ($n > 1$); $n = 1$ corresponds to the Bingham case. The mass conservation equation of the fluid layer may be written as

$$\frac{\partial h}{\partial t} = -\frac{\partial}{\partial x}(\bar{u} h), \tag{3.103}$$

where $\bar{u}(x, t)$ is the gap-averaged velocity, computed by double integration of Eq. (3.102) imposing symmetry and no-slip conditions at the wall. Introducing the

Fig. 3.11 Gravity current advancing in a fracture of constant width L_y (modified from Di Federico et al. 2017)

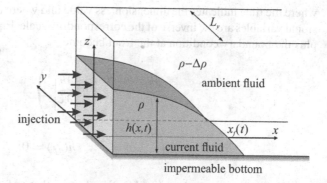

gap-averaged velocity computed in viscous-buoyancy balance means that we are neglecting inertial terms. As a first approach, we neglect yield stress and consider an Ostwald-de Waele fluid (power-law). An evolution equation for $h(x, t)$ alone is obtained, namely,

$$\frac{\partial h}{\partial t} = \left(\frac{L_y}{2}\right)^{(n+1)/n} \text{sgn}\left(\frac{\partial h}{\partial x}\right) \left(\frac{n}{2n+1}\right) \left(\frac{\Delta\rho\, g}{\mu_0}\right)^{1/n} \frac{\partial}{\partial x}\left(h\left|\frac{\partial h}{\partial x}\right|^{1/n}\right). \quad (3.104)$$

We can simplify the writing of the equation by introducing a velocity scale Ω:

$$\Omega = \left(\frac{L_y}{2}\right)^{(n+1)/n} \left(\frac{n}{2n+1}\right) \left(\frac{\Delta\rho\, g}{\mu_0}\right)^{1/n}, \quad (3.105)$$

and Eq. (3.104) becomes

$$\frac{\partial h}{\partial t} = \text{sgn}\left(\frac{\partial h}{\partial x}\right) \Omega \frac{\partial}{\partial x}\left(h\left|\frac{\partial h}{\partial x}\right|^{\frac{1}{n}}\right). \quad (3.106)$$

Suppose that the volume of fluid of the gravity current varies as

$$L_y \int_0^\infty h(x, t)\mathrm{d}x = Qt^\alpha, \quad (3.107)$$

where α, $Q \geq 0$. Then, the dimensions of the two parameters are $[\Omega] = L/T$, $[Q/L_y] = L^2/T^\alpha$. We also have a boundary condition corresponding to a null depth of the current at the front x_f; that is, $h(x_f) = 0$. It is convenient to express the dependent variable h and the independent variables x and t in nondimensional form as

$$\tilde{h} = h\left(\frac{Q}{L_y\Omega^\alpha}\right)^{1/(\alpha-2)}, \quad \tilde{x} = x\left(\frac{Q}{L_y\Omega^\alpha}\right)^{1/(\alpha-2)}, \quad \tilde{t} = t\left(\frac{Q}{L_y\Omega^2}\right)^{1/(\alpha-2)}, \quad (3.108)$$

where the tilde indicates the dimensionless value and where the coefficient of dimensional variables are the inverse of the corresponding scale. Equations (3.106)–(3.107) plus the boundary condition at the front become

$$\frac{\partial \tilde{h}}{\partial \tilde{t}} = -\frac{\partial}{\partial \tilde{x}} \left(\tilde{h} \left| \frac{\partial \tilde{h}}{\partial \tilde{x}} \right|^{\frac{1}{n}} \right), \tag{3.109}$$

$$\int_0^\infty \tilde{h}\, d\tilde{x} = \tilde{t}^\alpha, \quad \tilde{h}(\tilde{x}_f) = 0. \tag{3.110}$$

In the following, we neglect the tilde. We observe that so far with the adimensionalization process we still retain the initial number of variables.

Let us now see if we can identify a group of transformations for which the differential problem is invariant (see Sect. 3.3). We assume that the possible group is $x' = ax$, $t' = bt$, $h' = ch$, where the prime indicates the variable in the new system and a, b, c are real coefficients. Substituting into Eqs. (3.109)–(3.110) yields

$$\frac{\partial h'}{\partial t'} = -\frac{\partial}{\partial x'} \left(h' \left| \frac{\partial h'}{\partial x'} \right|^{\frac{1}{n}} \right) \rightarrow \frac{c}{b} \frac{\partial h}{\partial t} = -\frac{c^{1+1/n}}{a^{1+1/n}} \frac{\partial}{\partial x} \left(h \left| \frac{\partial h}{\partial x} \right|^{\frac{1}{n}} \right), \tag{3.111}$$

$$\int_0^\infty h'\, dx' = t'^\alpha, \rightarrow ac \int_0^\infty h\, dx = b^\alpha t^\alpha. \tag{3.112}$$

For the invariance, it is required that

$$\begin{cases} \dfrac{c}{b} = \dfrac{c^{1+1/n}}{a^{1+1/n}}, \\ ac = b^\alpha, \end{cases} \tag{3.113}$$

which admits the solution

$$b = a^{F_2}, \quad c = a^{F_1}, \quad \text{with} \quad F_1 = \frac{(n+1)\alpha - n}{n+\alpha}, \quad F_2 = \frac{n+1-F_1}{n}. \tag{3.114}$$

In summary, the differential equation (3.109), the integral constraint and the front condition in (3.110) are invariant under the following single-parameter group of transformations of space-time and depth of the current:

$$x' = ax, \quad t' = a^{F_2}t, \quad h' = a^{F_1}h. \tag{3.115}$$

We wonder if the transformation provides an advantage. At the present stage, it does not, but we can proceed and check if it is possible to reduce the number of variables by assuming that one of the three is fundamental, for example, time t. To this end, we look for the exponents of time that permit us to express the space x and the depth of the current h as

$$\frac{x}{t^r}, \quad \frac{h}{t^s}. \tag{3.116}$$

In dimensional analysis, the exponents must satisfy the principle of dimensional homogeneity. In the present case, the variables are already dimensionless, and the exponents must ensure that the two groups are still invariant in the transformation (3.115). The two groups are invariant if

$$\frac{x'}{t'^r} = \frac{ax}{a^{rF_2}t^r} \rightarrow r = 1/F_2 \equiv \frac{n+\alpha}{n+2}, \tag{3.117}$$

and

$$\frac{h'}{t'^s} = \frac{a^{F_1}h}{a^{sF_2}t^s} \rightarrow s = F_1/F_2 \equiv \frac{n(\alpha-1)+\alpha}{n+2}. \tag{3.118}$$

In a simpler way, we can consider time as an independent variable; from Eq. (3.115), we obtain the parameter $a \sim t^{1/F_2}$, and substituting into the first and the last term in Eq. (3.115) yields $x \sim t^{1/F_2}$ and $h \sim t^{F_1/F_2}$.

We now have a physical process whose mathematical counterpart is expressed as a function of two dimensionless groups, both invariant in the same group (3.115) wherein the differential problem is invariant. The typical equation is

$$f\left(\frac{h}{t^s}, \frac{x}{t^r}\right) = 0, \tag{3.119}$$

and for progress, we assume that it has the following structure:

$$\frac{h}{t^s} = f\left(\frac{x}{t^r}\right) \rightarrow h = t^s f(\eta), \quad \eta = x/t^r. \tag{3.120}$$

At this step, to avoid a variable boundary of the domain of integration, it is convenient to map the time-varying domain occupied by the current, $\eta \in [0, \eta_f]$ divided by η_f, which represents the front position in the self-similar variable. Hence,

$$h = \eta_f^{n+2} t^s f(\zeta), \quad \zeta = \eta/\eta_f, \quad \eta = x/t^r, \quad \eta_f = \left(\int_0^1 f \, d\zeta\right)^{-1/(n+2)}. \tag{3.121}$$

Substituting into Eq. (3.109) yields the following nonlinear ordinary differential equation:

$$\left(f \, |f'|^{1/n}\right)' + sf - rnf' = 0, \quad f(1) = 0, \quad \zeta \in [0, 1], \tag{3.122}$$

which for $\alpha = 0, 2$ admits an analytical solution with f represented by a parabola and a straight line, respectively. For different values of α, numerical integration is

(a)

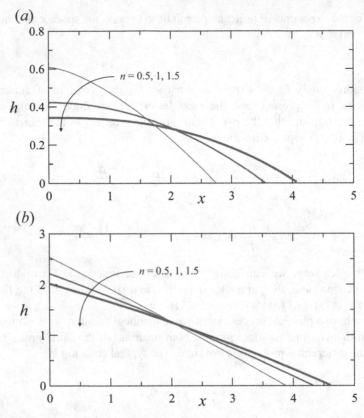

(b)

Fig. 3.12 Profiles of the gravity current at $t = 5$ for power-law fluids with different n, with (a) $\alpha = 0$ (a dam break, producing a parabolic shape); (b) $\alpha = 1$, constant inflow rate. Variables are nondimensional (modified from Di Federico et al. 2017)

required, achieved by expanding f for $\zeta \to 1$ to avoid singularities and to retrieve the value of the first derivative. By assuming that $f \approx a_0(1 - \zeta)^b$, substituting in (3.122) and balancing the lower-order terms, we obtain $b = 1$ and $a_0 = s^n$. Hence, it follows that

$$f|_{\zeta \to 1-\varepsilon} = s^n\varepsilon, \quad f'|_{\zeta \to 1-\varepsilon} = -s^n, \tag{3.123}$$

with ε a small quantity. This last aspect is particularly instructive because it indicates that the behaviour of the function describing the current profile is controlled by the conditions at the front, while what happens at the origin is irrelevant. This has been verified experimentally: the method of fluid injection does not affect the current profile at a suitable distance from the origin. Figure 3.12 shows the computed profiles for $\alpha = 0, 1$, corresponding to a dam break and a constant inflow rate, respectively.

We highlight that selecting a fundamental quantity different from time t brings us to a nonlinear ODE problem with the same solution.

This example shows the procedure: whenever a transformation group for a differential problem is obtained, the variables can be collected in power function combinations that are invariant in the same group of transformations, with a reduction equal to the number of parameters of the group transformation. We wonder if a differential problem such as (3.109)–(3.110), in the presence of a group of transformations wherein it is invariant, as with (3.115), always has self-similar solutions such as (3.121). The answer is yes, at least locally (see Birkhoff 1950).

The second important question is the range of validity of the self-similar solution. Following Barenblatt, we can assume that the solution is valid neither too early, when the initial and boundary conditions still exert their effects, nor too late, when the solution may become unstable due to perturbations. A detailed quantification is given for some gravity currents in Ball et al. (2017), Ball and Huppert (2019).

A third, even more important, question is as follows: is the self-similar solution realistic?

This last question is often addressed by conducting mathematical stability analyses of the solution, but the experimental validation becomes the true "dominus" of this analysis and, in general, of mathematical physics. Mathematical solutions, although elegant and satisfying all the criteria to meet stability, may not represent the physical reality that, in the final analysis, provides the true *resolution* of physical problems.

For the problem treated in (3.7), the answer is positive, and the self-similar solution has been verified by experiments within the limits of the experimental uncertainties. As an example, Fig. 3.13 shows the time evolution of a mound of power-law fluid with free drainage on the left in a V-shaped Hele-Shaw cell mimicking a porous medium with varying permeability and porosity, allowing a self-similar solution (Longo et al. 2015). The initial discrepancy progressively disappears and then appears again in the late stage of the process. See also the contribution by Di Federico et al. (2017) and references therein for the comparison between self-similar solutions and experiments.

Example 3.8 Let us extend the previous analysis to a Herschel–Bulkley fluid. By inserting the gap-averaged velocity into the mass conservation equation, we obtain

$$\frac{\partial h}{\partial t} = \left(\frac{L_y}{2}\right)^{(n+1)/n} \text{sgn}\left(\frac{\partial h}{\partial x}\right)\left(\frac{n}{2n+1}\right)\left(\frac{\Delta \rho \, g}{\mu_0}\right)^{1/n} \times$$

$$\frac{\partial}{\partial x}\left[h\left|\frac{\partial h}{\partial x}\right|^{1/n}\left(1 - \kappa\left|\frac{\partial h}{\partial x}\right|^{-1}\right)^{(n+1)/n}\left(1 + \left(\frac{n}{n+1}\right)\kappa\left|\frac{\partial h}{\partial x}\right|^{-1}\right)\right], \quad (3.124)$$

where $\kappa = 2\tau_p/(\Delta \rho \, g L_y)$ is a nondimensional number representing the ratio between yield stress and gravity-related stress. In dimensionless form, Eq. (3.124) can be written as

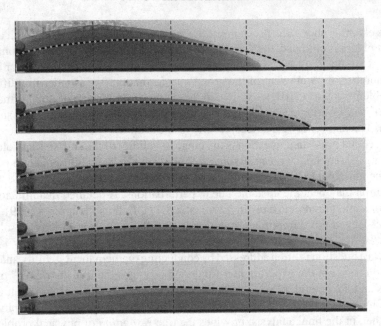

Fig. 3.13 Dipole of a power-law fluid. Photographs showing the current profiles at various times since release (10, 30, 50, 70 and 90 s) in a V-shaped Hele-Shaw cell. The curves with symbols represent the theoretical profile. The vertical dashed lines are 10 cm apart (modified from Longo et al. 2015)

$$
\frac{\partial h}{\partial t} = \mathrm{sgn}\left(\frac{\partial h}{\partial x}\right) \frac{\partial}{\partial x}\left[h\left|\frac{\partial h}{\partial x}\right|^{\frac{1}{n}} \left(1 - \kappa\left|\frac{\partial h}{\partial x}\right|^{-1}\right)^{\frac{n+1}{n}} \times \right.
$$
$$
\left. \left(1 + \left(\frac{n}{n+1}\right)\kappa\left|\frac{\partial h}{\partial x}\right|^{-1}\right)\right]. \qquad (3.125)
$$

The integral constraint (3.110) with the front condition of a null depth still hold.

The differential problem represented by Eqs. (3.125) and (3.110) is invariant under the group of transformations

$$
x' = ax, \quad t' = a^{2/\alpha}t, \quad h' = ah. \qquad (3.126)
$$

According to the Π-Theorem, selecting a single variable as independent, we can collect the three variables into two groups, for example

$$
\frac{x'}{t'} = \frac{ax}{a^{2/\alpha}t}, \quad \frac{h'}{t'} = \frac{ah}{a^{2/\alpha}t}, \qquad (3.127)
$$

which are invariant only if $\alpha = 2$. Hence, the solution is of the form

$$\frac{h}{t} = f\left(\frac{x}{t}\right) \rightarrow h = tf(\eta), \ \eta = x/t. \tag{3.128}$$

This result indicates that the differential problem (3.125)–(3.110) admits a self-similar solution only for the special case of volume changing $\propto t^2$ or equivalently of an inflow rate increasing linearly with time. We notice that for $\alpha = 2$ the scales expressed in Eq. (3.108) break down, but this problem can be easily circumvented by introducing new combinations of scales. At this point, we know the structure of the self-similar solution given by (3.128), and substituting into (3.125)–(3.110), we obtain a nonlinear ordinary differential problem for the function $f(\eta)$ in the independent variable η, which admits an analytical solution with a linear profile for f. Before integration, it is convenient to map the domain of the self-similar independent variable to $[0, 1]$, as shown for the case of a gravity current of a power-law fluid.

As a matter of evidence, the group of transformations arises from the structure of the equations of the differential problem. In particular, h and x must have the same time dependence otherwise in the following term in Eq. (3.125):

$$\left(1 - \kappa \left|\frac{\partial \tilde{h}}{\partial \tilde{x}}\right|^{-1}\right) \tag{3.129}$$

the partial derivative in space of h cannot appear with the unit, which is already invariant under any transformation.

We wonder whether, for $\alpha \neq 2$, the differential problem admits a self-similar solution. The answer is negative–it does not admit it–and this is evident because the presence of the yield stress introduces a new scale different from the scales already identified. The immediate question is why, for $\alpha = 2$, the presence of a new scale does not inhibit a self-similar solution.

When looking at the term in Eq. (3.129), it turns out that, in general, the transformation is

$$\kappa' \left|\frac{\partial h'}{\partial x'}\right|^{-1} = \kappa \left|\frac{\partial h\, a^{F_1}}{\partial a\, x}\right|^{-1} \rightarrow \kappa' = \kappa\, a^{F_1-1}, \tag{3.130}$$

and κ is invariant if $F_1 = 1 \rightarrow \alpha = 2$. Hence, the new scale due to the yield stress equals the other scale due to viscosity.

If we assume that κ is an additional variable, then the differential problem (3.125) and (3.110) plus the front condition of a null depth of the current is invariant under the group of transformations

$$x' = ax, \quad t' = a^{F_2}t, \quad h' = a^{F_1}h, \quad \kappa' = a^{1-F_1}\kappa. \tag{3.131}$$

In addition to the groups $x\,t^{-r}$ and $h\,t^{-s}$, with r and s given in (3.117)–(3.118), we have a new group $\kappa\,t^{-m}$, which is invariant if

$$\frac{\kappa'}{t'^m} = \frac{a^{1-F_1}\kappa}{a^{mF_2}t^m} \rightarrow m = \frac{(1-F_1)}{F_2} \equiv \frac{n(2-\alpha)}{2+n}. \tag{3.132}$$

At this point, we can seek a self-similar solution of the form

$$\frac{h}{t_s} = g\left(\frac{x}{t^r}, \frac{\kappa}{t^m}\right) \rightarrow h = t^s g(\eta, \zeta), \quad \eta = x/t^r, \quad \zeta = \kappa/t^m. \tag{3.133}$$

We do not have information about the structure of function g, but for small values of ζ, we can expand the function g about $\zeta = 0$ as:

$$g(\eta, \zeta) = g|_{(\eta,0)} + \left.\frac{\partial g}{\partial \zeta}\right|_{(\eta,0)} \zeta + O(\zeta^2), \tag{3.134}$$

and we can solve the differential problem with a perturbation method. This is the approach used in Di Federico et al. (2017), although it was achieved with a different reasoning.

Ultimately, the procedure of checking possible self-similar solutions of the first kind requires:

– the identification of the group of transformations that renders invariant the differential problem, including the initial and boundary conditions;
– the application of the Π-Theorem to recombine the variables into groups that are invariant with respect to that group of transformations.

Once the self-similar independent variable and the self-similar structure of the dependent variable have been identified, they are substituted into the original partial differential problem, obtaining an ordinary differential problem, usually a nonlinear problem.

3.3.3   The Derivation for Incomplete (Second-Kind) Similarity

Many other problems admit a self-similar solution, but sometimes the transformation group is not identifiable on the basis of the differential problem and on the criteria of dimensional analysis. Thermal diffusion processes with discontinuities in the diffusivity coefficient and filtration processes with discontinuities (Barenblatt and Sivashinskii 1969) belong to this category; other physical problems that often involve anomalous scale exponents are strong explosions and the collapse of particles or gases into a singular state; see Barenblatt et al. (1989) for a list of problems admitting self-similar solutions of the second kind.

The origin of second-kind self-similarities is the passage to the limit from the original problem to a degenerate problem admitting self-similar solutions: if the

Fig. 3.14 Schematic for a converging gravity current in a fracture of gap thickness $b = b_1 x^k$

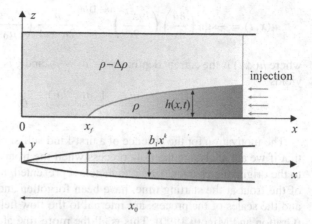

transition is not regular, dimensional considerations are not sufficient for defining all the scalings.

The type of scaling depends on the details of the underlying physics, and scaling generating first-kind and second-kind similarities can be observed in almost similar configurations. For example, the radius of a shock wave resulting from a strong explosion can be calculated from dimensional analysis, and the solution is self-similar of the first kind, while the implosion has anomalous scaling, and the solution is self-similar of the second kind. It also happens that some problems are amenable to a formal solution according to the criteria of a first-kind self-similarity, but the solution does not respect some constraints or comes into contradiction with the statements.

The solution of differential problems admitting a second-kind self similarity often results in a nonlinear eigenvalue problem with a continuous or a discrete set of eigenvalues and, in favourable cases, a single eigenvalue.

To clarify some aspects of the procedure, we detail the analysis for a converging gravity current, which brings us to a second-kind self-similar solution with a single eigenvalue (see Longo et al. 2021).

Example 3.9 We consider a gravity current advancing in a converging fracture toward the origin, starting at x_0 at $t = 0$ and reaching the origin $x = 0$ at $t = t_c$; see Fig. 3.14. Fluid rheology is a power-law that, in one dimension, reads

$$\tau = \mu_0 |\dot{\gamma}|^{n-1} \dot{\gamma} \tag{3.135}$$

where the tangential stress is τ and the shear rate is $\dot{\gamma}$, where the consistency index μ_0 represents a viscosity-like parameter, while the fluid behaviour index n indicates a shear-thinning ($n < 1$), shear-thickening ($n > 1$), or Newtonian ($n = 1$) fluid. The fluid advances in a horizontal channel with a gap thickness varying as $b(x) = b_1 x^k$ ($[b_1] = L^{1-k}$), and under several hypotheses, the gap-averaged horizontal velocity is

$$u(x,t) = -\text{sgn}\left(\frac{\partial h}{\partial x}\right)\left(\frac{b_1 x^k}{2}\right)^{(n+1)/n}\frac{n}{2n+1}\left(\frac{\Delta\rho\, g}{\mu_0}\right)^{1/n}\left|\frac{\partial h}{\partial x}\right|^{1/n}, \qquad (3.136)$$

where $h(x,t)$ is the current depth, $\Delta\rho \equiv \rho_c - \rho_a$ and g is gravity. Mass conservation yields

$$\frac{\partial h}{\partial t} + \frac{1}{x^k}\frac{\partial(x^k h u)}{\partial x} = 0. \qquad (3.137)$$

The motivation for the absence of a first-kind self-similar solution lies in the fact that if we are analysing the flow process when the front of the current is very close to the origin, the characteristic parameters represented, for example, by the position of the front at the starting time, have been forgotten and are no longer significant, and the scales of the process are internal to the flow field; hence, they are variable (Gratton and Minotti 1990). This is all the more true after closure, in the levelling phase.

It is advantageous to select x and $t_r = t_c - t$ as length and time scales, respectively, with t_c being the touch-down time. Defining U and H the dimensionless velocity and depth of the current, respectively, where

$$u(x,t) = \frac{x}{t_r}U(x,t_r), \qquad (3.138a)$$

$$h(x,t) = \left(\frac{2}{b_1}\right)^{n+1}\left(\frac{2n+1}{n}\right)^n\left(\frac{\mu_0}{\Delta\rho\, g}\right)\frac{x^{(n+1)(1-k)}}{t_r|t_r|^{n-1}}H(x,t_r), \qquad (3.138b)$$

and substituting into Eqs. (3.136)–(3.137) yields

$$U\,|U|^{n-1} + (n+1)(1-k)H + x\frac{\partial H}{\partial x} = 0, \qquad (3.139a)$$

$$t_r\frac{\partial H}{\partial t_r} - nH - HU(n+2-nk) - x\frac{\partial HU}{\partial x} = 0, \qquad (3.139b)$$

which is a system of two partial differential equations in H and U. We aim to find a group of transformations that leave invariant Eqs. (3.139a)–(3.139b):

$$U' = aU, \quad H' = bH, \quad x' = cx, \quad t_r' = \varpi\, t_r, \qquad (3.140)$$

where the prime indicates the variables in the transformed system. Substituting into Eqs. (3.139a)–(3.139b) yields

$$b = a^n, \quad b = ab, \qquad (3.141)$$

which admits only the trivial solution $a = b = 1$ and leaves undetermined both c and ϖ, which is of no help in advancing toward the solution of the problem. An attempt for progress is to check invariance in the following transformation group:

$$U' = U, \quad H' = H, \quad x' = cx, \quad t_r' = c^{1/\delta}t_r, \tag{3.142}$$

where δ is unknown (in the first-kind self-similar solutions, the transformation group is completely defined). The formal solution of Eqs. (3.139a)–(3.139b),

$$f(U, H, x, t_r) = 0, \tag{3.143}$$

is invariant under the group of transformations (3.142) and can be expressed with only three arguments:

$$\tilde{f}\left(U, H, \frac{x}{t_r^\delta}\right) = 0, \tag{3.144}$$

all invariants within the same group of transformations (3.142) and with x/t_r^δ the candidate similarity variable.

For an easy computation, we assume a slightly different similarity variable, with $\xi = x\chi^{-1}t_r^{-1}|t_r|^{1-\delta}$, where the exponent δ is an unknown eigenvalue, where χ is a dimensional prefactor with dimension $[\chi] = LT^{-\delta}$ and where the absolute value is inserted to change the sign of t_r.

Inserting the similarity variable into Eqs. (3.139a)–(3.139b) gives

$$U|U|^{n-1} + \xi H' + (n+1)(1-k)H = 0, \tag{3.145a}$$
$$\delta\xi H' + nH + \xi(HU)' + (n+2-nk)HU = 0, \tag{3.145b}$$

where the prime indicates the derivative with respect to ξ and where the variable t_r does not appear, or to be more precise, it does not appear independently but only in combination with the variable x. This is exactly what happens when applying Buckingham's Theorem: the fundamental quantities no longer appear on their own but always in combination with each other. Eliminating ξ from the two equations results in

$$\begin{cases} \dfrac{dH}{dU} = \dfrac{H[(n+1)(1-k)H + U|U|^{n-1}]}{H[(k+1)U - (n+1)(1-k)\delta + n] - (U+\delta)U|U|^{n-1}}, & (3.146a) \\[3mm] \dfrac{d\ln\xi}{dH} = -\dfrac{1}{U|U|^{n-1} + (n+1)(1-k)H}, & (3.146b) \end{cases}$$

representing a set of autonomous planar ODEs, with boundary conditions represented by points in the phase space. The possible solutions of the differential problem connect two singular points, which are defined as simultaneous zeros of the numerator and denominator, including also a singular point where the denominator tends to infinity. There are four singular points, namely,

$$O : (H, U) \equiv (0, 0),$$

$$A : (H, U) \equiv (0, -\delta),$$

$$B : (H, U) \equiv \left(\left[\frac{n}{2 + n(1 - k)} \right]^n \frac{1}{(n + 1)(1 - k)}, -\frac{n}{2 + n(1 - k)} \right),$$

$$C : (H, U) \equiv \left(-\infty, \frac{(n + 1)(1 - k)\delta - n}{k + 1} \right),$$

and by reasoning on the meaning of each of them, we conclude that the pre-closure phase, when $t_r > 0$ and the current is approaching the origin, is described by a curve connecting O and A; the post-closure (levelling) phase, when $t_r < 0$ and the current has filled all the channel, is described by a curve connecting O and C. The computation of these curves requires a trial and error procedure starting near the origin and computing $H(U)$ with a first-attempt eigenvalue. The correct eigenvalue yields a curve reaching exactly point A, the front of the current. A visual representation in the phase space is shown in Fig. 3.15, with a critical eigenvalue $\delta_c = 1.815\,357$ for the case of $k = 0.6$ and $n = 0.7$.

The post-closure phase connects points O and C and is computed starting from C and reaching O. Iteration is not required since the critical eigenvalue has already been computed.

After computing $H(U)$, it is possible to integrate Eq. (3.146b) to evaluate the coordinate ξ for the pre- and postclosure phases.

As usual, experimental validation is requested, which clearly demonstrates whether the self-similar solution is stable and has a physical sense. Figure 3.16 shows the profiles of the current taken at different times in the post-closure phase, confirming the reliability of the procedure and of the results.

In summary, we can say that whenever dimensional analysis does not help in finding the transformation group, the global solution of the problem allows the identification of the anomalous scaling: the transformation itself is unknown and is determined through the overall solution of the differential problem. In the previous example, we adopted phase plane analysis; in other cases, different techniques were used. The approach permits significant simplifications and helps in solving the differential problems, often allowing analytical solutions.

There are several other problems in which self-similar solutions of the second kind have proven successful, even with the presence of multiple eigenvalues (Papageorgiou 1995) or with a continuous spectrum of eigenvalues (Barenblatt and Zel'Dovich 1972).

Summarising Concepts

• After identification of the relevant dimensionless groups, the next step requires identification of the structure of the functional relationship between the variables

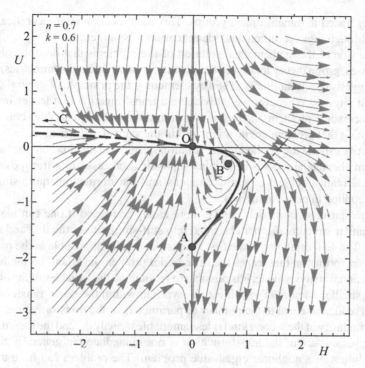

Fig. 3.15 Converging gravity current in a fracture of gap thickness $b = b_1 x^k$. Phase portrait of (3.146a) for $n = 0.7$ (shear-thinning fluid) and $k = 0.6$, with $\delta_c = 1.815\,357$. The continuous curve refers to the preclosure phase, the dashed curve refers to the postclosure (levelling) phase, the thin red horizontal line indicates the asymptote in the levelling phase, and the dashed-dotted red curves are the approximate solutions about points O and A

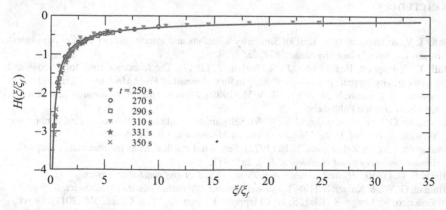

Fig. 3.16 Experimental profiles of the current in converging channel flow with $k = 0.6$ during levelling (symbols) compared to the theoretical self-similar solution (bold line). Newtonian fluid $n = 1$, $t_c = 229$ s, $\delta_c = 1.787\,4$ (modified from Longo et al. 2021)

already traced to dimensionless groups. This identification on a theoretical basis goes through logic, intuition, and identification of symmetries.

- On the side of the mathematical solution of the problems that describe physical processes, group theory facilitates the identification of internal symmetries, possibly suggesting self-similarity. The key element is the principle of general covariance in physics. Whenever we individuate a transformation that leaves invariant a mathematical problem, the best and in-minimum-number variables completely describing that problem are invariant within the same transformation.

- Similarity criteria are applied to the solution of mathematical problems; incidentally, mathematical models describing a physical process are a transposition of physical reality, subject to approximations and transformations quite similar to those applied in physical models.

- A particular class of models can be more easily addressed if one can identify a self-similar structure, further classified into self-similarity of the first and second kinds. The self-similarity of the first kind is potentially applicable in the presence of a transformation that leaves the differential problem invariant: on the basis of dimensional analysis alone, it is possible to recombine the variables of the problem into a smaller number of variables, all invariant within the same transformation, with a reduction equal to the number of parameters of the transformation.

- Self-similarity of the second kind is less amenable to analysis, and the identification of the parameters of the transformation is not immediate but generally requires the solution of a nonlinear eigenvalue problem. The problem can have a single solution, a discrete number of solutions, or a continuous spectrum of solutions.

References

Ball, T. V., & Huppert, H. E. (2019). Similarity solutions and viscous gravity current adjustment times. *Journal of Fluid Mechanics, 874*, 285–298.

Ball, T. V., Huppert, H. E., Lister, J., & Neufeld, J. (2017). The relaxation time for viscous and porous gravity currents following a change in flux. *Journal of Fluid Mechanics, 821*, 330–342.

Barenblatt, G. I., Entov, V. M., & Ryzhik, V. M. (1989). *Theory of fluid flows through natural rocks.* Kluwer Academic Publishers.

Barenblatt, G. I., & Sivashinskii, G. I. (1969). Self-similar solutions of the second kind in nonlinear filtration. *Journal of Applied Mathematics and Mechanics, 33*(5), 836–845.

Barenblatt, G. I., & Zel'Dovich, Y. B. (1972). Self-similar solutions as intermediate asymptotics. *Annual Review of Fluid Mechanics, 4*(1), 285–312.

Birkhoff, G. (1950). *Hydrodynamics: A study in logic, fact and similitude.* Dover.

Bluman, G. W., & Kumei, S. (1989). *Symmetries and differential equations.* Springer.

Di Federico, V., Longo, S., King, S. E., Chiapponi, L., Petrolo, D., & Ciriello, V. (2017). Gravity-driven flow of Herschel-Bulkley fluid in a fracture and in a 2D porous medium. *Journal of Fluid Mechanics, 821*, 59–84.

Gratton, J., & Minotti, F. (1990). Self-similar viscous gravity currents: Phase-plane formalism. *Journal of Fluid Mechanics, 210*, 155–182.

Hornung, H. G. (2006). *Dimensional analysis: Examples of the use of symmetry.* Dover Publications Inc.

Longo, S., Di Federico, V., & Chiapponi, L. (2015). A dipole solution for power-law gravity currents in porous formations. *Journal of Fluid Mechanics, 778,* 534–551.

Longo, S., Chiapponi, L., Petrolo, S., Lenci, A., & Di Federico, V. (2021). Converging gravity currents of power-law fluid. *Journal of Fluid Mechanics, 918,* A5, 1–30.

Papageorgiou, D. T. (1995). On the breakup of viscous liquid threads. *Physics of Fluids, 7*(7), 1529–1544.

Raszillier, H., & Durst, F. (1991). Coriolis-effect in mass flow metering. *Archive of Applied Mechanics, 61*(3), 192–214.

Raszillier, H., & Raszillier, V. (1991). Dimensional and symmetry analysis of Coriolis mass flowmeters. *Flow Measurement and Instrumentation, 2*(3), 180–184.

Szirtes, T. (2007). *Applied dimensional analysis and modeling.* Butterworth-Heinemann.

Chapter 4
The Theory of Similarity and Applications to Models

> *... in the sciences, the authority of thousands of opinions is not worth as much as one tiny spark of reason in an individual man...*
>
> *Galileo Galilei*

The concept of similarity is widely used in many fields of geometry and mathematics; for application purposes, it is necessary to extend and to specify it according to the field of interest. In our case, the concept of similarity is closely related to the theory of physical models, with applications also in the interpretation of complex systems. In some cases, it is advantageous to distort the model, to amplify one dimension with respect to the others to facilitate measurements or reduce uncertainties, or to ensure the balance of certain quantities.

4.1 Similarities

A *physical model* is a material reproduction, on a geometric scale, of a *prototype*: it can be a structure, a system, or a device on which to carry out experiments, which allows low-cost changes to be made to optimise performances and results.

The physical model is required when we need experimental data to be extrapolated at full scale or when we wish to obtain empirical relationships between the variables involved in the physical process in the presence of complex analytical relationships that are difficult to solve, inaccurate or simply unknown. The physical model is suggested when the prototype is too small or too large, when the prototype is not accessible, when the quantities to be measured take on values that are too large or too small, when the execution of the measurements would take too much or too little time.

S. G. Longo, *Principles and Applications of Dimensional Analysis and Similarity*, Mathematical Engineering, https://doi.org/10.1007/978-3-030-79217-6_4

The *theory of similarity* provides the necessary support to design models and to extrapolate in full scale the measures taken and the results obtained. This process of extrapolation is not error-free, with uncertainties generically attributed to *scale effects*.

Physical models are almost always at a reduced geometric scale, but there are also examples of models that are realised with an enlarged geometric scale or without a change in scale. For example, to interpret the mechanics of insect flight, physical models have been made with a highly enlarged geometric scale to find space to place sensors and to reduce the speed and frequency of wing beats. Sometimes, for reasons of similarity or economy (for instance, in some real systems where the material or fluid used, such as hydrogen, high-temperature steam, oil, is dangerous or expensive or difficult to handle), the physical models use a different material or fluid from the real one. In most cases, the same fluid is used in the model and in the prototype.

There are some physical processes for which physical modelling is not applicable, such as the study of fracture propagation, the study of plastic flow (creep), shrinkage, and adhesion effects. These are processes strongly conditioned by the scale effects and that do not tolerate the changes of the conditions of the prototype, and the models in such cases can give completely wrong indications.

4.1.1 Geometric Similarity

As we demonstrated in Sect. 1.2.4, two systems of units based on the same fundamental quantities (belonging by definition to the same class) are linked by the relationships between the selected units of measurement. For two systems of the class M, L, T,

$$\begin{cases} M'' = r_M M', \\ L'' = r_L L', \\ T'' = r_T T'. \end{cases} \tag{4.1}$$

The ratios r_M, r_L, r_T are numbers and are called *scale ratios* or *scales*. The conventional symbol for the ratio of length scales is λ. By definition, dimensionless groups (or numbers) are invariant under transformations (4.1). Each scale can be considered as a conversion factor from a unit of measurement to another unit of measurement of the same variable: for example, the length conversion factor between the British Imperial system and the International System is $r_L = 1$ in/0.0254 m. A scale can also be considered a factor of proportionality between measures of the same variable in the same system of units of measurement, but in two different spaces. In this latter case, we define the two spaces as the *model space* and the *prototype space* (or *real space*).

If the two spaces are linked only by the length scale

$$\lambda = \frac{L''}{L'},$$ (4.2)

the objects in the two spaces are considered *geometrically similar*, having the same shape and differing only in size. Following the schematic in Fig. 4.1, a *centre of similarity* O can then be identified such that

$$\frac{\overline{OA'}}{\overline{OA''}} = \frac{\overline{OB'}}{\overline{OB''}} = \frac{\overline{OC'}}{\overline{OC''}}.$$ (4.3)

If the two objects occupy the same semispace with respect to the similarity centre O and satisfy the conditions of similarity, they are considered *homothetics*, with the area and volume scaling as λ^2 and λ^3, respectively.

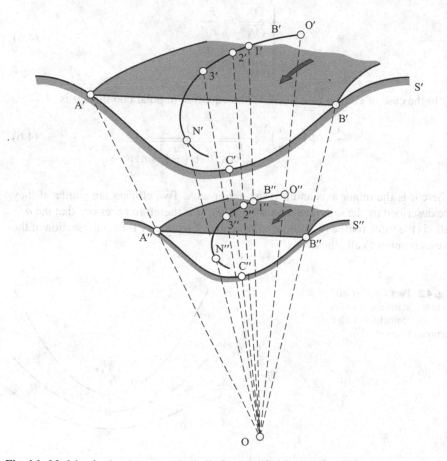

Fig. 4.1 Models of a river in geometric similarity (modified from Yalin 1971)

In some cases, the geometric similarity can be immediately verified: two spheres of different diameters or two cubes of different lengths of the edge are geometrically similar. For other curves, flat or solid figures, the verification is not immediate. For example, we wish to check if two parabolas are similar. It is convenient to express the equation of the parabola in polar coordinates, $\rho = \Psi(\theta)$, where ρ is the vector ray and θ is the angle (anomaly). With the origin of the vector ray in the focus, the equation of a parabola has the expression

$$\frac{\rho}{F} = \Phi(\theta) = \frac{2(1 - \cos\theta)}{\sin^2\theta}, \tag{4.4}$$

where F is the distance of the focus from the vertex. Two parabolas are similar if the function $\Phi(\theta)$ is the same. Since no parameters appear in the function $\Phi(\theta)$ of Eq. (4.4), it follows that all parabolas are geometrically similar and that the centre of similarity is the common focus (see Fig. 4.2):

$$\begin{cases} \dfrac{\overline{OA'}}{\overline{OA''}} = \dfrac{F'}{F''}, \\[2mm] \dfrac{\overline{OA'}}{\overline{OA'''}} = \dfrac{F'}{F'''}. \end{cases} \tag{4.5}$$

In the case of ellipses (see Fig. 4.3), the equation in polar coordinates is

$$\frac{\rho}{a} = \Phi\left(\theta, \frac{a}{b}\right) = \frac{1}{\sqrt{\cos^2\theta + \left(\dfrac{a}{b}\right)^2 \sin^2\theta}}, \tag{4.6}$$

where a is the minor axis and b is the major axis. Two ellipses are similar if they are described by the same function $\Phi(\theta, a/b)$. It is therefore necessary that the a/b ratio is the same for the two ellipses; the centre of similarity is the intersection of the axes, common to all ellipses:

Fig. 4.2 Parabolas are all similar and have a centre of similarity coincident with the common focus

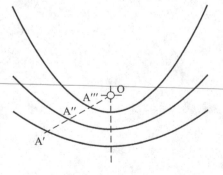

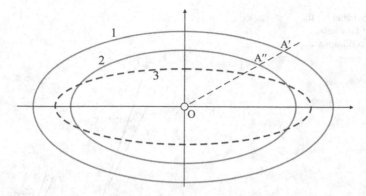

Fig. 4.3 Geometrically similar ellipses 1 and 2, and dissimilar ellipses, 1 and 3 or 2 and 3

$$\frac{\overline{OA'}}{\overline{OA''}} = \frac{a'}{a''} \equiv \frac{b'}{b''}. \tag{4.7}$$

Finally, *fractals* are geometric entities characterised by a particular form of geometric self-similarity, defined as *internal homothety*.

4.1.2 Kinematic Similarity

Let us consider a model and a prototype subject to a physical process that modifies over time the geometric position of a single object or of the whole. The prerequisite for kinematic similarity is that the trajectories of the homologous moving parts are geometrically similar. We wish to analyse the conditions required for complete kinematic similarity.

Let us consider the schematic in Fig. 4.4 with the point P_1 in the space *(1)* and the point P_2 in the space *(2)*, and let us indicate with $\mathbf{r}_1$ and with $\mathbf{r}_2$ their position vectors at time t_1 and t_2, respectively. We have selected the origin of the time so that, in the hypothesis that the time scale is different for the two spaces, at time t_1 the vector $\mathbf{r}_1$ is parallel and equiverse to the vector $\mathbf{r}_2$ at time t_2. In a time interval dt_1, the vector $\mathbf{r}_1$ becomes $\mathbf{r}_1 + d\mathbf{r}_1$, and in a time interval dt_2, the vector $\mathbf{r}_2$ becomes $\mathbf{r}_2 + d\mathbf{r}_2$. To guarantee that the position vectors remain parallel and equiverse in all subsequent times, it is necessary that the increments $d\mathbf{r}_1$ and $d\mathbf{r}_2$ are parallel and equiverse (which requires the similarity of the trajectories) and that the ratio between their magnitude is equal to the geometric scale λ. If the time scale is r_T, we can write

$$\begin{cases} d\mathbf{r}_2 = \lambda \, d\mathbf{r}_1, \\ dt_2 = r_T \, dt_1, \end{cases} \tag{4.8}$$

Fig. 4.4 Scheme for the
analysis of kinematic
similarity conditions

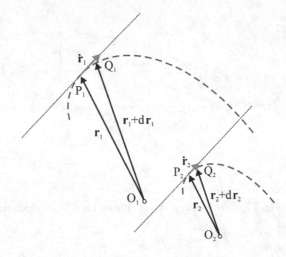

and dividing both sides yields

$$\frac{\mathrm{d}\mathbf{r}_2}{\mathrm{d}t_2} = \frac{\lambda}{r_T}\frac{\mathrm{d}\mathbf{r}_1}{\mathrm{d}t_1} \rightarrow \dot{\mathbf{r}}_2 = \frac{\lambda}{r_T}\dot{\mathbf{r}}_1, \tag{4.9}$$

where the dot indicates a time derivative. Differentiating again with respect to the
time yields

$$\ddot{\mathbf{r}}_2 = \frac{\lambda}{r_T^2}\ddot{\mathbf{r}}_1. \tag{4.10}$$

Therefore, the kinematic similarity imposes a constraint between the geometric
scale λ and the time scale r_T, which are involved in all kinematic quantities. The
polygons of the velocity vectors in the model and in the prototype are geometrically
similar. The velocity, acceleration and volumetric flow rate (just to mention a few of
the most frequently used kinematic quantities) scale as follows:

$$\begin{cases} r_V = \dfrac{L''/T''}{L'/T'} = \dfrac{\lambda}{r_T}, \\[2mm] r_a = \dfrac{L''/T''^2}{L'/T'^2} = \dfrac{\lambda}{r_T^2}, \\[2mm] r_Q = \dfrac{L''^3/T''}{L'^3/T'} = \dfrac{\lambda^3}{r_T}. \end{cases} \tag{4.11}$$

4.1.3 Dynamic Similarity

The concept of dynamic similarity considers the forces acting in the model and in the prototype. If the relationship between the forces acting on a body in the model and in the prototype is fixed, that relationship must govern the relationship between all the forces acting in the model and in the prototype on that body and on all the corresponding bodies. Consequently, the polygons of the forces in the model and prototype are geometrically similar. The prerequisite for dynamic similarity is the existence of kinematic similarity.

We first analyse the frictionless unconstrained motion of point masses.

Let us consider a point mass that moves in space *(1)* and its counterpart moving in space *(2)*, and let us indicate with P_1 the point in space *(1)* that corresponds to P_2 in space *(2)*, where P_1 and P_2 belong to the trajectories of the two particles in the two spaces. We wish to calculate the conditions necessary for similarity of the trajectories and for kinematic similarity of the particles.

We assume that the position vectors are related to each other as

$$\overline{O_2P_2} = \lambda\,\overline{O_1P_1}, \tag{4.12}$$

and we postulate that the forces per unit of mass acting on the two particles in their respective spaces, defined as *specific mass forces*, are represented by two parallel and equiverse vectors $\mathbf{f}_1$ and $\mathbf{f}_2$ at each corresponding time, and with magnitudes such that

$$|\mathbf{f}_2| = r_f\,|\mathbf{f}_1|, \tag{4.13}$$

where r_f assumes a constant value within the same physical process. The two specific forces are applied on the two particles in the homologous instants and in the homologous positions occupied by these particles in the two spaces. Again, we postulate that the velocities $\mathbf{V}_1$ and $\mathbf{V}_2$ of the two particles behave like the specific forces of mass (they are, therefore, parallel and equiverse), with the ratio between their magnitude, at the time when the particles occupy the points P_1 and P_2, equal to

$$|\mathbf{V}_2| = r_V\,|\mathbf{V}_1|. \tag{4.14}$$

We wish to check under which conditions the value of r_V remains unchanged (and the geometrical relations between the two vectors also remain unchanged) when the particles move from P_1 to Q_1 and from P_2 to Q_2.

From the condition of proportionality of the specific forces of mass follows an equal proportionality of the accelerations. In addition, the ratio between the times required to travel from P_2 to Q_2 and from P_1 to Q_1 is equal to

$$\frac{\Delta t_{P_2 \to Q_2}}{\Delta t_{P_1 \to Q_1}} = \frac{\dfrac{\overline{P_2 Q_2}}{|\mathbf{V}_2|}}{\dfrac{\overline{P_1 Q_1}}{|\mathbf{V}_1|}} \equiv \frac{\lambda}{r_V}, \tag{4.15}$$

since $\overline{P_2 Q_2} = \overline{P_2 O_2} + \overline{O_2 Q_2} = \lambda \overline{P_1 O_1} + \lambda \overline{O_1 Q_1} = \lambda \overline{P_1 Q_1}$.

The velocity increments are two parallel and equiverse vectors, with the ratio between the magnitudes equal to

$$\frac{|\Delta \mathbf{V}|_{P_2 \to Q_2}}{|\Delta \mathbf{V}|_{P_1 \to Q_1}} = \frac{\Delta t_{P_2 \to Q_2} |\mathbf{f}_2|}{\Delta t_{P_1 \to Q_1} |\mathbf{f}_1|} \equiv \frac{\lambda r_f}{r_V}. \tag{4.16}$$

For r_V to remain unchanged, it is necessary that

$$r_V = \frac{\lambda r_f}{r_V} \to \frac{r_V^2}{\lambda r_f} = 1. \tag{4.17}$$

This requires that

$$\frac{V_1^2}{l_1 f_1} = \frac{V_2^2}{l_2 f_2}, \tag{4.18}$$

and the ratio

$$\frac{V^2}{lf} \tag{4.19}$$

assumes the same value in the two spaces. The ratio (4.19) is called the *Reech number* (Reech 1852) (or Froude number). We observe that to ensure dynamic similarity, in addition to having the same Reech number in the model and prototype, it is also necessary to ensure parallelism and the same orientation for homologous vectors in the model and prototype.

Ultimately, in conditions of dynamic similarity, the trajectories of two point masses are geometrically similar, and the motion is kinematically similar if:

- the specific mass forces acting on the particles, when they occupy corresponding points in the two paths, are parallel, equiverse and in an invariant proportion;
- the geometric scale, the velocity scale and the specific mass force scale satisfy Eq. (4.17).

If these conditions are met, the velocity vectors at any corresponding time are parallel, are applied in the corresponding point and have an invariant ratio of magnitudes, with a scale of the travel times equal to λ/r_V.

This result can be obtained more rigorously by using the equations of motion of point masses. The equations of motion for two homologous point masses in the two spaces are:

$$\begin{cases} m_1 \dfrac{d^2 x_1}{dt^2} = F_{1x}, \\[2mm] m_1 \dfrac{d^2 y_1}{dt^2} = F_{1y}, \\[2mm] m_1 \dfrac{d^2 z_1}{dt^2} = F_{1z}, \end{cases} \tag{4.20}$$

$$\begin{cases} m_2 \dfrac{d^2 x_2}{dt^2} = F_{2x}, \\[2mm] m_2 \dfrac{d^2 y_2}{dt^2} = F_{2y}, \\[2mm] m_2 \dfrac{d^2 z_2}{dt^2} = F_{2z}, \end{cases} \tag{4.21}$$

and can be re-written as:

$$\begin{cases} u_{1x} \dfrac{du_{1x}}{dx} = f_{1x}(\mathbf{r}_1), \\[2mm] u_{1y} \dfrac{du_{1y}}{dy} = f_{1y}(\mathbf{r}_1), \\[2mm] u_{1z} \dfrac{du_{1z}}{dz} = f_{1z}(\mathbf{r}_1), \end{cases} \tag{4.22}$$

$$\begin{cases} u_{2x} \dfrac{du_{2x}}{dx} = f_{2x}(\mathbf{r}_2), \\[2mm] u_{2y} \dfrac{du_{2y}}{dy} = f_{2y}(\mathbf{r}_2), \\[2mm] u_{2z} \dfrac{du_{2z}}{dz} = f_{2z}(\mathbf{r}_2), \end{cases} \tag{4.23}$$

where f_1 and f_2 are the specific mass forces, calculated at the homologous points identified by the position vectors $\mathbf{r}_1$ and $\mathbf{r}_2$ occupied by the particles in homologous instants. In fact, for the x-component,

$$\frac{d^2 x}{dt^2} = \frac{du_x}{dt} = \frac{du_x}{dx}\frac{dx}{dt} = u_x \frac{du_x}{dx}. \tag{4.24}$$

Since by hypothesis the specific mass forces must be parallel and equiverse, we have

$$\begin{cases} f_{2x}(\mathbf{r}_2) = r_f\, f_{1x}(\mathbf{r}_1) \equiv r_f\, f_{1x}\left(\dfrac{\mathbf{r}_2}{\lambda}\right), \\[2mm] f_{2y}(\mathbf{r}_2) = r_f\, f_{1y}(\mathbf{r}_1) \equiv r_f\, f_{1y}\left(\dfrac{\mathbf{r}_2}{\lambda}\right), \\[2mm] f_{2z}(\mathbf{r}_2) = r_f\, f_{1z}(\mathbf{r}_1) \equiv r_f\, f_{1z}\left(\dfrac{\mathbf{r}_2}{\lambda}\right), \end{cases} \tag{4.25}$$

where r_f is the ratio of specific mass forces and λ is the geometric scale.

The kinematic similarity is satisfied as long as the following requirement is met:

$$
\begin{cases}
\mathbf{r}_2 = \lambda\,\mathbf{r}_1 \to x_2 = \lambda\,x_1, \quad y_2 = \lambda\,y_1, \quad z_2 = \lambda\,z_1, \\
u_{2x} = r_V\,u_{1x}, \quad u_{2y} = r_V\,u_{1y}, \quad u_{2z} = r_V\,u_{1z}.
\end{cases}
\tag{4.26}
$$

Substituting Eqs. (4.25)–(4.26) into Eq. (4.23) yields

$$
\begin{cases}
\dfrac{r_V^2}{\lambda\,r_f}\,u_{1x}\,\dfrac{du_{1x}}{dx_1} = f_{1x}(\mathbf{r}_1), \\[2mm]
\dfrac{r_V^2}{\lambda\,r_f}\,u_{1y}\,\dfrac{du_{1y}}{dy_1} = f_{1y}(\mathbf{r}_1), \\[2mm]
\dfrac{r_V^2}{\lambda\,r_f}\,u_{1z}\,\dfrac{du_{1z}}{dz_1} = f_{1z}(\mathbf{r}_1),
\end{cases}
\tag{4.27}
$$

which become identical to Eq. (4.22) if

$$
\frac{r_V^2}{\lambda\,r_f} = 1.
\tag{4.28}
$$

Therefore, to ensure dynamic similarity, it is necessary that the motion of homologous point masses has the same Reech number.

If dynamic friction is present, to guarantee the parallelism of the total specific force on the contact surface (normal component plus tangential component due to friction) in the model and in the prototype, it is necessary that the friction coefficient assumes the same value in the model and in the prototype.

Example 4.1 Let us consider the motion of a planet around the Sun, and let us indicate with r_0 the average distance between the two celestial bodies.

Let us assume the initial position of the planet coincides with r_0 and that its velocity is a vector tangent to the orbit with magnitude V_0. The only force considered is the gravitational attraction, which, at the initial time, is equal to

$$
f_0 = \frac{k\,M_s}{r_0^2},
\tag{4.29}
$$

where k is the gravitational constant and M_s is the mass of the Sun. The condition of dynamic similarity of the orbits of the planets requires that

$$
\begin{cases}
r_V^2 = r_f\,\lambda \\
r_f = \dfrac{1}{\lambda^2}
\end{cases}
\to r_V^2 = \frac{1}{\lambda}, \qquad \therefore \tag{4.30}
$$

Fig. 4.5 Experimental verification of Kepler's third law for the solar system

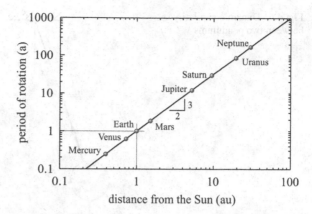

which is equivalent to $V_0^2 r_0 = \text{const}$. The time scale is $r_T = \lambda / r_V$ hence $r_T^2 = \lambda^3$. This is Kepler's third law, which states that the period of rotation of the planets around the Sun varies according to the power 3/2 of the average distance from the Sun, as shown in Fig. 4.5.

Example 4.2 We wish to calculate the period of oscillation of a pendulum, which schematically represents the constrained motion of a point mass.

Let us suppose that we move the pendulum from the equilibrium configuration of an angle α_0 and then release it with zero initial velocity. We can use the similarity between two pendulums of different lengths, see Fig. 4.6, with masses subject to acceleration of gravity and to centripetal acceleration, the only two specific forces of mass (we neglect the fictitious forces due to the non-inertial terrestrial reference system). At the beginning, with circular trajectories, all the kinematic and geometrical prerequisites of the dynamic similarity are satisfied, and it is sufficient to verify that the Reech number is the same for both pendulums to ensure dynamic similarity:

$$\frac{V_{1,0}^2}{g\,l_1} = \frac{V_{2,0}^2}{g\,l_2} \rightarrow r_V = \sqrt{\lambda}. \tag{4.31}$$

The ratio of the centripetal acceleration in the two pendulums must be equal to

$$\frac{V_1^2/l_1}{V_2^2/l_2} = 1, \tag{4.32}$$

which is always satisfied. In general,

$$\frac{V^2}{g\,l} = \text{const} \rightarrow V = C_1\sqrt{g\,l} \rightarrow t = C_2\sqrt{\frac{l}{g}}, \tag{4.33}$$

Fig. 4.6 Dynamic similarity
between two pendulums

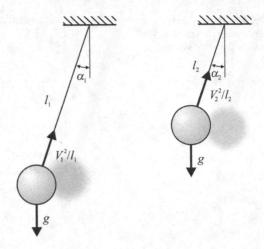

where t is the period of oscillation and the coefficient C_2 is a function of the initial
angle α_0,

$$C_2 = f(\alpha_0). \tag{4.34}$$

Integrating the motion equation, we calculate

$$f(\alpha_0) = 4K\left(\sin^2\frac{\alpha_0}{2}\right), \tag{4.35}$$

where $K(\ldots)$ is the elliptical integral of the first kind. Expanding in series the function
$K(\ldots)$ for $\alpha_0 \rightarrow 0$ yields $f(\alpha_0) \approx 2\pi$. Substituting in Eq. (4.33), we obtain the classic
formula of the period of infinitesimal oscillations of a pendulum. We observe the
absence of the mass in the expression of the period.

The analysis can be extended to pendulums with more complex trajectories. For
instance, if the wire is forced to rest on a cycloidal support, the trajectory of the
pendulum becomes a cycloid (see Fig. 4.7). By integrating the motion equation, the
period of oscillation is still formally expressed by Eq. (4.33) but is not a function of
the initial angle value α_0 and hence is independent of the amplitude of the oscillations.
This is the cycloidal pendulum, presented by Huygens (1629–1645) in the treatise
Horologium oscillatorium sive de motu pendulorum, 1673, and analysed by Newton
in his *Principia Mathematica, 1687*, as the equivalent of the oscillation of a liquid
in a U-shaped tube.

For a cycloidal pendulum and for any oscillation amplitude, the period of oscillations is

$$t = 2\pi\sqrt{\frac{4a}{g}}. \tag{4.36}$$

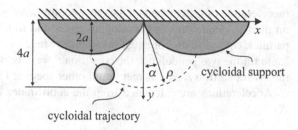

Fig. 4.7 Cycloidal pendulum

cycloidal trajectory

The parametric equation of the trajectory in the coordinate system $x - y$ in Fig. 4.7 is

$$\begin{cases} x = a\,(n + \sin n), \\ y = a\,(3 + \cos n), \end{cases} \tag{4.37}$$

where n is the parameter, and can be expressed in parametric polar coordinates as

$$\begin{cases} \dfrac{\rho}{a} = \sqrt{(n + \sin n)^2 + (3 + \cos n)^2}, \\ \alpha = \tan^{-1}\left(\dfrac{n + \sin n}{3 + \cos n}\right). \end{cases} \tag{4.38}$$

This equation always fulfils the conditions of geometric similarity, regardless of the value of a (see Sect. 4.1.1).

We now wish to verify that there is also kinematic and dynamic similarity.

Let us consider two cycloidal pendulums that describe geometrically similar trajectories; see Fig. 4.8.

If the velocities of the two pendulums at the initial time are parallel vectors with a ratio of their magnitude r_V, then r_V must be invariant at all times, and the ratio of the specific forces of mass must satisfy the condition $r_V^2 = \lambda\,r_f$. Since we have already calculated that $r_t = \sqrt{\lambda/r_f}$, we must also have $r_f = 1$; i.e., the specific mass forces

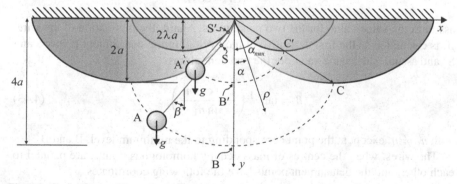

Fig. 4.8 Cycloidal pendulums in similarity

(acceleration of gravity and all the other components associated with motion) acting on the two pendulums in homologous times and in homologous positions must be parallel, equiverse and of equal magnitude. Since the acceleration of gravity is the same for the two pendulums, the condition $r_g = 1$ is already met, and it is sufficient that the condition $r_f = 1$ is met for all other specific forces of mass.

Accelerations are calculated from the coordinates of the centres of mass A and A':

$$\begin{cases} x_A = a\,(n + \sin n), \\ y_A = 2a + a\,(1 + \cos n), \end{cases} \qquad \begin{cases} x_{A'} = \lambda\,a\,(n' + \sin n'), \\ y_{A'} = 2\lambda\,a + \lambda\,a\,(1 + \cos n'). \end{cases} \tag{4.39}$$

Differentiating twice with respect to time yields

$$\begin{cases} \ddot{x}_A = a\,\dfrac{d^2 n}{dt^2}\,(1 + \cos n) - a\left(\dfrac{dn}{dt}\right)^2 \sin n, \\[2mm] \ddot{y}_A = -a\,\dfrac{d^2 n}{dt^2}\,\sin n - a\left(\dfrac{dn}{dt}\right)^2 \cos n, \end{cases} \tag{4.40}$$

and

$$\begin{cases} \ddot{x}_{A'} = \lambda\,a\,\dfrac{d^2 n'}{dt'^2}\,(1 + \cos n') - \lambda\,a\left(\dfrac{dn'}{dt'}\right)^2 \sin n', \\[2mm] \ddot{y}_{A'} = -\lambda\,a\,\dfrac{d^2 n'}{dt'^2}\,\sin n' - \lambda\,a\left(\dfrac{dn'}{dt'}\right)^2 \cos n'. \end{cases} \tag{4.41}$$

Since $t' = \sqrt{\lambda}\,t$, comparing the expressions of the acceleration for the two pendulums confirms that the accelerations are parallel and equiverse only if $n = n'$. As a consequence, the angle α of the position vector passing through the points of detachment of the wires from the cycloids, S and S', which is equal to

$$\alpha = \tan^{-1}\left(\frac{n - \sin n}{1 - \cos n}\right), \tag{4.42}$$

assumes the same value for the two pendulums. Again, the inclination of the wire, β, is calculated as the first derivative of the cycloid in the detachment points S and S' and assumes the same value for the two pendulums,

$$\beta = \tan^{-1}\left(\frac{1 - \cos n_1}{\sin n_1}\right), \tag{4.43}$$

with $n_1 \neq n$, except at the point corresponding to the minimum level, B and B'.

The wires, when the centres of mass occupy homologous points, are parallel to each other, and the detachment points have the following coordinates:

$$\begin{cases} x_S = a\,(n - \sin n), \\ y_S = a\,(1 - \cos n), \end{cases} \quad \begin{cases} x_{S'} = \lambda\,a\,(n - \sin n), \\ y_{S'} = \lambda\,a\,(1 - \cos n). \end{cases} \tag{4.44}$$

Starting from the vertical direction, if a pendulum arrives at A at time t_A, a similar pendulum will arrive at A$'$ at time $t_{A'} = \sqrt{\lambda}\, t_A$.

Using the equation of the coordinates of the pendulum centre of mass, it can be demonstrated that the velocities of the two pendulums, in homologous points, are parallel and satisfy a ratio equal to $\sqrt{\lambda}$. Alternatively, by applying energy conservation and assuming a zero velocity at the maximum elongation of the pendulums, the magnitude of the velocity, as a function of the parameters a and n, is equal to

$$V = \sqrt{2a\,g\,(1 + \cos n)}, \quad V' = \sqrt{2\lambda\,a\,g\,(1 + \cos n)}. \qquad \therefore \quad (4.45)$$

Hence, the two pendulums are in dynamic similarity.

4.1.4 Dynamic Similarity for Interacting Material Particle Systems

To ensure dynamic similarity for interacting particle systems, the conditions already analysed for an isolated particle in free or constrained motion must be met for each particle. Therefore, the conditions of dynamic similarity are:

– geometric similarity and initial velocities (vectors) in an invariant ratio;
– specific forces of mass (vectors) in an invariant ratio;
– identical ratios between the masses of pairs of corresponding particles for all pairs.

This last condition derives from the necessity to guarantee that all the forces, including those deriving from the interaction, are parallel, equiverse and in an invariant ratio of their magnitude in the two similar systems; this can happen only if the nature of the interactions is identical (for example, the same coefficient of friction for collisions between particles) and if the masses of the particles are in an invariant ratio (that is, the same ratio for all the pairs of similar particles). If this were not the case, an interaction such as the collision between pairs of particles characterised by a different mass ratio, in the model and in the prototype, would result in accelerations in a non-invariant ratio. This condition can be generalised by also involving the masses and imposing that all the forces, including those deriving from the interaction between the particles, are vectorially in an invariant ratio. Consequently, the Reech number relative to the forces of interaction F must satisfy the equality

$$\frac{m_1\,V_1^2}{l_1\,F_1} = \frac{m_2\,V_2^2}{l_2\,F_2}, \tag{4.46}$$

where m_1 and m_2 are the masses in the two spaces. We observe that F_1 and F_2 are forces, while f_1 and f_2 in Eq. (4.18) are specific mass forces, with the dimension of an acceleration.

4.1.5 Dynamic Similarity for Rigid Bodies

Rigid bodies can be described as particle systems with the constraint that the distances between any two particles in the system are invariant over time, since they do not deform under the action of applied forces.

The conditions for dynamic similarity are:

- geometrical similarity of the two bodies;
- constant ratio of density near corresponding points for all pairs of points;
- identical initial orientation of the vectors of position, velocity, acceleration, and force;
- invariant ratio of the initial velocity (in the vectorial sense) of corresponding points and for all pairs of points;
- invariant ratio of the forces (in the vectorial sense) acting near corresponding points and for all pairs of points;
- equality of the Reech number in the form $m_1 V_1^2/(l_1 F_1) = m_2 V_2^2/(l_2 F_2)$, where m_1 and m_2 are the masses of the two bodies, V_1 and V_2 are their speeds, and F_1 and F_2 are the applied forces.

The condition regarding the density ratio can be replaced by the condition of collinearity of the principal axes of inertia and by a suitable ratio between the moments of inertia.

The initial velocity condition can be replaced by a similar relationship between the velocities of the centres of mass and the rotational rate.

Finally, the condition regarding forces can be replaced by a similar relationship between the resulting forces and moments applied to the two bodies.

We now check the sufficiency and the consequences of all these itemised conditions.

The equations of motion for a rigid body, in orthogonal Cartesian coordinates, are:

$$\begin{cases} m\dfrac{d^2 x}{dt^2} = F_x, \quad m\dfrac{d^2 y}{dt^2} = F_y, \quad m\dfrac{d^2 z}{dt^2} = F_z, \\ mR_x^2\dfrac{d^2\theta}{dt^2} = M_x, \quad mR_y^2\dfrac{d^2\phi}{dt^2} = M_y, \quad mR_z^2\dfrac{d^2\psi}{dt^2} = M_z, \end{cases} \quad (4.47)$$

where m is the mass; R_x, R_y and R_z are the radii of gyration for rotation with respect to the axis x, y and z, respectively; F and M are the acting forces and moments, respectively; and x, y, z, θ, ϕ and ψ are the Lagrangian parameters of the motion. The equations can be re-written as

$$\begin{cases} u_x \dfrac{du_x}{dx} = f_x, \quad u_y \dfrac{du_y}{dy} = f_y, \quad u_z \dfrac{du_z}{dz} = f_z, \\[2mm] \omega_x \dfrac{d\omega_x}{d\theta} = n_x, \quad \omega_y \dfrac{d\omega_y}{d\phi} = n_y, \quad \omega_z \dfrac{d\omega_z}{d\psi} = n_z, \end{cases} \tag{4.48}$$

where f are the specific mass forces, ω is the angular velocity, and $n_{(\ldots)} = M_{(\ldots)}/(m R_{(\ldots)}^2)$ are the moments per unit moment of inertia.

If we consider two rigid bodies in two different spaces (1) and (2) and we assume that the physical processes they are involved in are in dynamic similarity, following the system (4.48), for the first body yields:

$$\begin{cases} u_{1x} \dfrac{du_{1x}}{dx_1} = f_{1x}(\mathbf{r}_1, \, \boldsymbol{\chi}_1), \\[2mm] \ldots, \\[2mm] \omega_{1x} \dfrac{d\omega_{1x}}{d\theta_1} = n_{1x}(\mathbf{r}_1, \, \boldsymbol{\chi}_1), \\[2mm] \ldots, \end{cases} \tag{4.49}$$

where $\mathbf{r}_1$ is the position vector of the centre of mass and $\boldsymbol{\chi}_1$ is the vector that detects the orientation, since we assume that the specific forces and specific moments depend on the 6 Lagrangian parameters of the body. For the second body,

$$\begin{cases} u_{2x} \dfrac{du_{2x}}{dx_2} = f_{2x}(\mathbf{r}_2, \, \boldsymbol{\chi}_2), \\[2mm] \ldots, \\[2mm] \omega_{2x} \dfrac{d\omega_{2x}}{d\theta_2} = n_{2x}(\mathbf{r}_2, \, \boldsymbol{\chi}_2), \\[2mm] \ldots. \end{cases} \tag{4.50}$$

Since by hypothesis the specific mass forces must be parallel and equiverse, it results

$$f_{2x}(\mathbf{r}_2, \, \boldsymbol{\chi}_2) = r_f f_{1x}(\mathbf{r}_1, \, \boldsymbol{\chi}_1) \equiv r_f f_{1x}\left(\dfrac{\mathbf{r}_2}{\lambda}, \, \boldsymbol{\chi}_2\right), \tag{4.51}$$

where r_f is the ratio of specific mass forces and λ is the length scale.

We observe that the two vectors $\boldsymbol{\chi}_1$ and $\boldsymbol{\chi}_2$ must be coincident, unlike the position vectors $\mathbf{r}_1$ and $\mathbf{r}_2$, which, although parallel and equiverse, have a magnitude with a scale equal to the length scale.

The specific moments per unit of moment of inertia must be parallel and equiverse; hence,

$$n_{2x}(\mathbf{r}_2, \, \boldsymbol{\chi}_2) = \dfrac{r_f}{\lambda} n_{1x}(\mathbf{r}_1, \, \boldsymbol{\chi}_1) \equiv \dfrac{r_f}{\lambda} n_{1x}\left(\dfrac{\mathbf{r}_2}{\lambda}, \, \boldsymbol{\chi}_2\right). \tag{4.52}$$

The equations for the second body are re-written as

$$
\begin{cases}
u_{2x} \dfrac{du_{2x}}{dx_2} = r_f f_{1x}\left(\dfrac{\mathbf{r}_2}{\lambda}, \ \mathbf{\chi}_2\right), \\
\cdots, \\
\omega_{2x} \dfrac{d\omega_{2x}}{d\theta_2} = \dfrac{r_f}{\lambda} n_{1x}\left(\dfrac{\mathbf{r}_2}{\lambda}, \ \mathbf{\chi}_2\right), \\
\cdots.
\end{cases}
\tag{4.53}
$$

Kinematic similarity is satisfied if:

$$
\begin{cases}
\mathbf{r}_2 = \lambda\, \mathbf{r}_1 \rightarrow x_2 = \lambda\, x_1, \quad y_2 = \lambda\, y_1, \quad z_2 = \lambda\, z_1, \\
\mathbf{\chi}_2 = \mathbf{\chi}_1 \rightarrow \theta_2 = \theta_1, \quad \phi_2 = \phi_1, \quad \psi_2 = \psi_1, \\
u_{2x} = r_V\, u_{1x}, \quad u_{2y} = r_V\, u_{1y}, \quad u_{2z} = r_V\, u_{1z}, \\
\omega_{2x} = \dfrac{r_V}{\lambda}\, \omega_{1x}, \quad \omega_{2y} = \dfrac{r_V}{\lambda}\, \omega_{1y}, \quad \omega_{2z} = \dfrac{r_V}{\lambda}\, \omega_{1z}.
\end{cases}
\tag{4.54}
$$

Inserting Eq. (4.54) into Eq. (4.53) yields

$$
\begin{cases}
\dfrac{r_V^2}{\lambda\, r_f}\, u_{1x} \dfrac{du_{1x}}{dx_1} = f_{1x}(\mathbf{r}_1, \ \mathbf{\chi}_1), \\
\cdots, \\
\dfrac{r_V^2}{\lambda\, r_f}\, \omega_{1x} \dfrac{d\omega_{1x}}{d\theta_1} = n_{1x}(\mathbf{r}_1, \ \mathbf{\chi}_1), \\
\cdots,
\end{cases}
\tag{4.55}
$$

which become identical to Eq. (4.49) if

$$
\frac{r_V^2}{\lambda\, r_f} = 1.
\tag{4.56}
$$

Hence, the dynamic similarity of the two rigid bodies requires a constant Reech number.

In summary, the case of rigid bodies is treated similarly to the case of noninteracting discrete material particle systems. The extension to interacting rigid bodies is immediate and yields the invariance of the Reech number in the form

$$
\frac{r_m\, r_V^2}{\lambda\, r_F} = 1,
\tag{4.57}
$$

in which the mass forces and the mass appear separately, not just as specific forces of mass.

4.1.6 Affine Transformations of Trajectories and Conditions of Distorted Similarity

Thus far, we have checked that the condition of dynamic similarity requires that the Reech number be the same in the model and in the prototype, with isotropic scales. We wish to check which condition of similarity is possible if the scales of velocity and of the mass force and the geometric scale assume different values for two different axes, a condition defined as *distorted similarity*.

If the mass forces in one direction do not depend on the motions in the other directions (if, therefore, the mass force is aligned with the displacement vector), it is possible to guarantee distorted similarity. Moreover, if the time scale is unique in all directions, the trajectories will be connected by an affine transformation:

$$\begin{cases} x_2 = \lambda_{11} x_1 + \lambda_{12} y_1 + \lambda_{13} z_1, \\ y_2 = \lambda_{21} x_1 + \lambda_{22} y_1 + \lambda_{23} z_1, \\ z_2 = \lambda_{31} x_1 + \lambda_{32} y_1 + \lambda_{33} z_1, \end{cases} \tag{4.58}$$

where the subscript for the coordinates x, y, z refers to the two similar spaces and λ_{ij}, $(i, j = 1, 2, 3)$ are the coefficients of the transformation. In compact form, we have

$$\mathbf{r}_2 = \mathbf{\Lambda} \cdot \mathbf{r}_1, \tag{4.59}$$

where $\mathbf{\Lambda}$ is the matrix of transformation. The scale ratio between the velocities can be expressed as a second-order tensor:

$$\mathbf{R}_V = \begin{bmatrix} \dfrac{V_{2x}}{V_{1x}} & \dfrac{V_{2x}}{V_{1y}} & \dfrac{V_{2x}}{V_{1z}} \\[2ex] \dfrac{V_{2y}}{V_{1x}} & \dfrac{V_{2y}}{V_{1y}} & \dfrac{V_{2y}}{V_{1z}} \\[2ex] \dfrac{V_{2z}}{V_{1x}} & \dfrac{V_{2z}}{V_{1y}} & \dfrac{V_{2z}}{V_{1z}} \end{bmatrix}, \tag{4.60}$$

which is always considered diagonal as

$$\mathbf{R}_V = \begin{bmatrix} \dfrac{V_{2x}}{V_{1x}} & 0 & 0 \\[2ex] 0 & \dfrac{V_{2y}}{V_{1y}} & 0 \\[2ex] 0 & 0 & \dfrac{V_{2z}}{V_{1z}} \end{bmatrix}, \tag{4.61}$$

and, for undistorted similarity, is isotropic:

$$\mathbf{R}_V = \frac{V_2}{V_1} \begin{bmatrix} 1 & 0 & 0 \\ 0 & 1 & 0 \\ 0 & 0 & 1 \end{bmatrix}. \tag{4.62}$$

In a similar way, the ratio of specific mass forces can also be expressed as a second-order tensor $\mathbf{R}_f$, which must be diagonal for the hypothesis of collinearity between forces and displacements, while the geometric scale is already generalised by the transformation matrix $\mathbf{\Lambda}$.

By performing again the computations, the kinematic similarity condition requires that

$$\mathbf{R}_V = \frac{1}{r_T} \mathbf{\Lambda}, \tag{4.63}$$

where the ratio of times is a scalar. The condition of dynamic similarity requires that

$$\mathbf{R}_V \cdot \mathbf{R}_V = \mathbf{\Lambda} \cdot \mathbf{R}_f, \tag{4.64}$$

where the product between the tensors is rows by columns. The nondiagonal terms of the tensor $\mathbf{R}_V$ must be null since $\mathbf{\Lambda} \cdot \mathbf{R}_f$ is diagonal, while for the three diagonal terms,

$$\frac{r_{V_x}^2}{\lambda_x r_{f_x}} = \frac{r_{V_y}^2}{\lambda_y r_{f_y}} = \frac{r_{V_z}^2}{\lambda_z r_{f_z}} = 1, \tag{4.65}$$

where we have assumed $\lambda_x \equiv \lambda_{11}, \ \lambda_y \equiv \lambda_{22}, \ \lambda_z \equiv \lambda_{33}$. In addition, isotropy of the time-scale ratio yields to:

$$r_T = \frac{\lambda_x}{r_{V_x}} = \frac{\lambda_y}{r_{V_y}} = \frac{\lambda_z}{r_{V_z}} \rightarrow \frac{r_{V_x}}{r_{f_x}} = \frac{r_{V_y}}{r_{f_y}} = \frac{r_{V_z}}{r_{f_z}}. \tag{4.66}$$

4.1.7   The Constitutive Similarity and the Other Criteria of Similarity

There are other similarity criteria that apply to the constitutive equations. In constitutive similarity, the materials in the model and in the prototype are characterised by homologous rheological properties. The rheological properties of materials are defined by a mechanical constitutive equation that relates the stress state to the strain (or to the strain rate in the case of fluids). A constitutive equation must satisfy some principles (see Tanner 2000):

- it must be independent from the system of units of measurement and can be expressed in dimensionless form;
- it must be independent from the system of coordinates and can be expressed in tensorial form;
- it must satisfy the second principle of thermodynamics;
- it must be independent from the reference system, that is it must respect the principle of *material objectivity*.

In many cases of practical interest, the tensorial formulation of the constitutive equation is complex or does not exist. In these cases, there are empirical or semiempirical equations that, for example, for an elastic body correlate the stress state to the deformation state as a function of temperature (in the phenomena of creep and relaxation), as a function of the space (possible inhomogeneity) and orientation (possible anisotropy). The interaction of stresses acting in two distinct directions, described through Poisson's ratio, and the interaction between two neighbouring points, by means of the spatial gradient of the stress, are often considered.

For example, for an elastic body in uniaxial geometry, an empirical expression of the constitutive equation is (Harris et al. 1962):

$$
\sigma\left(x,\, \varepsilon,\, t,\, \theta,\, \dot{\varepsilon}\right) = \sigma_0\left(x,\, \varepsilon_0,\, t_0,\, \theta_0,\, \dot{\varepsilon}_0\right) + \sum_{n=1}^{\infty} \frac{(\varepsilon - \varepsilon_0)^n}{n!} \left\{ \frac{\partial^n \sigma}{\partial \varepsilon^n}\bigg|_{\varepsilon=\varepsilon_0} + \right.
$$

$$
\sum_{j=1}^{\infty} \frac{(t - t_0)^j}{j!} \frac{\partial^{n+j}\sigma}{\partial \varepsilon^n \partial t^j}\bigg|_{t=t_0} + \sum_{l=1}^{\infty} \frac{(\theta - \theta_0)^l}{l!} \frac{\partial^{n+l}\sigma}{\partial \varepsilon^n \partial \theta^l}\bigg|_{\theta=\theta_0} +
$$

$$
\left. \sum_{k=1}^{\infty} \frac{(\dot{\varepsilon} - \dot{\varepsilon}_0)^k}{k!} \frac{\partial^{n+k}\sigma}{\partial \varepsilon^n \partial \dot{\varepsilon}^k}\bigg|_{\dot{\varepsilon}=\dot{\varepsilon}_0} \right\}\Bigg|_{\sigma=\sigma_0},
$$

$$
\tag{4.67}
$$

where ε is the strain, t is the time, θ is the temperature, and $\dot{\varepsilon}$ is the strain rate. The maximum strain is expressed as

$$
\varepsilon_{max}\left(x,\, t,\, \theta,\, \dot{\varepsilon}\right) = \varepsilon_{max}\left(x,\, t_0,\, \theta_0,\, \dot{\varepsilon}_0\right) + \sum_{j=1}^{\infty} \frac{(t - t_0)^j}{j!} \frac{\partial^j \varepsilon_{max}}{\partial t^j}\bigg|_{t=t_0} +
$$

$$
\sum_{l=1}^{\infty} \frac{(\theta - \theta_0)^l}{l!} \frac{\partial^l \varepsilon_{max}}{\partial \theta^l}\bigg|_{\theta=\theta_0} + \sum_{k=1}^{\infty} \frac{(\dot{\varepsilon} - \dot{\varepsilon}_0)^k}{k!} \frac{\partial^k \varepsilon_{max}}{\partial \dot{\varepsilon}^k}\bigg|_{\dot{\varepsilon}=\dot{\varepsilon}_0}. \tag{4.68}
$$

If the material has memory, the equations must be written for both the loading and unloading phases. The conditions of constitutive similarity for Eq. (4.67) are:

$$\begin{cases} r_{\left(\frac{\partial^n \sigma}{\partial \varepsilon^n}\right)} = r_\sigma, \\[2mm] r_{\left(\frac{\partial^j}{\partial t^j} \frac{\partial^n \sigma}{\partial \varepsilon^n}\right)} = r_\sigma \, r_t^{-j}, \\[2mm] r_{\left(\frac{\partial \varepsilon}{\partial t}\right)} = r_t^{-1}, \\[2mm] r_{\left(\frac{\partial^l}{\partial \theta^l} \frac{\partial^n \sigma}{\partial \varepsilon^n}\right)} = r_\sigma \, r_\theta^{-l}, \\[2mm] r_{\left(\frac{\partial^k}{\partial \dot\varepsilon^k} \frac{\partial^n \sigma}{\partial \varepsilon^n}\right)} = r_\sigma \, r_t^{k}. \end{cases} \tag{4.69}$$

As a consequence of Eq. (4.68), the other conditions are:

$$\begin{cases} r_{\left(\frac{\partial^j \varepsilon_{max}}{\partial t^j}\right)} = r_t^{-j}, \\[2mm] r_{\left(\frac{\partial^l \varepsilon_{max}}{\partial \theta^l}\right)} = r_\theta^{-l}, \\[2mm] r_{\left(\frac{\partial^k \varepsilon_{max}}{\partial \dot\varepsilon^k}\right)} = r_t^{k}. \end{cases} \tag{4.70}$$

The similarity conditions must be satisfied for all points and for each n, j, k and l.

Example 4.3 Let us consider a material in the prototype with the following empirical description of the rheological behaviour:

$$\begin{cases} \sigma_p = \left(a + b\,t + c\,\theta^2 + d\,\dot\varepsilon\right)\varepsilon + (e + g\,\theta)\,\varepsilon^2, \\[2mm] \varepsilon_{p,max} = a_1 + b_1\,t + c_1\,\theta + d_1\,\dot\varepsilon. \end{cases} \tag{4.71}$$

We wish to calculate the coefficients of a material in constitutive similarity. Using the definitions of the coefficients of Eqs. (4.67)–(4.68), we calculate:

$$\begin{cases} \dfrac{\partial \sigma}{\partial \varepsilon}(t_0,\,\theta_0,\,\dot\varepsilon_0) = a, & \dfrac{\partial^2 \sigma}{\partial \varepsilon \partial t} = b, \\[3mm] \dfrac{\partial^3 \sigma}{\partial \varepsilon \partial \theta^2} = 2\,c, & \dfrac{\partial^2 \sigma}{\partial \varepsilon \partial \dot\varepsilon} = d, \\[3mm] \dfrac{\partial^2 \sigma}{\partial \varepsilon^2}(t_0,\,\theta_0,\,\dot\varepsilon_0) = 2\,e, & \dfrac{\partial^3 \sigma}{\partial \varepsilon^2 \partial \theta} = g, \\[3mm] \dfrac{\partial \varepsilon_{max}}{\partial t} = b_1, & \dfrac{\partial \varepsilon_{max}}{\partial \theta} = c_1, \\[3mm] \dfrac{\partial \varepsilon_{max}}{\partial \dot\varepsilon} = d_1. \end{cases} \tag{4.72}$$

On the basis of the conditions of similarity, the model yields:

$$\begin{cases} \dfrac{\partial \sigma}{\partial \varepsilon}(t_0,\ \theta_0,\ \dot\varepsilon_0) = r_\sigma\, a, & \dfrac{\partial^2 \sigma}{\partial \varepsilon \partial t} = r_\sigma\, r_t^{-1}\, b, \\[2mm] \dfrac{\partial^3 \sigma}{\partial \varepsilon \partial \theta^2} = 2\, r_\sigma\, r_\theta^{-2}\, c, & \dfrac{\partial^2 \sigma}{\partial \varepsilon \partial \dot\varepsilon} = r_\sigma\, r_t\, d, \\[2mm] \dfrac{\partial^2 \sigma}{\partial \varepsilon^2}(t_0,\ \theta_0,\ \dot\varepsilon_0) = 2\, r_\sigma\, e, & \dfrac{\partial^3 \sigma}{\partial \varepsilon^2 \partial \theta} = r_\sigma\, r_\theta^{-1}\, g, \\[2mm] \dfrac{\partial \varepsilon_{max}}{\partial t} = r_t^{-1}\, b_1, & \dfrac{\partial \varepsilon_{max}}{\partial \theta} = r_\theta^{-1}\, c_1, \\[2mm] \dfrac{\partial \varepsilon_{max}}{\partial \dot\varepsilon} = r_t\, d_1. \end{cases} \qquad (4.73)$$

Therefore, a material in similarity must have the following constitutive equation:

$$\begin{cases} \sigma_m = \left(r_\sigma\, a + r_\sigma\, r_t^{-1}\, b\, t + r_\sigma\, r_\theta^{-2}\, c\, \theta^2 + r_\sigma\, r_t\, d\, \dot\varepsilon \right) \varepsilon + \\[2mm] \quad \left(r_\sigma\, e + r_\sigma\, r_\theta^{-1}\, g\, \theta \right) \varepsilon^2, \\[2mm] \varepsilon_{m,max} = a_1 + r_t^{-1}\, b_1\, t + r_\theta^{-1}\, c_1\, \theta + r_t\, d_1\, \dot\varepsilon. \end{cases} \qquad \therefore \ (4.74)$$

The high number of coefficients involved and the impossibility of producing programmable materials, with rheological characteristics that can be established and imposed, discourages the adoption of strict criteria of constitutive similarity. In many cases, it is possible to neglect the dependence on temperature and time, and it is also possible to consider only the first-order contribution of ε. Thus, constitutive similarity requires only that $\sigma_m = r_\sigma\, a\, \varepsilon$, $\varepsilon_{m,max} = \varepsilon_{p,max}$. Since it must be $r_\varepsilon = 1$, it must also result $r_\sigma = r_a$.

If the material is a *viscous fluid* (otherwise defined as a Stokes fluid or Reiner-Rivlin fluid) using the Cayley-Hamilton Theorem, the constitutive equation has the following structure:

$$T_{ij} = A\, \delta_{ij} + B\, D_{ij} + C\, \left(D_{ij}\right)^2, \qquad (4.75)$$

where $T_{ij} \equiv \mathbf{T}$ is the stress tensor, $D_{ij} \equiv \mathbf{D}$ is the strain rate tensor, δ_{ij} is the tensor of Kronecker, A, B and C are functions of the principal invariants of the tensor of the strain rates, written as I_1, I_2 and I_3. The structure of the three functions is different for different categories of materials, necessitating the definition of *material functions*. It can be demonstrated that the two functions A and B cannot be constants if C is not zero. The case where C is null and $\mathbf{T} = f(\mathbf{D})$ is linear describes a Newtonian fluid, with the following constitutive equation:

$$\mathbf{T} = (-p + \xi\, \nabla \mathbf{V})\, \mathbf{I} + 2\, \mu\, \mathbf{D}, \qquad (4.76)$$

where ξ is the bulk viscosity and μ is the dynamic viscosity.

The conditions of constitutive similarity for a Newtonian fluid are:

$$r_\tau = r_p = r_\xi\, \frac{r_V}{\lambda} = r_\mu\, \frac{r_V}{\lambda}. \qquad (4.77)$$

For an ideal gas, $3\lambda + 2\mu = 0$, and one condition is superfluous. If the flow field is isochoric, the bulk viscosity is not relevant, and the corresponding condition is superfluous. Finally, if the pressure scales as $r_p = r_\rho\, r_V^2$, constitutive similarity is equivalent to Reynolds similarity:

$$\frac{r_\rho\, r_V\, \lambda}{r_\mu} = 1. \tag{4.78}$$

4.2 The Condition of Similarity on the Basis of Dimensional Analysis

Two physical processes involving the same variables x_i and described by the same homogeneous function $f(x_i) = 0$, are defined to exhibit *similarity* if the ratios $r_{x_i} = x_i''/x_i'$ between the measure of the i-th variable, read for the second process (x_i''), and the measure of the same variable, read for the first process (x_i'), are known.

We now demonstrate that a homogeneous function that allows similarity without any constraint on the ratios r_{x_i} is a power function of the type

$$f(x_i) = C\, x_1^{\delta_1}\, x_2^{\delta_2} \cdots x_n^{\delta_n} = 0. \tag{4.79}$$

The condition of similarity implies that

$$f\left(x_i''\right) = f\left(x_i'\right) = 0 \rightarrow f\left(r_{x_i}\, x_i'\right) = f\left(x_i'\right) = 0. \tag{4.80}$$

As a property of homogeneous functions (see Appendix A), we can collect scale ratios with a common factor,

$$f\left(r_{x_i}\, x_i'\right) = \Phi\left(r_{x_i}\right) f\left(x_i'\right) = 0 \tag{4.81}$$

and differentiating with respect to r_{x_1} yields

$$\frac{\partial f\left(r_{x_i}\, x_i'\right)}{\partial r_{x_1}} = \frac{\partial \Phi}{\partial r_{x_1}} f\left(x_i'\right); \tag{4.82}$$

hence,

$$x_1'\, \frac{\partial f\left(r_{x_i}\, x_i'\right)}{\partial \left(r_{x_1}\, x_1'\right)} = \frac{\partial \Phi}{\partial r_{x_1}} f\left(x_i'\right). \tag{4.83}$$

This expression must be valid for any value of $r_{x_1},\, r_{x_2},\, \ldots,\, r_{x_n}$, including the value $r_{x_i} = 1$, $(i = 1,\, 2,\, \ldots,\, n)$. Therefore,

$$x_1' \frac{\partial f(x_i')}{\partial x_1'} = \frac{\partial \Phi}{\partial r_{x_1}}\bigg|_{r_{x_i}=1} f(x_i'). \tag{4.84}$$

The derivative of the function Φ is calculated for a unit value of its argument and is a number that we indicate with δ_1. Following separation of variables, Eq. (4.84) is integrated:

$$\ln f = \delta_1 \ln x_1' + \text{const} \rightarrow f(x_i') = C_1 x_1'^{\delta_1}, \tag{4.85}$$

where C_1 is a dimensionless coefficient. Repeating the procedure for all other variables yields

$$f(x_i) = C x_1^{\delta_1} x_2^{\delta_2} \cdots x_n^{\delta_n}. \qquad \text{(q.e.d.)} \tag{4.86}$$

Unfortunately, there is no physical phenomenon governed by a homogeneous power function, and the scales r_{x_i} cannot be arbitrary. The simplest way to impose the condition of similarity is to express the homogeneous function that describes the phenomenon under analysis as a function of dimensionless groups using Buckingham's Theorem:

$$f(x_1, x_2, \ldots, x_n) = 0 \rightarrow \tilde{f}(\Pi_1, \Pi_2, \ldots, \Pi_{n-k}) = 0. \tag{4.87}$$

The condition of similarity is satisfied if the corresponding dimensionless groups, which in the two physical processes have the expression

$$\Pi_i'' = \frac{x_{k+i}''}{\left(x_1''\right)^{\alpha_i} \left(x_2''\right)^{\beta_i} \cdots \left(x_k''\right)^{\delta_i}}, \quad \Pi_i' = \frac{x_{k+i}'}{\left(x_1'\right)^{\alpha_i} \left(x_2'\right)^{\beta_i} \cdots \left(x_k'\right)^{\delta_i}},$$

$$(i = 1, 2, \ldots, n-k), \tag{4.88}$$

have the same numerical value, or

$$\Pi_i'' = \Pi_i', \quad (i = 1, 2, \ldots, n-k). \tag{4.89}$$

Since we have

$$\Pi_i'' = \frac{x_{k+i}''}{\left(x_1''\right)^{\alpha_i} \left(x_2''\right)^{\beta_i} \cdots \left(x_k''\right)^{\delta_i}} = \frac{r_{x_{k+i}} x_{k+i}'}{\left(r_{x_1} x_1'\right)^{\alpha_i} \left(r_{x_2} x_2'\right)^{\beta_i} \cdots \left(r_{x_k} x_k'\right)^{\delta_i}} =$$

$$\frac{r_{x_{k+i}}}{r_{x1}^{\alpha_i} r_{x2}^{\beta_i} \cdots r_{xk}^{\delta_i}} \frac{x_{k+i}'}{\left(x_1'\right)^{\alpha} \left(x_2'\right)^{\beta_i} \cdots \left(x_k'\right)^{\delta_i}} = \frac{r_{x_{k+i}}}{r_{x1}^{\alpha_i} r_{x2}^{\beta_i} \cdots r_{xk}^{\delta_i}} \Pi_i',$$

$$(i = 1, 2, \ldots, n-k), \tag{4.90}$$

the set of Eq. (4.88) yields the following $(n-k)$ equations:

$$\frac{r_{x_{k+i}}}{r_{x1}^{\alpha_i} r_{x2}^{\beta_i} \cdots r_{xk}^{\delta_i}} = 1, \quad (i = 1, 2, \ldots, n-k). \tag{4.91}$$

Hence, the two phenomena are similar if the dimensionless groups, calculated by introducing the scale ratios of the variables in place of the variables, assume a value of unity.

The procedure results in a system of $(n - k)$ equations in the n unknowns represented by the n scale ratios r_{x_i}, which admits ∞^k solutions and leaves k degrees of freedom in selecting the scales: we can fix k scales and then are able to compute the residual $(n - k)$ scales. However, additional constraints are often added that reduce the number of degrees of freedom, and sometimes, the system of equations ends up admitting only the trivial solution $r_{x_i} = 1$.

Example 4.4   Let us analyse a structure statically loaded in the elastic regime. The quantities involved are

- the geometric variables that define the structure, such as a length l and the ratios of the other geometric dimensions with respect to l, indicated with $h_1, h_2, \ldots, h_n$;
- the mechanical characteristics of the material of the structure, such as Young's modulus E and Poisson's ratio ν;
- the load condition, indicated with a concentrated load P and the ratios of the other concentrated loads to P, i.e., $s_1, s_2, \ldots, s_m$;
- the point of application of the concentrated loads, $p_0, p_1, \ldots, p_m$, expressed in a dimensionless form as a ratio to the reference dimension l;
- the angle of inclination of these loads, $\theta_0, \theta_1, \ldots, \theta_m$.

We wish to describe the dependence of the normal stress σ (a governed variable) acting near the point (x, y, z). The typical equation is

$$\sigma = f(x, y, z, E, l, P, \nu,$$
$$h_1, \ldots, h_n, s_1, \ldots, s_m, p_0, p_1, \ldots, p_m, \theta_0, \theta_1, \ldots, \theta_m). \quad (4.92)$$

Without losing generality, for simplicity, we analyse the normal stress acting in a specific section of abscissa x due to the action of a single concentrated orthogonal load. The typical equation simplifies as

$$\sigma = f(x, E, l, P, \nu). \quad (4.93)$$

The dimensional matrix of the six variables is

$$
\begin{array}{c|cccccc}
 & \sigma & x & E & l & P & \nu \\
\hline
M & 1 & 0 & 1 & 0 & 1 & 0 \\
L & -1 & 1 & -1 & 1 & 1 & 0 \\
T & -2 & 0 & -2 & 0 & -2 & 0
\end{array}
\quad (4.94)
$$

and has rank 2 (the first and last rows are a linear combination). Selecting 2 independent variables, for example, E and l, we can define $(6 - 2) = 4$ dimensionless groups, for instance, the following:

$$\Pi_1 = \frac{\sigma}{E}, \quad \Pi_2 = \frac{x}{l}, \quad \Pi_3 = \frac{P}{E\,l^2}, \quad \Pi_4 = \nu, \tag{4.95}$$

where ν is already dimensionless. The functional relationship between dimensionless groups is

$$\frac{\sigma}{E} = \tilde{f}\left(\frac{x}{l}, \frac{P}{E\,l^2}, \nu\right). \tag{4.96}$$

In a similar way, it is possible to define the functional relation for other variables, such as the deflection δ:

$$\frac{\delta}{l} = \tilde{f_1}\left(\frac{x}{l}, \frac{P}{E\,l^2}, \nu\right). \tag{4.97}$$

We observe that these expressions are very general and are valid even in the case of large deformations, provided that the criteria of completeness of the set of variables involved are met: for nonlinear elastic materials or in a plastic regime, it is necessary to add the other variables that intervene in the process.

The criterion of similarity requires that:

$$\begin{cases} \Pi_1' = \Pi_1'' \rightarrow r_\sigma = r_E, \\ \Pi_2' = \Pi_2'' \rightarrow r_x = \lambda, \\ \Pi_3' = \Pi_3'' \rightarrow r_P = r_E\,\lambda^2, \\ \nu' = \nu'' \rightarrow r_\nu = 1. \end{cases} \tag{4.98}$$

The unknowns are the 6 scales, constrained by 4 independent equations. For example, by selecting a geometric scale λ and the ratio of Young's modulus r_E, we can calculate all the other scales. There is the constraint that Poisson's ratio must assume the same value in the model and in the prototype.

4.3 The Condition of Similarity on the Basis of Direct Analysis

The similarity condition on the basis of dimensional analysis, described in Sect. 4.2, has the advantage of generality but leaves a large degree of indeterminacy.

Less indeterminacy is left if we already know the equation(s) describing the physical process. In this case, it is sufficient to apply the dimensional homogeneity criterion to obtain a system of equations in the variables r_{x_i}. For example, if the process can be expressed with an equality of two power functions, necessarily having the same dimension:

$$P_1(x_i) = P_2(x_i), \tag{4.99}$$

similarity requires that:

$$
\begin{cases} P_1(x_i') = P_2(x_i') \\ P_1(x_i'') = P_2(x_i'') \end{cases} \rightarrow
$$

$$
P_1(r_{x_i} x_i') = P_2(r_{x_i} x_i') \rightarrow P_1(r_{x_i})\, P_1(x_i') = P_2(r_{x_i})\, P_2(x_i'). \tag{4.100}
$$

These conditions are satisfied providing that

$$
P_1(r_{x_i}) = P_2(r_{x_i}). \tag{4.101}
$$

As an example, let us suppose we wish to evaluate the similarity rules for the flow rate in a channel, and suppose we adopt the Chézy formula with the Gauckler-Strickler coefficient:

$$
U = k\, R^{1/6} \sqrt{R\, i_f}. \tag{4.102}
$$

Direct analysis indicates that the following equation must be satisfied:

$$
r_U = r_k\, \lambda^{2/3}\, r_{i_f}^{1/2}. \tag{4.103}
$$

Equation (4.103) leaves 2 degrees of freedom in the choice of the 3 unknown scales r_U, r_k and λ (we assume, for simplicity, $r_{i_f} = 1$): with 2 scales arbitrarily selected, the third scale is computed.

If the power functions contain functions of dimensionless groups, the condition of similarity also requires the equality of the dimensionless groups in the model and in the prototype.

For example, we wish to analyse the pressure drop in a circular pipeline on the basis of Darcy's law:

$$
J = f\!\left(\mathrm{Re},\ \frac{\varepsilon}{D}\right) \frac{U^2}{2g}\frac{1}{D}, \tag{4.104}
$$

where J is the energy drop per unit weight and per unit length, $f(\ldots)$ is the friction factor (a function of the Reynolds number and of the relative roughness ε/D), U is the average velocity in the pipeline, g is the acceleration of gravity and D is the diameter of the pipeline. Direct analysis indicates that

$$
\begin{cases} r_J = \dfrac{r_U^2}{r_g}\dfrac{1}{\lambda}, \\[2ex] \dfrac{r_\rho\, r_U\, \lambda}{r_\mu} = 1, \\[2ex] r_\varepsilon = \lambda, \end{cases} \tag{4.105}
$$

which leaves 3 degrees of freedom in the choice of 4 unknown ratios. The second and third equations impose an equal Reynolds number and relative roughness, respectively, in the model and the prototype.

In the general case, if the physical process is described by an equation such as

$$\sum_{j=1}^{N} \varsigma_j P_j(x_i) = 0,$$ (4.106)

where ς_j are dimensionless terms, for instance, constants or transcendental functions with dimensionless arguments, i.e.,

$$\varsigma_j = \varsigma_j(\Pi_k(x_i)), \quad (k = 1, 2, \ldots, m),$$ (4.107)

and P_j are the terms of the equation, the condition of similarity requires that

$$\begin{cases} \sum_{j=1}^{N} \varsigma_j(\Pi_k(x_i)) P_j(x_i) = 0, \\ \sum_{j=1}^{N} \varsigma_j(\Pi_k(r_{x_i} x_i)) P_j(r_{x_i} x_i) = 0. \end{cases}$$ (4.108)

The second equation in (4.108) can be re-written as

$$\sum_{j=1}^{N} \varsigma_j(\Pi_k(r_{x_i}) \Pi_k(x_i)) P_j(r_{x_i}) P_j(x_i) = 0,$$ (4.109)

and the system of equations is satisfied if

$$\begin{cases} \Pi_k(r_{x_i}) = 1, \quad (k = 1, 2, \ldots, m), \\ P_1(r_{x_i}) = P_2(r_{x_i}) = \ldots = P_N(r_{x_i}). \end{cases}$$ (4.110)

In summary, we can write $(m + N - 1)$ equations for the scales; m corresponds to the dimensionless groups involved, and $(N - 1)$ corresponds to the condition of homogeneity imposed on the N terms of the equation that defines the process.

Example 4.5 Theory indicates that the wavelength l of sea gravity waves is

$$l = \frac{g t^2}{2\pi} \tanh\left(\frac{2\pi h}{l}\right),$$ (4.111)

where t is the period of the wave, h is the local depth, and g is the acceleration of gravity. We wish to calculate the scales if the scale for the depth is λ.

Direct analysis indicates that

$$\begin{cases} r_l = r_g\, r_t^2 \equiv r_t^2, \\ r_l = \lambda, \end{cases} \qquad (4.112)$$

hence,

$$\begin{cases} r_t = \sqrt{\lambda}, \\ r_l = \lambda. \end{cases} \qquad \therefore \quad (4.113)$$

In a small-scale model with $\lambda = 1/25$, the time ratio is $r_t = 1/5$, and the wavelength ratio is $r_l = 1/25$.

The similarity conditions based on the criteria of the direct analysis are valid only for the interval of validity of the equation used to describe the physical process: in the example of sea gravity waves, the calculated scales are valid until the surface tension becomes dominant, when a different expression of the wavelength replaces Eq. (4.111).

4.4   An Extension of the Concept of Similarity: Some Scale Laws in Biology

Similarity is a tool used not only to extrapolate the measurements made in the model to the prototype, but also to analyse the behaviour of some complex systems that have scaling-invariant properties.

If we observe the profile of the ribs or the canopy of a tree, the geometry of the contour (in the first case) or of the branches (in the second case) appears similar or identical when comparing what is observed at different levels of detail. This property is also referred to as *internal homothety*. Already Galilei in his *Discorsi e dimostrazioni matematiche intorno a due nuove scienze*, 1683, wondered about the effect of a dilation of the same factor for all the linear dimensions of a living being, concluding that such a giant animal could not support its own weight above a certain size limit. Therefore, the skeleton of large animals must necessarily occupy a larger portion of the body volume than that of smaller animals. Bones should be of a much stronger material in the giant, or they should not scale isometrically (see Fig. 4.9). In fact, the strength of bones and tendons varies according to λ^2, while the weight scales with λ^3 and the average stress scales with $\lambda^3/\lambda^2 = \lambda$. This is why when the maximum limit of allowable stress is exceeded, the cross-section of the bones should scale by more than λ^2. For animals living in water, a significant contribution to the reduction of loads comes from the buoyant force of Archimedes: whales have, in proportion, smaller bones than they would have if they lived on land. This is confirmed by the

Fig. 4.9 Bones reproduced on a distorted geometric scale (from Galileo Galilei, *Discorsi e dimostrazioni matematiche intorno a due nuove scienze*, 1683)

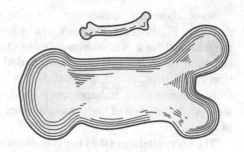

fact that in the case of beaching, without the buoyant force of Archimedes, death occurs by suffocation and crushing of the animal under its own weight. The almost neutral buoyancy of most animals living in water allows much larger sizes than for animals living outside water.

As another example, it is known that children move with greater agility on stones or on irregular surfaces than adults. This is because the body weight varies according to λ^3, while the footprint area of the feet varies according to λ^2; therefore, the contact pressure varies according to λ and is lower in children than in adults. The contact pressure is so much reduced at small λ that it also compensates for the more delicate biological tissues of children.

Biological systems and living organisms are of such complexity that deterministic analyses are almost unaffordable, and a possible quantitative analysis uses scale laws to correlate macroscopic processes with cellular or molecular processes. Some of these scale laws and the analysis that correlates them are so well founded that they correctly interpret many variables in a range of more than 20 orders of magnitude. For example, it is possible to predict the dependence on body mass (i) of the basal metabolic rate (the power needed to ensure life), (ii) of certain characteristic times (the lifespan, the period of heart pulsation), and (iii) of many geometric characteristics (the length of the aorta, the height of a tree).

A relation is defined *allometric*, as opposed to *isometric*, if it is a parametrisation of the measure of some variable (relative, for example, to the physiology of an organism), in relation to the measure of some other variable (usually a global dimension, such as mass or volume), with a non-linear proportionality. For instance, the heart rate varies according to the body mass to the power of $-1/4, f \sim M^{-1/4}$; the lifespan is $\sim M^{1/4}$ and, therefore, the smallest organisms die earlier; again, the radius and length of both the aorta and the branches of the trees are respectively $\sim M^{3/8}$ and $\sim M^{1/4}$.

The best known allometric relation correlates the basal metabolic rate to the body mass with an exponent of $3/4 \pm 0.10$, valid over a range of 6 orders of magnitude (*Kleiber's law*, Kleiber 1947). The same exponent can be measured, albeit with different forms of normalisation, even at the molecular level, for instance, in mitochondria, extending its range to 27 orders of magnitude. According to Kleiber's law, an increase in mass leads to an increase in the efficiency of living organisms. By comparison, in internal combustion engines, power is isometric with mass.

Some allometric relationships combine to provide invariants or isometric relationships. For instance, the heart rate by lifespan is invariant, as if organisms were programmed for a maximum number of cycles: to be performed at high frequency and short times in small organisms, and at low frequency and for longer times in larger organisms.

It is an important and apparently anomalous fact that most dependent variables are a power function of body mass, or other quantities, with an exponential multiple of $\frac{1}{4}$.

The interpretation of the known scale laws refers to the mechanisms of transport of nutrients and wastes, which are based on some structures partly invariant and partly variant with the scale. The systems of nutrient distribution and waste collection can be described as a branched network in which the size of the vessels decreases and structured with slightly different characteristics in plants and mammals. In plants, the vessels are in parallel, and there are no large vessels corresponding to the aorta; here, the exchange mechanism and the nature of the fluids are different (liquids or gases), as is the pumping system (pulsating compression such as the heart in the cardiovascular circle, pulsating bellows in the respiratory system, osmotic and vapour pressure such as in plants).

It seems useful to present here one of these models to highlight the great potential of the analysis of the similarity of complex systems. We observe that many systems of engineering interest are complex in the same sense as biological systems.

Let us consider a model representative of the circulatory system in a mammal in cases where:

(a) a network is necessary to supply the organism, and this network has a branched hierarchical structure that reaches the whole organism;
(b) the terminal parts of the network, where the exchange of nutrients and wastes takes place, are scale invariant, that is, they have dimensions that do not depend on the size of the organism;
(c) the system has evolved in such a way as to optimise efficiency.

For hypothesis (a), the vessels of the network are classified in hierarchical levels, where the level 0 corresponds to the maximum size (the aorta) and the level N corresponds to the minimum size (the capillaries); see Fig. 4.10.

The volumetric flow rate Q_k for vessels at the k level for $k = 0, \ldots, N$ is

$$Q_k = \pi r_k^2 \bar{u}_k, \tag{4.114}$$

where r_k is the radius and $\bar{u}_k$ is the cross-section average fluid velocity. The subscript refers the variable to the hierarchical level k. Due to the branched structure, there is no bypass between noncontiguous levels, and at the same hierarchical level, the vessels function in parallel. Applying the conservation of the volumetric flow rate yields

$$Q = R_k Q_k = R_k \pi r_k^2 \bar{u}_k \equiv R_N \pi r_N^2 \bar{u}_N, \tag{4.115}$$

Fig. 4.10 Scheme of a blood vessel network (modified from Antonets et al. 1991)

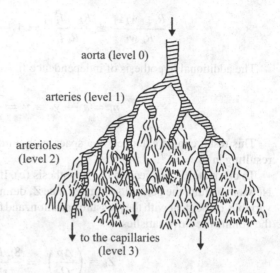

aorta (level 0)

arteries (level 1)

arterioles (level 2)

to the capillaries (level 3)

where R_k the number of vessels at the level k. Since the nutrients needed for basal metabolism are transported by the fluid flowing in the vessels with a linear proportionality (doubling the volumetric flow rate doubles the mass flow rate of the nutrients), if the basal metabolic rate is $B \sim M^a$, then $Q \sim M^a$ with the same exponent, equal to 3/4 according to Kleiber. Moreover, for hypothesis (b), the radius of the capillary vessels and the average velocity of the fluid flowing in the capillaries are scale invariant. The result is $R_N \sim M^a$. For $a = 3/4$, this relationship indicates that the number of capillaries grows less than isometrically. Since the number of cells grows isometrically (the characteristic size of cells in an organism does not vary significantly), this implies a greater efficiency of the capillaries for larger organisms: each capillary serves a greater number of cells.

To describe the network, it is necessary to identify how the radius and length of the vessels vary with the rank. We introduce the following scale factors:

$$\beta_k = \frac{r_{k+1}}{r_k}, \quad \gamma_k = \frac{l_{k+1}}{l_k}, \quad n_k = \frac{R_{k+1}}{R_k}, \tag{4.116}$$

where l_k is the length of the vessel at rank k. To ensure that all cells are fed, it is necessary that the network extends to the whole organism and completely fills the space (*space-filling* hypothesis). The capillary vessels directly feed the cells, and if w_N is the volume of cells served by each capillary, we must have $W = R_N w_N$, where W is the volume of the organism. In a network with many branches, this concept can be extended to all ranks, and we find that $W = R_k w_k$, where w_k is the volume of cells served by each vessel of k rank. Since the vessels have a much smaller radius than the length, $w_k \sim l_k^3$ for sufficiently large k. At each level, the relation $W \approx R_k w_k \sim R_k l_k^3$ holds. Ultimately,

$$\frac{R_{k+1} w_{k+1}}{R_k w_k} \equiv \frac{R_{k+1} l_{k+1}^3}{R_k l_k^3} = 1 \rightarrow n_k \gamma_k^3 = 1. \tag{4.117}$$

The additional hypothesis of independence from the level k requires that

$$n_k = n \rightarrow \gamma_k \equiv \gamma = \frac{1}{n^3}. \tag{4.118}$$

This result can be extended to a space with a number of dimensions equal to d, resulting in $\gamma_k = n^{-d}$.

To calculate β_k, we need to use hypothesis (c). If we consider a laminar flow of a Newtonian fluid, the generalised impedance Z, defined as the ratio between the stress variable, coincident with the pressure variation, and the flow variable, coincident with the volumetric flow, satisfies

$$Z_k \equiv \left(\frac{\Delta p}{Q}\right)_k = \frac{8\mu l_k}{\pi r_k^4}, \tag{4.119}$$

where Δp is the pressure variation and μ is the dynamic viscosity. Due to the structure of the network, vessels of the same rank are in parallel, while vessels of different ranks are in cascade; therefore, the network impedance is equal to

$$Z \equiv \left(\frac{\Delta p}{Q}\right) = \sum_{k=0}^{N} \frac{8\mu l_k}{\pi r_k^4} \frac{1}{R_k}. \tag{4.120}$$

The greatest contribution is that of the capillaries, which results in

$$Z \sim R_N^{-1} \sim M^{-a}, \tag{4.121}$$

which indicates that flow resistance is reduced for a larger network. From this point of view, a larger organism is also more efficient. We observe that the overpressure in the aorta is equal to

$$\Delta p = Q Z \tag{4.122}$$

and is independent of the mass of the organism, since the flow rate is proportional to M^a and the impedance is proportional to M^{-a}. In fact, the aortic overpressure is the same for a whale (with an aorta diameter of 60 cm) and for a mouse (with an aorta diameter of less than 0.2 mm). The average velocity of the flow in the aorta is also identical in the two mammals.

Adopting hypothesis (c), we impose that with an assigned metabolic rate and a circulating volume of fluid W_b, the power of the heart muscle should be minimised in the presence of a *space-filling* network. Using the Lagrange multipliers, we impose that a linear combination of (i) the power of the heart muscle, (ii) the total volume of fluid in circulation, (iii) the volume of cells employed by the network at each rank and (iv) the mass assumes a minimum value:

$$F(r_k,\ l_k,\ n_k) = P(r_k,\ l_k,\ n_k,\ M) +$$

$$\lambda\, W_b(r_k,\ l_k,\ n_k,\ M) + \sum_{k=0}^{N} \lambda_k\, R_k\, l_k^3 + \lambda_M\, M, \qquad (4.123)$$

where λ, λ_k and λ_M are the Lagrange multipliers. Minimising the power P is equivalent to minimising the impedances. Imposing that

$$\frac{\partial F}{\partial r_k} = \frac{\partial F}{\partial l_k} = \frac{\partial F}{\partial n_k} = 0, \qquad (4.124)$$

we calculate $\beta_k = n^{1/3}$ with $n_k = n$ at every rank k. This is equivalent to an increase in the area of the total cross-section of the vessels as the rank increases and, consequently, to a reduction in the mean velocity of the flow from the aorta to the capillaries.

Using Eq. (4.115) and given $n^N = R_N$, the ratio of the average velocity in the capillaries to the average velocity in the aorta is equal to

$$\frac{\bar{u}_N}{\bar{u}_0} = \frac{\bar{u}_N}{\bar{u}_{N-1}} \frac{\bar{u}_{N-1}}{\bar{u}_{N-2}} \cdots \frac{\bar{u}_1}{\bar{u}_0} =$$

$$\left(\frac{R_{N-1}}{R_N} \frac{r_{N-1}^2}{r_N^2} \right) \left(\frac{R_{N-2}}{R_{N-1}} \frac{r_{N-2}^2}{r_{N-1}^2} \right) \cdots \left(\frac{R_0}{R_1} \frac{r_0^2}{r_1^2} \right) =$$

$$\frac{1}{\left(n\,\beta_N^{-2} \right)} \frac{1}{\left(n\,\beta_{N-1}^{-2} \right)} \cdots \frac{1}{\left(n\,\beta_1^{-2} \right)} = \frac{1}{\left(n^N\,\beta^{-2N} \right)} \equiv \frac{1}{R_N^{1/3}}. \qquad (4.125)$$

For humans, the number of capillary vessels is $R_N \approx 10^{10}$ and results in $\bar{u}_N/\bar{u}_0 \approx 10^{-3}$; hence, the average velocity of the flow in the capillaries is equal to one thousandth of the average velocity in the aorta. We observe that a capillary velocity equal to the aorta velocity would make diffusive exchange with the vessel walls impossible. Table 4.1 lists some parameters of blood vessels in humans.

4.4.1 A Derivation of the Exponent of Kleiber's Law

To better understand this type of analysis, we attempt to derive Kleiber's exponent.

We consider the total volume of fluid in the network of vessels, W_b, as a function of the volume W_N in the capillaries, with

$$W_b = \sum_{k=0}^{N} W_k = W_N \left[1 + \frac{1}{\beta^2\,\gamma\,n} + \frac{1}{\left(\beta^2\,\gamma\,n \right)^2} + \cdots + \frac{1}{\left(\beta^2\,\gamma\,n \right)^N} \right]. \qquad (4.126)$$

Table 4.1 Some geometric characteristics and the number of blood vessels in humans

	Radius (cm)	R	Area of the total cross-section (cm^2)	Wall thickness (cm)	Length (cm)
Aorta	1.25	1	4.9	0.2	50
Arteries	0.2	159	20	0.1	50
Arterioles	$1.5 \cdot 10^{-3}$	$5.7 \cdot 10^7$	400	$2 \cdot 10^{-3}$	1
Capillaries	$3 \cdot 10^{-4}$	$1.6 \cdot 10^{10}$	4500	$1 \cdot 10^{-4}$	0.1
Venules	$1 \cdot 10^{-3}$	$1.3 \cdot 10^9$	4000	$2 \cdot 10^{-4}$	0.2
Veins	0.25	200	40	0.05	2.5
Vena cava	1.5	1	18	0.15	50

The term multiplying the volume W_N is a geometric series:

$$1 + \frac{1}{\beta^2 \gamma n} + \frac{1}{(\beta^2 \gamma n)^2} + \cdots + \frac{1}{(\beta^2 \gamma n)^N} \equiv$$

$$\frac{(\beta^2 \gamma n)^{-(N+1)} - 1}{(\beta^2 \gamma n)^{-1} - 1} = \frac{1}{(\beta^2 \gamma n)^N} \frac{1 - (\beta^2 \gamma n)^{N+1}}{1 - \beta^2 \gamma n}$$

$$\tag{4.127}$$

and approximately results in (West 1999)

$$W_b \approx \frac{W_N (\beta^2 \gamma)^{-N}}{1 - \beta^2 \gamma n}. \tag{4.128}$$

The minimum energy principle results in $W_b \sim M$, and since $W_N \sim M^0$, we also obtain $(\beta^2 \gamma)^{-N} \sim M$. Assuming for the basal metabolic rate B an allometric relation $B \sim M^a$ and considering that the number of capillaries is $R_N \sim M^a$ yields

$$R_N \equiv n^N \sim M^a \rightarrow N \ln n = a \ln M \rightarrow N \ln n =$$

$$a \ln (\beta^2 \gamma)^{-N} \rightarrow a = -\frac{\ln n}{\ln (\beta^2 \gamma)}. \tag{4.129}$$

If the hypothesis leading to Eq. (4.118), which results in $\gamma = n^{-1/3}$, is fulfilled, and if $\beta = n^{-1/3}$, then $a = 1$ is calculated, a different value than indicated in Kleiber's law.

In fact, the correct derivation of Kleiber's law requires the analysis of the hydraulic behaviour of the network, including the pulsating nature of the flow (this is not the case in plant lymphatic networks).

A linearised solution indicates that a pressure wave propagates in a fluid in a pipe with elastic thin walls with a speed equal to

$$\left(\frac{c}{c_0}\right)^2 = -\frac{J_2\left(i^{3/2}\alpha\right)}{J_0\left(i^{3/2}\alpha\right)},$$

(4.130)

where $c_0 = \sqrt{E\delta/(2\rho_m r)}$ is the Korteweg–Moens velocity, calculated for an elastic wave in a fluid. E is Young's modulus of the wall material, r is the radius, δ is the thickness and ρ_m is the density of the walls. The symbol $\alpha = r(\omega\rho/\mu)^{1/2}$ is the Womersley number, $\omega = 2\pi/T$ is the pulsation, T is the wave period, ρ is the density of the fluid, μ is the dynamic viscosity, J_0 and J_2 are functions of Bessel of order 0 and 2, respectively, and $i^2 = -1$. The impedance is equal to

$$Z = \frac{c_0^2\,\rho}{\pi r^2\,c}.$$

(4.131)

Usually, c and Z are complex functions of ω, and the pressure wave dampens and is dispersive. For large vessels, α assumes high values, and viscosity plays a marginal role; hence, we calculate $c \approx c_0$ and $Z \approx c_0\,\rho_m/\pi r^2$, and the wave propagates without attenuation (the speed and impedance, in fact, are real). For sufficiently small α, the speed and impedance are complex, the latter $\sim r^{-2}$ compared to $\sim r^{-4}$ of the nonpulsating case. The minimum energy condition leads to the conservation condition of the total cross-sectional area at each network level, i.e., $\beta_k = n^{-1/2}$. This condition ensures impedance matching between large vessels, which limits the reflected energy and optimises the transmitted energy. For increasing rank k, $\alpha \to 0$ and the viscosity dominate the flow field, with $c \to 0$. The oscillations are strongly damped, and the average component of the flow dominates, according to Poiseuille's law, with an impedance $\sim r^{-4}$. Hence, for the smaller vessels we obtain $\beta_k = n^{-1/3}$.

In summary, if the flow in the network is pulsating, biological evolution has adapted the impedance between the contiguous levels, and β_k is no longer invariant for the whole network but is a step function that takes the values $\beta_< = n^{-1/2}$ for $k < k_{crit}$ and $\beta_> = n^{-1/3}$ for $k > k_{crit}$.

When taking into account the transition of β, the volume of the fluid has a more cumbersome expression than Eq. (4.128) (West et al. 1997), that is,

$$W_b = \frac{W_N}{\left(\beta_>^2\,\gamma\right)^N}\left\{\left(\frac{\beta_>}{\beta_<}\right)^{2k_{crit}}\frac{1 - \left(n\,\beta_<^2\,\gamma\right)^{k_{crit}}}{1 - \left(n\,\beta_<^2\,\gamma\right)}\right.$$
$$\left. + \left[\frac{1 - \left(n\,\beta_>^2\,\gamma\right)^N}{1 - \left(n\,\beta_>^2\,\gamma\right)} - \frac{1 - \left(n\,\beta_>^2\,\gamma\right)^{k_{crit}}}{1 - \left(n\,\beta_>^2\,\gamma\right)}\right]\right\}.$$

(4.132)

Table 4.2 Allometric exponents of the $y \sim M^{\chi}$ relationship for some variables of the cardiovascular system in mammals (modified from West et al. 1997)

Variable y	Theoretical exponent χ	Experimental exponent χ
Aortic radius r_0	$3/8 = 0.375$	0.36
Aortic pressure Δp_0	$0 = 0.00$	0.032
Aortic flow velocity u_0	$0 = 0.00$	0.07
Volume of blood W_b	$1 = 1.00$	1.00
Recirculation time	$1/4 = 0.25$	0.25
Heart volume	$1 = 1.00$	1.03
Heart rate ω	$-1/4 = -0.25$	-0.25
Heart power	$3/4 = 0.75$	0.74
Number of capillaries R_N	$3/4 = 0.75$	–
Womersley number α	$1/4 = 0.25$	0.25
Capillary density	$-1/12 = -0.083$	-0.095
Total impedance Z	$-3/4 = -0.75$	-0.76
Basal metabolic rate B	$3/4 = 0.75$	0.75

The value of k_{crit} of the transition coincides with the rank where the impedance of the pulsating system and the impedance of Poiseuille become of the same order of magnitude. It can be demonstrated that $k_{crit} \sim N \sim \ln M$. The expression in Eq. (4.132) is dominated by the first contribution, corresponding to the large vessels and $W_b \sim n^{(N+1/3)} k_{crit} \sim n^{4/3} N$. This results in $a = 3/4$, according to the law of Kleiber. This analysis also explains the deviations from Kleiber's law for small animals, where the β transition takes place almost immediately and the contribution of large vessels to the W_b volume of blood is comparable with that of small vessels.

Table 4.2 lists the allometric exponents for some variables of the cardiovascular system in mammals.

Summarising Concepts

● The concept of a group and invariance are susceptible to applicative extensions. We recall that the power function structure of derived variables expressed as a function of fundamental quantities is the consequence of invariance when passing from one system of units to another of the same class; Buckingham's Theorem can be demonstrated within group theory and invoking the principle of covariance. Similarly, we can imagine that model and prototype reproduce a physical process in such a way that this process is invariant.

- From an analytical perspective, model and prototype invariance requires a transformation involving scale relations. The transformation allows the identification of the similarity criteria. The three basic similarities are geometric, kinematic, and dynamic.
- Identification of the transformation, or more simply the similarity criteria, requires knowledge of the mathematical description of the physical process, defined either by a functional relation involving dimensionless groups, or described in terms of known algebraic or differential equations, including initial and boundary conditions. In the former case, similarity criteria are derived 'on the basis of dimensional analysis'; in the latter case, similarity criteria are derived 'on the basis of direct analysis'. Both procedures have pros and cons.
- The criteria of similarity find a powerful application in the interpretation of complex natural systems, with numerous variables that appear evolved in such a way as to minimise some function.

References

Antonets, V. A., Antonets, M. A., & Shereshevsky, I. A. (1991). The statistical cluster dynamics in the dendroid transfer systems. *Fractals in the Fundamental and Applied Sciences*, 59–71.

Harris, H. G., Pahl, P. J., & Sharma, S. D. (1962). *Dynamic studies of structures by means of models*. MIT.

Kleiber, M. (1947). Body size and metabolic rate. *Physiological Reviews, 27*, 511–541.

Reech, F. (1852). *Cours de Mécanique d'après la Nature Genéralement Flexible et Elastique des Corps*. Carilian-Goeury.

Tanner, R. I. (2000). *Engineering Rheology*. Oxford University Press.

West, G. B. (1999). The origin of universal scaling laws in biology. *Physica A: Statistical Mechanics and its Applications, 263*(1), 104–113. In *Proceedings of the 20th IUPAP International Conference on Statistical Physics*.

West, G. B., Brown, J. H., & Enquist, B. J. (1997). A general model for the origin of allometric scaling laws in biology. *Science, 276*(5309), 122–126.

Yalin, M. S. (1971). *Theory of hydraulic models*. Macmillan Publishers Ltd.

Part II
The Applications

Chapter 5
Applications in Fluid Mechanics and Hydraulics

Starting with the experiments by Smeaton (1759) on water wheels, fluid mechanics has favoured the applications of physical modelling. The great variety of problems, otherwise unsolvable, resulted in great development in the techniques of hydraulic physical modelling, both for internal and external flows. The history of models in hydraulic engineering is widely documented in the classic book by Rouse and Ince (1963), and the numerous applications are detailed in the monographs by Yalin (1971) and by Ivicsics (1980), Novak and Čábelka (1981), Hughes (1993), all exclusively devoted to hydraulic engineering problems.

5.1 The Dimensionless Groups of Interest in Fluid Mechanics

As mentioned many times, the selection of the dimensionless groups is preferably on the basis of their physical meaning. The proper choice is of great help in the analysis of the typical equation that describes the physical process and during the interpretation of the experimental results. The first step to identify such groups is to render dimensionless the fundamental equations that govern the physical process and boundary conditions.

5.1.1 The Linear Momentum Balance Equation

Suppose we wish to analyse the isochoric flow field of a Newtonian fluid in the presence of gravity, described by the Navier–Stokes equation,

S. G. Longo, *Principles and Applications of Dimensional Analysis and Similarity*, Mathematical Engineering, https://doi.org/10.1007/978-3-030-79217-6_5

177

$$\underbrace{\frac{\partial \mathbf{u}}{\partial t}}_{\substack{local \\ inertia}} + \underbrace{\mathbf{u} \cdot \nabla \mathbf{u}}_{\substack{convective \\ inertia}} - \underbrace{\mathbf{f}}_{\substack{gravity \\ action}} + \underbrace{\frac{\nabla p}{\rho}}_{\substack{pressure \\ action}} - \underbrace{\frac{\mu}{\rho} \nabla^2 \mathbf{u}}_{\substack{viscosity \\ action}} = 0 \qquad (5.1)$$

which, for the i-th component, assumes the following form in scalar notation:

$$\frac{\partial u_i}{\partial t} + u_k \frac{\partial u_i}{\partial x_k} - g_i + \frac{1}{\rho} \frac{\partial p}{\partial x_i} - \frac{\mu}{\rho} \nabla^2 u_i = 0. \qquad (5.2)$$

Following Einstein notation, a subscript repeated twice within the same term implies the sum of all values. The symbol ∇^2 is the operator of Laplace (Laplacian), which, in orthogonal Cartesian coordinates, has the expression

$$\nabla^2 \equiv \frac{\partial^2}{\partial x_1^2} + \frac{\partial^2}{\partial x_2^2} + \frac{\partial^2}{\partial x_3^2} \equiv \frac{\partial^2}{\partial x^2} + \frac{\partial^2}{\partial y^2} + \frac{\partial^2}{\partial z^2}. \qquad (5.3)$$

We can proceed to non-dimensionalise the problem by choosing scales that appear to be significant. Without going into the details of the choice, which will be adequate only after having gained insights into the phenomenon, we indicate these scales with the symbols u_0, l_0, t_0, p_0, representing the velocity scale, the length scale, the time scale and the pressure scale, respectively. We also assume that the temperature is invariant, and the dynamic viscosity is constant and equal to μ_0. The dimensionless variables, indicated by the symbol $\sim$, are:

$$\begin{cases} \tilde{u} = \dfrac{u}{u_0}, \\[2mm] \tilde{x} = \dfrac{x}{l_0}, \\[2mm] \tilde{t} = \dfrac{t}{t_0}, \\[2mm] \tilde{p} = \dfrac{p}{p_0}. \end{cases} \qquad (5.4)$$

Equation (5.2) in dimensionless variables reads:

$$\frac{u_0}{t_0} \frac{\partial \tilde{u}_i}{\partial \tilde{t}} + \frac{u_0^2}{l_0} \tilde{u}_k \frac{\partial \tilde{u}_i}{\partial \tilde{x}_k} - g_i + \frac{p_0}{\rho_0 l_0} \frac{\partial \tilde{p}}{\partial \tilde{x}_i} - \frac{\mu_0}{\rho_0} \frac{u_0}{l_0^2} \tilde{\nabla}^2 \tilde{u}_i = 0. \qquad (5.5)$$

Dividing all terms by the dimensional coefficient u_0^2/l_0 of the convective inertia term yields:

$$\left(\frac{l_0}{u_0 t_0}\right)\frac{\partial \tilde{u}_i}{\partial \tilde{t}} + \tilde{u}_k\frac{\partial \tilde{u}_i}{\partial \tilde{x}_k} - \left(\frac{l_0 g_i}{u_0^2}\right)$$

$$+ \left(\frac{p_0}{\rho_0 u_0^2}\right)\frac{\partial \tilde{p}}{\partial \tilde{x}_i} - \left(\frac{\mu_0}{\rho_0 l_0 u_0}\right)\tilde{\nabla}^2 \tilde{u}_i = 0, \quad (5.6)$$

where the monomials in parentheses are dimensionless.

The first monomial

$$St = \frac{l_0}{u_0 t_0} \quad (5.7)$$

is the *Strouhal number* and represents the ratio between local inertia and convective inertia; it is irrelevant in steady-state flow.

The second monomial

$$\frac{l_0 g_i}{u_0^2} = \frac{1}{Fr^2} \rightarrow Fr = \frac{u_0}{\sqrt{g\, l_0}} \quad (5.8)$$

is the *Froude number* computed by assuming that the i-direction is parallel to gravity. The Froude number is the ratio between convective inertia and the action of gravity; its definition is not uniform in the literature and is sometimes expressed as $Fr = u_0^2/(g\, l_0)$.

The third monomial

$$Eu = \frac{p_0}{\rho_0 u_0^2} \quad (5.9)$$

is the *Euler number*. If pressure scales with $\rho_0 u_0^2$, the Euler number assumes a unit value.

The last monomial

$$\frac{\mu_0}{\rho_0 l_0 u_0} = \frac{1}{Re} \rightarrow Re = \frac{\rho_0 l_0 u_0}{\mu_0} \quad (5.10)$$

is the inverse of the Reynolds number. The Reynolds number is the ratio of convective inertia and the action of viscosity.

Equation (5.6) assumes the new expression as a linear combination of non-dimensional terms:

$$St\frac{\partial \tilde{u}_i}{\partial \tilde{t}} + \tilde{u}_k\frac{\partial \tilde{u}_i}{\partial \tilde{x}_k} - \frac{1}{Fr^2} + Eu\frac{\partial \tilde{p}}{\partial \tilde{x}_i} - \frac{1}{Re}\tilde{\nabla}^2 \tilde{u}_i = 0. \quad (5.11)$$

Depending on the category and regime of flow, some terms are very small compared to others and can be neglected, which considerably simplifies the analysis. For example, for $Re \rightarrow \infty$, the viscous term becomes negligible except near the boundaries. The Euler equation holds (missing the viscous term in Eq. (5.6)), valid for ideal fluids (zero viscosity) and approximating turbulent flows of viscous fluids far from boundaries. Euler equation cannot satisfy the no-slip condition at the boundaries, and a boundary layer develops where the full Navier–Stokes equation is again necessary. In addition, much care is necessary to apply the approximation since, in some cases, the boundary layer is not confined and occupies the entire flow domain, like in pipes and in hydraulic jumps, or is well extended in the flow domain due, e.g., to separation. In this case, Euler approximation cannot be used.

Similarly, for Fr $\to \infty$ the contribution of the gravity to the balance in the vertical direction is negligible (in directions orthogonal to gravity the contribution is obviously always null), and inertia controls the flow field.

Adopting the same procedure as the other balance or conservation equations, we can select other dimensionless groups with a specific physical meaning. The most frequent dimensionless groups in fluid mechanics are:

$$\text{Re} = \frac{\rho\,u\,l}{\mu} \equiv \frac{\rho\,u^2\,l^2}{\mu\,u\,l} = \frac{convective\ inertia\ force}{viscous\ force} \qquad \text{(Reynolds)},$$

$$\text{M} = \frac{u}{\sqrt{\dfrac{\varepsilon}{\rho}}} \equiv \sqrt{\frac{\rho\,u^2\,l^2}{\varepsilon\,l^2}} = \sqrt{\frac{convective\ inertia\ force}{elastic\ force}} \qquad \text{(Mach)},$$

$$\text{We} = \frac{u\,\sqrt{\rho\,l}}{\sqrt{\sigma}} \equiv \sqrt{\frac{\rho\,u^2\,l^2}{\sigma\,l}} = \sqrt{\frac{convective\ inertia\ force}{surface\ tension\ force}} \qquad \text{(Weber)}, \quad (5.12)$$

$$\text{Fr} = \frac{u}{\sqrt{g\,l}} \equiv \sqrt{\frac{\rho\,u^2\,l^2}{\rho\,g\,l^3}} = \sqrt{\frac{convective\ inertia\ force}{gravity\ force}} \qquad \text{(Froude)},$$

$$\text{St} = \frac{l}{u\,t} \equiv \frac{\rho\,u\,l^3\,t^{-1}}{\rho\,u^2\,l^2} = \frac{local\ inertia\ force}{convective\ inertia\ force} \qquad \text{(Strohual)},$$

$$\text{Eu} = \frac{\Delta p}{\rho\,u^2} \equiv \frac{\Delta p\,l^2}{\rho\,u^2\,l^2} = \frac{pressure\ force}{convective\ inertia\ force} \qquad \text{(Euler)}.$$

Convective inertia is adopted as a reference to relate all other contributions because convective inertia is the most characteristic term of fluid flow motion. It is necessary here to pay particular attention to the consequences of the choice of the way to relate pressure and time.

If pressure is related to dynamic pressure, the Euler number is unity and disappears from the balance equation. This is equivalent to assuming that the forces of pressure and convective inertia are comparable, which is not always correct.

If time scales as l_0/u_0, the Strohual number is unity, and therefore, it is assumed that local inertia and convective inertia are comparable, which is also not always correct.

Note that Mach number is better known as $\text{M} = u/c$, where $c = \sqrt{\varepsilon/\rho}$ is the speed of sound.

Wishing to make the comparison with forces different from the convective one, we can proceed by composing the classic dimensionless groups in monomial form or in power function form. For example, we calculate:

$$\frac{pressure\ force}{viscous\ force} = \text{Eu} \cdot \text{Re},$$

$$\frac{pressure\ force}{gravity\ force} = \text{Eu} \cdot \text{Fr}, \tag{5.13}$$

$$\frac{viscous\ force}{gravity\ force} = \frac{\text{Fr}}{\text{Re}}.$$

The last ratio is often defined in a slightly different way, that is,

$$\text{Ga} = \frac{\text{Re}^2}{\text{Fr}} = \frac{g\,l_0^3}{\nu^2}, \tag{5.14}$$

and is named the *Galileo number*, which does not contain a velocity scale.

If the effects of buoyancy are important, the Galileo number changes to

$$\text{Ar} = \frac{\Delta\rho}{\rho}\,\text{Ga} = \frac{\Delta\rho}{\rho}\,\frac{g\,l_0^3}{\nu^2} \tag{5.15}$$

and it is named the *Archimedes number*.

Some characteristic numbers arise from the need to adapt the analysis to particular flow fields. For example, in cylindrical circular duct curves, to take into account secondary circulations, called Dean's vortices, the Reynolds number changes to

$$\text{Dn} = \text{Re}\sqrt{\frac{r}{r_c}}, \tag{5.16}$$

where r is the duct radius, r_c is the curvature radius, and the new number is called the *Dean number*. Similarly, in situations where pressure can be reduced to vapour tension, the Euler number changes, referring to the pressure change to the vapour tension, and is called *number of cavitation* or *Thoma number*:

$$\text{Th} = \frac{p - p_{vap}}{\rho\,u_0^2}, \tag{5.17}$$

where p_{vap} is the vapour partial pressure.

For flows in non-inertial frames, some ratios between fictitious (inertial) forces and other characteristic forces are defined. Therefore, for example, the *Ekman number* is defined as

$$\text{Ek} = \frac{viscous\ force}{Coriolis\ force} = \frac{\nu}{\Omega\,l_0^2}, \tag{5.18}$$

where Ω is the angular velocity of the non-inertial frame of reference and the *Rossby number*

$$\text{Ro} = \frac{inertia}{Coriolis\ force} = \frac{u_0}{\Omega\,l_0}. \tag{5.19}$$

Some numbers, especially for non-Newtonian behaviour fluids, refer to the properties of the fluid and not of the flow field. For Herschel-Bulkley fluids, the *Bingham number* is the ratio between the yield stress and the viscous stress,

$$\text{Bm} = \frac{yield\ stress}{viscous\ stress} = \frac{\tau_y\, l_0}{\mu_p\, u_0}, \tag{5.20}$$

where τ_y is the yield stress and μ_p is the apparent viscosity.

The *Deborah number* is

$$\text{De} = \frac{T_r}{T_f}, \tag{5.21}$$

where T_r is the relaxation time and T_f is the time scale of the flow field. The relaxation time for water is equal to $T_r \approx 10^{-12}$ s, and for mineral lubricating oil, it is equal to $T_r \approx 10^{-6}$ s; it is equal to a few seconds for polymers and tends to be infinite for solids. For De $\ll 1$, the fluid behaves like a solid. Therefore, we can walk on water if the support action lasts less than 10^{-12}s. In fact, all solids have a very large but not infinite relaxation time. So, for example, the glass in cathedral glass windows is thicker at the bottom because, over hundreds of years and under the action of gravity, the glass flowed downward.

If the fluid is compressible, it is subject to variations in density expressed as a function of pressure variation as $\Delta\rho/\rho = \Delta p/\varepsilon$, where ε is the bulk modulus of the fluid. If the pressure varies as $\Delta p \propto \rho\, u_0^2$, the variation in relative density is important if $\Delta\rho/\rho > 1$, that is, if $u_0 > (\varepsilon/\rho)^{1/2} \equiv c$, where c is the speed of sound that becomes the natural scale for velocity. The ratio $M = u_0/c$ is the *Mach number*, also known as *Cauchy number* with $\text{Ch} = M^2$.

The model of a continuous medium is no longer valid when the average free path l_p of the molecules is comparable with the geometric scale l_0 of the domain, for example, for a very rarefied gas in limited domains. The ratio $\text{Kn} = l_p/l_0$ is the *Knudsen number*, which assumes very high values for low-pressure gas in porous media or microchannels. The Knudsen number is also expressed as $\text{Kn} = M/\text{Re}$.

5.1.2 Boundary Conditions

The kinematic boundary condition at the boundary of the fluid domain is

$$\frac{DF}{Dt} = 0 \rightarrow \frac{\partial F}{\partial t} + \mathbf{u} \cdot \nabla F = 0, \tag{5.22}$$

where the equation $F(x,\ y,\ z,\ t) = 0$ describes the boundary. If the boundary is stationary, it results in $\mathbf{u} \cdot \nabla F \equiv \mathbf{u} \cdot \mathbf{n} = 0$; therefore, the normal velocity component at the boundary is zero. This is the condition of no penetration, while the no-slip condition requires that the tangential velocity is locally null. The result of both conditions is $\mathbf{u} = 0$ at the boundary, and since there are no terms to relate, we cannot gain useful information.

In practise, some particular cases require the abandonment of the no-slip hypothesis, with the introduction of a parallel velocity at the boundary related to the velocity gradient to define a characteristic length equal to

$$\beta = \frac{u_s}{\left.\dfrac{du}{dy}\right|_{y=0}}, \tag{5.23}$$

where u_s is the slip velocity and $du/dy \equiv \dot{\gamma}$ is the velocity gradient at the boundary ($y = 0$).

In liquids, the characteristic scale length is on the order of 0.1 μm, and experimentally, this results in

$$\begin{cases} \beta = A\,\dot{\gamma}^{B} \\ u_s = A\,\dot{\gamma}^{B+1} \end{cases}, \tag{5.24}$$

where $B \approx 1/2$ and A is a dimensionless coefficient.

In gases, the slip velocity is expressed by Maxwell's relation (Maxwell 1878),

$$u_s = \frac{2 - \sigma_v}{\sigma_v}\,\mathrm{Kn}\, l_0\, \dot{\gamma}, \tag{5.25}$$

where σ_v is the ratio between the number of molecules hitting the frontier that are not mirror reflected and the total number, l_0 is the scale length of the flow and $\dot{\gamma}$ is the velocity gradient at the wall.

The dynamic condition at a rigid wall is of no specific interest. The analysis is more interesting if the boundary delimits domains of fluids of different natures. In this case, the condition on the normal stress component at the interface results in Laplace's equation

$$\Delta p = \sigma \left(\frac{1}{R_1} + \frac{1}{R_2} \right), \tag{5.26}$$

where σ is the surface tension and R_1 and R_2 are the main curvature radii. In dimensionless form, we obtain:

$$\left(\frac{p_0 l_0}{\sigma} \right) \Delta \tilde{p} = \left(\frac{1}{\tilde{R}_1} + \frac{1}{\tilde{R}_2} \right). \tag{5.27}$$

The monomial in parentheses on the left-hand side is the *Laplace number*, which takes a different form depending on the choice of the pressure scale; the geometric scale must necessarily be representative of the curvature of the interface and can be, for example, the radius of a water drop or air bubble.

If the pressure relates to convective inertia, the *Weber number* is obtained,

$$\mathrm{We} = \frac{\rho \, l_0 \, u_0^2}{\sigma} = \frac{convective \; inertia \; force}{surface \; tension \; force}. \tag{5.28}$$

The Weber number is involved in the study of water drops or gas bubbles and in the presence of curvature of fluid trajectories, but always with an interface with other fluids. For example, it is involved in the physical process that allows some insects to move on water.

If pressure varies with local inertia, it results in

$$\frac{\rho \, l_0^2 \, u_0}{\sigma \, t_0} = \frac{local \; inertia \; force}{capillarity \; force}. \tag{5.29}$$

If we consider the effects of the fluid viscosity, we obtain

$$\mathrm{Ca} = \frac{\mu \, u_0}{\sigma} = \frac{viscous \; force}{capillarity \; force}, \tag{5.30}$$

where Ca is the *capillarity number*, which is involved in all small-scale phenomena and in the presence of interfaces between liquids and gases, such as coalescence and adhesion.

Finally, if the pressure varies with gravity, we have

$$\mathrm{Bo} = \frac{\rho \, g \, l_0^2}{\sigma} = \frac{gravity \; (or \; buoyancy) \, force}{capillarity \; force}, \tag{5.31}$$

where Bo is the *Bond number* that pertains, for example, in the study of the behaviour of a drop at rest on a flat horizontal surface. The shape of the drop depends on the action of gravity and surface tension, that is, the number of Bond. The condition $\mathrm{Bo} = 1$ allows us to calculate the capillary length scale $l_c = [\sigma/(\rho g)]^{1/2}$.

The second dynamic condition at the interface requires continuity of the tangential stresses if the surface tension is spatially homogeneous. In the presence of a space gradient of such stress (due, for example, to gradients of temperature or concentration of a component, if the liquid is a mixture), the condition is

$$\tau_A - \tau_B - \nabla_s \sigma = 0, \tag{5.32}$$

where τ_A and τ_B are the tangential stresses at the interface in the domain occupied by the fluid A and B, respectively, and $\nabla_s \sigma$ is the component of the space gradient of surface tension projected onto the interface. Thus, a space gradient of surface tension is equivalent to a tangential stress, which generates a mass flow from regions with a lower value to a higher value of σ. A dimensionless group representative of this physical process is the *Marangoni number*, defined as

$$\mathrm{Ma} = \frac{\tau_{\mathrm{Ma}}}{\tau}, \tag{5.33}$$

where τ_{Ma} is the Marangoni tangential stress and τ is the tangential stress due, for example, to the shear rate of a viscous fluid. For a flat interface, $\tau_{\mathrm{Ma}} \approx \Delta\sigma/l_0$ and

$$\mathrm{Ma} = \frac{\Delta\sigma}{\mu\, u_0}, \tag{5.34}$$

by assuming that $\tau \propto \mu\, u_0/l_0$.

The change in surface tension can also be a consequence of the temperature gradient. In this case,

$$\mathrm{Ma} = \frac{l_0}{\mu k}\frac{d\sigma}{d\theta}\,\Delta\theta \quad \text{(thermocapillarity)}, \tag{5.35}$$

where k is the thermal diffusivity and $\Delta\theta$ is the temperature difference. If the change in surface tension is caused by the concentration gradient of a surfactant, then it results in

$$\mathrm{Ma} = \frac{l_0}{\mu\kappa}\frac{d\sigma}{dC}\,\Delta C \quad \text{(action of a surfactant)}, \tag{5.36}$$

where κ is the diffusivity of the surfactant and ΔC is its concentration variation.

The Marangoni effect stabilises soap bubbles and is also the origin of the small arches that form in glasses of wine and alcoholic drinks in general: in a wine film leaning against the wall, the evaporation of alcohol is more intense in the upper part, where the film is thinner. A lower alcohol concentration corresponds to a higher surface tension, which draws fluid from below. The liquid literally climbs the wall of the glass and accumulates at the top until it falls again when the action of gravity balances the tension of Marangoni (see Fig. 5.1). For the analysis of this effect, the most representative dimensional number is $\mathrm{Ma} \cdot \mathrm{Ca}$, with Ma calculated on the basis of the concentration gradient of the alcohol.

Example 5.1 A classic application of dimensional analysis in fluid mechanics is the identification of the formal solution of the flow field in the wall boundary layer. The wall boundary layer is a subdomain of the flow field in which the Navier–Stokes equations are still valid but with considerable simplifications thanks to the special

Fig. 5.1 Wine climbs due to the Marangoni effect. The imbalance of surface tension is due to the alcohol concentration gradient

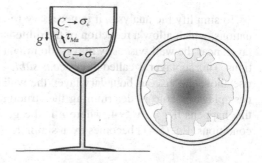

geometry, which permits the elimination of some terms on the basis of the analysis
of their order of magnitude.

The physical process that describes the velocity profile is defined by the velocity
u at the distance y from the wall, measured orthogonally to the wall, and depends on
the geometric roughness ε, the dynamic viscosity μ, the density of the fluid ρ and
the tangential stress at the wall τ_b. It also depends on the geometric characteristics of
the external flow, represented by a length scale l. The typical equation can be written
as follows:

$$u = f(y, \ \varepsilon, \ l, \ \mu, \ \rho, \ \tau_b). \tag{5.37}$$

Space of dimensions is 3-dimensional, and by virtue of Buckingham's Theorem,
the physical process is represented as a function of $(7 - 3) = 4$ dimensionless groups.
If we select u, y and ρ as fundamental quantities, the most immediate dimensional
groups are:

$$\frac{\varepsilon}{y}, \ \frac{l}{y}, \ \frac{\mu}{\rho u y}, \ \frac{\tau_b}{\rho u^2}. \tag{5.38}$$

We can attribute precise physical meanings to the various terms. For example, τ_b/ρ
has the dimensions of a velocity squared such that, since it involves the characteristics
of the flow field near the wall, is conventionally called *friction velocity* and indicated
with u_*. On the basis of this new velocity scale, the following 4 dimensionless groups
are traditionally defined:

$$\frac{u}{u_*}, \ \frac{y u_*}{\nu}, \ \frac{y}{\varepsilon}, \ \frac{y}{l}, \tag{5.39}$$

and the typical equation becomes

$$\frac{u}{u_*} = \tilde{f} \left(\frac{y u_*}{\nu}, \ \frac{y}{\varepsilon}, \ \frac{y}{l} \right). \tag{5.40}$$

If the boundary layer is sufficiently extended in length (such as, for example, on
an infinite plate at a great distance from the leading edge), the l length macroscale
is irrelevant, and Eq. (5.40) reduces to

$$\frac{u}{u_*} = \tilde{f} \left(\frac{y u_*}{\nu}, \ \frac{y}{\varepsilon} \right). \tag{5.41}$$

To simplify the analysis, it is necessary to consider the possible asymptotic situations, which allow a reduction of the dimensionless groups. Within the boundary
layer, near the wall, viscosity plays a dominant role and dampens turbulent fluctuations. This sublayer is called the *viscous sublayer*. If the surface roughness does not
rise above the viscous boundary layer, the wall is hydraulically smooth, and roughness plays no role in determining the structure of the flow field. Experimentally,
this happens if $\varepsilon u_*/\nu \lesssim 4$, where ε is the geometric scale of roughness. In these
conditions, Eq. (5.41) becomes even simpler:

$$\frac{u}{u_*} = \tilde{f}\left(\frac{y\,u_*}{\nu}\right). \tag{5.42}$$

Dimensional analysis does not offer further tools to detect the expression of the $\tilde{f}$ function. The function has been theoretically identified by Prandtl based on a turbulence model, and results in

$$\frac{u}{u_*} = \frac{1}{\kappa}\ln\left(\frac{y\,u_*}{\nu}\right) + C_1, \tag{5.43}$$

where κ is the von Kármán constant, equal to 0.4, and C_1 is a constant, experimentally equal to 5.0 for velocity profiles in a duct of circular cross-section. In the viscous sublayer, the tangential stress is only viscous, and since the fluid is Newtonian, it is $\tau = \mu\,\partial u/\partial y$; assuming that the tangential stress is uniform along the vertical direction and equal to the wall tangential stress, $\tau = \tau_b = \rho\,u_*^2$, results in

$$\tau_b = \rho\,u_*^2 = \mu\,\frac{\partial u}{\partial y} \rightarrow \frac{u}{u_*} = \frac{y\,u_*}{\nu}. \tag{5.44}$$

Therefore, the velocity profile in the viscous boundary layer is linear, with a conventional thickness represented by the intersection between the logarithmic velocity profile, specific to the outer region (and calculated by other means), and the linear profile is equal to $y = 11.8\nu/u_*$ (Nikuradse thickness) (Fig. 5.2).

A second asymptotic condition occurs when the thickness of the viscous boundary layer is much less than the scale of roughness ε; when $(\varepsilon\,u_*)/\nu \gtrsim 70$, see Fig. 5.3. In this case, the viscosity no longer plays any role in the structure of the flow field, and Eq. (5.41) reduces to:

$$\frac{u}{u_*} = \tilde{f}\left(\frac{y}{\varepsilon}\right). \tag{5.45}$$

Fig. 5.2 Velocity profile in the turbulent boundary layer for hydraulically smooth walls

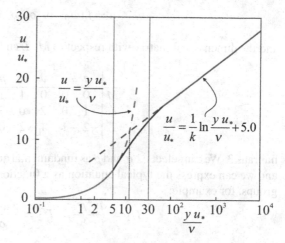

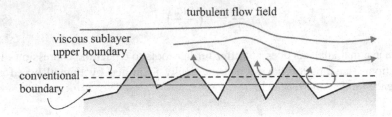

Fig. 5.3 Flow field near a rough wall

Again, dimensional analysis is of no further help to define the structure of the function $\tilde{f}$. Prandtl and von Kármán derived, on a theoretical basis, the following velocity profile:

$$\frac{u}{u_*} = \frac{1}{\kappa} \ln\left(\frac{y}{\varepsilon}\right) + C_1, \qquad \therefore \quad (5.46)$$

where C_1 is an integration constant experimentally equal to 8.5 for fully developed turbulent flow in a rough duct of circular cross-section.

Example 5.2 We wish to analyse the process of breaking the interface between a liquid and a gas as a consequence of an acceleration impressed on the fluid container, with possible expulsion of drops of liquid in the gas.

The physical process can be studied experimentally by placing liquid in a container fixed on a shaking table undergoing vertical sinusoidal oscillation with amplitude z_0 and angular frequency ω_0. The variables involved are the kinematic viscosity of the liquid ν, the surface tension at the interface σ and the density of the liquid ρ. Instead of the amplitude of the oscillations, it is convenient to consider the impressed acceleration $a = z_0 \omega_0^2$. The typical equation of the process is

$$a = f(\omega_0, \ \nu, \ \sigma, \ \rho), \qquad (5.47)$$

and the dimensional matrix with respect to M, L and T

$$
\begin{array}{c|ccccc}
 & a & \omega_0 & \nu & \sigma & \rho \\
\hline
M & 0 & 0 & 0 & 1 & 1 \\
L & 1 & 0 & 2 & 0 & -3 \\
T & -2 & -1 & -1 & -2 & 0 \\
\end{array}
\qquad (5.48)
$$

has rank 3. We can select ν, σ and ρ as fundamental quantities (they are independent), and we can express the typical equation as a function of $(5 - 3) = 2$ dimensionless groups, for example:

$$\Pi_1 = \frac{a\nu^4\rho^3}{\sigma^3}, \qquad \Pi_2 = \frac{\omega_0\nu^3\rho^2}{\sigma^2}. \qquad (5.49)$$

Hence, we can write:

Fig. 5.4 Experimental
results to identify threshold
conditions for the formation
of liquid droplets in a gas
(modified from Goodridge
et al. 1997)

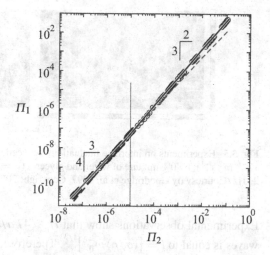

$$\Pi_1 = \tilde{f}(\Pi_2) \rightarrow \quad \frac{a v^4 \rho^3}{\sigma^3} = \tilde{f}\left(\frac{\omega_0 v^3 \rho^2}{\sigma^2}\right). \tag{5.50}$$

Figure 5.4 shows the results of experiments by Goodridge et al. (1997) with the
data, plotted on a log-log scale, revealing a double slope of the interpolating line: for
$\Pi_2 < 10^{-5}$, we obtain $\Pi_1 \propto \Pi_2^{4/3}$, equivalent to

$$a = c_1 \left(\frac{\sigma}{\rho}\right)^{1/3} \omega_0^{4/3}, \tag{5.51}$$

with a regime controlled by surface tension; for $\Pi_2 > 10^{-5}$, we obtain $\Pi_1 \propto \Pi_2^{3/2}$,
equivalent to

$$a = c_2 \, v^{1/2} \, \omega_0^{3/2}, \tag{5.52}$$

with a regime controlled by viscosity; the two numerical coefficients, calculated by
least squares interpolation, are $c_1 = 0.261$ and $c_2 = 1.306$. Figure 5.5 shows that the
interface looks substantially different between the two regimes. This is a case in
which the form of the typical equation and the exponents derive from the analysis
of the diagrams of the experimental tests, with the values of the measured quantities
represented on the basis of the indications of the dimensional analysis. We wish now
to underpin the origin of the two functions.

In the regime controlled by the surface tension, we can assume that droplets occur
when the H height of the capillary waves becomes comparable with the l wavelength.
The height is proportional to the imposed acceleration:

$$H \propto \frac{a}{\omega_0^2}. \tag{5.53}$$

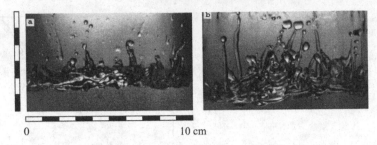

0   10 cm

Fig. 5.5 Experiments on interface fluctuations between a liquid and a gas: **a** distilled water ($\nu = 10^{-6}$ m^2 s^{-1}); **b** 80% mixture of water and glycerol ($\nu = 43 \cdot 10^{-6}$ m^2 s^{-1}). Excitation frequency 20 Hz (Courtesy by Goodridge et al. 1997, Copyright 2007 by the American Physical Society)

Experimental observations show that $H \approx 47 \, a/\omega_0^2$, and the length of the capillary waves is equal to $l = \left[(\sigma/\rho) \, \omega_0^{-2} \right]^{1/3}$. Therefore,

$$a \propto \left(\frac{\sigma}{\rho} \right)^{1/3} \omega_0^{4/3}. \tag{5.54}$$

In the viscosity-controlled regime, we can assume that a droplet occurs when the input power equals the power dissipated by the viscosity. The input power per unit mass P_i depends on acceleration and frequency, has dimensions $L^2 \, T^{-3}$ and is proportional to a^2/ω_0. The power dissipated per unit of mass is proportional to the kinematic viscosity ν and to the velocity gradient; dimensionally, we have

$$P_o \propto \nu \left(\frac{V}{l} \right)^2, \tag{5.55}$$

where V is a velocity scale.

A possible velocity scale is $H \, \omega$, and a length scale is the length of capillary waves l. Hence,

$$P_o \propto \nu \left(\frac{H \, \omega}{l} \right)^2. \tag{5.56}$$

Assuming that $H \approx l$ and equating input and output power yields

$$a \propto \omega_0^{3/2} \nu^{1/2}, \qquad\qquad \therefore \tag{5.57}$$

coincident with Eq. (5.52).

From this example, the invaluable support offered by dimensional analysis is evident, both for the preliminary data processing and for the physical interpretation of the results.

Example 5.3 We now analyse the motion of a body in a compressible fluid with heat transfer and friction. We wish to calculate the drag force on the body, which is a function of the geometry, represented by the longitudinal dimension l and transversal dimension d, the velocity of the current U, the pressure p, the surface temperature of the body θ_w, and the properties of the gas ρ, μ, k, c_v, and R, respectively representing the density, the dynamic viscosity, the thermal conductivity, the specific heat at constant volume, and the gas constant. The typical equation is

$$F = f(l, d, U, p, \theta_w, \rho, \mu, k, c_v, R), \tag{5.58}$$

and the dimensional matrix

$$
\begin{array}{c|ccccccccccc}
 & F & l & d & U & p & \theta_w & \rho & \mu & k & c_v & R \\
\hline
M & 1 & 0 & 0 & 0 & 1 & 0 & 1 & 1 & 1 & 0 & 0 \\
L & 1 & 1 & 1 & 1 & -1 & 0 & -3 & -1 & 1 & 2 & 2 \\
T & -2 & 0 & 0 & -1 & -2 & 0 & 0 & -1 & -3 & -2 & -2 \\
\Theta & 0 & 0 & 0 & 0 & 0 & 1 & 0 & 0 & -1 & -1 & -1
\end{array}
\tag{5.59}
$$

has rank 4. According to Buckingham's Theorem, it is possible to express the physical process as a function of $(11 - 4) = 7$ dimensionless groups. The groups usually selected are:

$$\frac{F}{\rho\, U^2\, l^2} = \tilde{\Phi}\left(\frac{d}{l}, \frac{\rho\, l\, U}{\mu}, \frac{U}{\sqrt{\gamma\, p/\rho}}, \frac{c_v + R}{c_v}, \frac{R\,\theta_w}{U^2}, \frac{\mu\,(c_v + R)}{k}\right), \tag{5.60}$$

or,

$$\frac{F}{\rho\, U^2\, l^2} = \tilde{\Phi}\left(\frac{d}{l}, \text{Re}, \text{M}, \gamma, \frac{R\,\theta_w}{U^2}, \text{Pr}\right), \tag{5.61}$$

where $\gamma = (c_v + R)/c_v \equiv c_p/c_v$ from Mayer's relation and c_p is the specific heat at constant pressure. For a specific gas, both γ and the Prandtl number can be considered constant; hence,

$$\frac{F}{\rho\, U^2\, l^2} = \tilde{\Phi}\left(\frac{d}{l}, \text{Re}, \text{M}, \frac{R\,\theta_w}{U^2}\right), \tag{5.62}$$

and, following conventional notation,

$$F = \frac{1}{2}\, \rho\, U^2\, l^2\, C_D\left(\frac{d}{l}, \text{Re}, \text{M}, \frac{R\,\theta_w}{U^2}\right), \tag{5.63}$$

where C_D is the *drag coefficient*, a function of the shape factor, Reynolds number, Mach number and of the last dimensional group missing a specific name.

If we wish to realise a physical model, the condition of similarity requires that, in addition to the geometric similarity, the similarity of Reynolds and the similarity of Mach are satisfied; i.e., $r_{d/l} = r_{\text{Re}} = r_{\text{M}} = r_{C_D} = 1$. The system of equations in the unknown scales reads:

$$\begin{cases} r_d = r_l = \lambda, \\ r_\rho\, r_U\, \lambda = r_\mu, \\ r_U = r_\gamma^{1/2}\, r_p^{1/2}\, r_\rho^{-1/2}, \\ r_R\, r_{\theta_w} = r_U^2. \end{cases} \qquad (5.64)$$

Using the same gas, in the model and prototype, $r_\gamma = 1$. In addition, we require

$$\frac{r_p^{1/2}\, r_\rho^{1/2}}{r_\mu} = \frac{1}{\lambda}. \qquad \therefore \quad (5.65)$$

The experiments for the determination of the drag coefficient are conducted in wind tunnels; see Sect. 9. For small geometric scale models, it is necessary to increase the pressure to values that are often incompatible with the forces and deformations of the structural elements in the model. Alternatively, one can take advantage of the fact that the ratio $\sqrt{r_\rho}/r_\mu$ increases by lowering the temperature (see Fig. 9.3), and it is therefore possible, by cooling the gas, to limit the pressure increase in the model. The wind tunnels equipped to perform this operation are called *cryogenic tunnels* (see Sect. 9.2).

Example 5.4 We wish to calculate the flow rate scale for a rectangular sharp-crested weir with lateral contraction (see Fig. 5.6). The geometrical variables are the width of the crest b, the water level with respect to the crest h_m, the width of the arrival channel B, the crest elevation d, and the distance of the water level measurement section from the weir section L_h. The characteristics of the fluid are the density ρ, the dynamic viscosity μ and the surface tension σ. Further variables are the acceleration of gravity g and the volumetric flow Q. The physical process can be expressed with the typical equation

$$Q = f(b,\ h_m,\ B,\ d,\ L_h,\ \rho,\ \mu,\ \sigma,\ g). \qquad (5.66)$$

This is a function of 10 variables, 3 of which are independent, such as the triad d, ρ and g. We apply Buckingham's Theorem to obtain the new function

$$\frac{Q}{d^2\sqrt{g\,d}} = \tilde{f}\left(\frac{b}{d},\ \frac{h_m}{d},\ \frac{B}{d},\ \frac{L_h}{d},\ \frac{\mu}{\rho d\sqrt{g\,d}},\ \frac{\sigma}{\rho d^2}\right). \qquad (5.67)$$

Many of the dimensionless groups in Eq. (5.67) are without physical meaning and can be more conveniently rewritten as a function of other groups. The procedure for the selection of the most suitable dimensionless groups involves the execution of a series of experiments and an in-depth knowledge of the physical process. For the present analysis, an appropriate selection of dimensionless groups brings us to the following typical equation:

Fig. 5.6 Sharp-crested weir
with. lateral contraction
(modified from Longo and
Petti 2006)

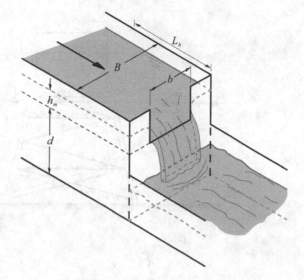

$$\frac{Q}{b\,h_m\,\sqrt{g\,h_m}} = \Phi_1\left(\frac{h_m}{d}, \ \frac{h_m\,\sqrt{g\,\rho\,h_m}}{\mu}, \ \frac{h_m\,\sqrt{g\,\rho}}{\sqrt{\sigma}}, \ \frac{b}{B}, \ \frac{b}{d}, \ \frac{L_h}{d}\right) \equiv$$

$$\Phi_1\left(\frac{h_m}{d}, \ \text{Re}, \ \text{We}, \ \frac{b}{B}, \ \frac{b}{d}, \ \frac{L_h}{d}\right). \tag{5.68}$$

The dimensional matrix of the new set of groups

	b	h_m	B	d	L_h	ρ	μ	σ	g	Q
Π_1	-1	$-3/2$	0	0	0	0	0	0	$-1/2$	1
Π_2	0	1	0	-1	0	0	0	0	0	0
Π_3	0	$3/2$	0	0	0	1	-1	0	$1/2$	0
Π_4	0	$1/2$	0	0	0	$1/2$	0	$-1/2$	$1/2$	0
Π_5	1	0	-1	0	0	0	0	0	0	0
Π_6	1	0	0	-1	0	0	0	0	0	0
Π_7	0	0	0	-1	1	0	0	0	0	0

$$(5.69)$$

has rank 7, equal to the number of rows (see Sect. 1.4.2.2); hence, the groups are
independent and are a complete set.

If the fluid is water and in the presence of standard Earth gravity, the values of μ,
σ, ρ and g are constant. The number of sufficient dimensionless groups is equal to
$(n-k) - (n_f - k_f)$, where $n_f = 4$ is the number of constant variables and k_f their
rank (see Sect. 1.4.7). The rank of the dimensional matrix of μ, σ, ρ and g,

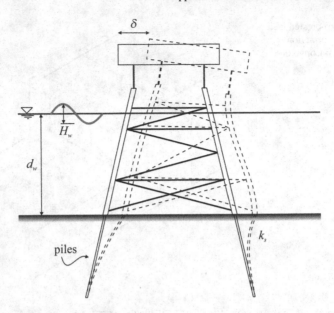

Fig. 5.7 Schematic of an offshore platform

$$
\begin{array}{c|cccc}
 & \mu & \sigma & \rho & g \\
\hline
M & 1 & 1 & 1 & 0 \\
L & -1 & 0 & -3 & 1 \\
T & -1 & -2 & 0 & -2
\end{array}' \tag{5.70}
$$

is 3. Therefore, only one dimensional group can be eliminated, and the typical equation becomes

$$
\frac{Q}{b\,h_m\,\sqrt{g\,h_m}} = \Phi_2\left(\frac{h_m}{d},\ \frac{h_m\,\rho\,\sigma}{\mu^2},\ \frac{b}{B},\ \frac{b}{d},\ \frac{L_h}{d} \right). \qquad \therefore \tag{5.71}
$$

The Φ_2 function is commonly referred to as the weir efflux coefficient.

Example 5.5 We wish to analyse the deformation of an offshore metal truss platform with a pile foundation, subject to the action of sea waves (see Fig. 5.7).

The variables involved are related to the structure, ground, and sea waves. If we are interested in structure deformation, the typical equation is

$$
\delta = f(l,\ g,\ \rho,\ M_s,\ E_s,\ I_{ms},\ k_s,\ u_w,\ H_w,\ d_w,\ \mu,\ t), \tag{5.72}
$$

where δ is a deformation (e.g., the maximum horizontal displacement), l is a characteristic geometric dimension of the structure (for instance, its height), g is the acceleration of gravity, ρ is the density of water, M_s is the total mass, E_s is Young's modulus of the structure material, I_{ms} is the moment of inertia, k_s is the elastic constant of the ground, u_w is the velocity of water particles, H_w is the wave height, d_w is the local water depth, μ is the dynamic viscosity of water, and t is the period of the incident wave. The dimensional matrix, as a function of M, L and T, is

$$
\begin{array}{c|ccc|ccccccccccc}
 & l & g & \rho & \delta & M_s & E_s & I_{ms} & k_s & u_w & H_w & d_w & \mu & t \\
\hline
M & 0 & 0 & 1 & 0 & 1 & 1 & 1 & 1 & 0 & 0 & 0 & 1 & 0 \\
L & 1 & 1 & -3 & 1 & 0 & -1 & 2 & 0 & 1 & 1 & 1 & -1 & 0 \\
T & 0 & -2 & 0 & 0 & 0 & -2 & 0 & -2 & -1 & 0 & 0 & -1 & 1 \\
\end{array}
\tag{5.73}
$$

and has rank 3. We can extract the following minor from the matrix in Eq. (5.73):

$$
\mathbf{A} = \begin{bmatrix} 0 & 0 & 1 \\ 1 & 1 & -3 \\ 0 & -2 & 0 \end{bmatrix},
\tag{5.74}
$$

which is non-singular. As a consequence, the three dimensions l, g and ρ are independent and can represent a basis. The residual matrix is

$$
\mathbf{B} = \begin{bmatrix} 0 & 1 & 1 & 1 & 1 & 0 & 0 & 0 & 1 & 0 \\ 1 & 0 & -1 & 2 & 0 & 1 & 1 & 1 & -1 & 0 \\ 0 & 0 & -2 & 0 & -2 & -1 & 0 & 0 & -1 & 1 \end{bmatrix}.
\tag{5.75}
$$

The matrix of the dimensional exponents of the other variables with respect to the fundamental quantities is calculated as

$$
\mathbf{C} = \mathbf{A}^{-1} \cdot \mathbf{B},
\tag{5.76}
$$

with the following result:

$$
\mathbf{C} = \begin{bmatrix} 1 & 3 & 1 & 5 & 2 & 0.5 & 1 & 1 & 1.5 & 0.5 \\ 0 & 0 & 1 & 0 & 1 & 0.5 & 0 & 0 & 0.5 & -0.5 \\ 0 & 1 & 1 & 1 & 1 & 0 & 0 & 0 & 1 & 0 \end{bmatrix}.
\tag{5.77}
$$

The dimensionless groups are:

$$
\Pi_1 = \frac{\delta}{l}, \quad \Pi_2 = \frac{M_s}{\rho\, l^3}, \quad \Pi_3 = \frac{E_s}{\rho\, g\, l}, \quad \Pi_4 = \frac{I_{ms}}{\rho\, l^5},
$$

$$
\Pi_5 = \frac{k_s}{\rho\, g\, l^2}, \quad \Pi_6 = \frac{u_w}{\sqrt{g\, l}}, \quad \Pi_7 = \frac{H_w}{l}, \quad \Pi_8 = \frac{d_w}{l},
\tag{5.78}
$$

$$
\Pi_9 = \frac{\mu}{\rho\, l\, \sqrt{g\, l}}, \quad \Pi_{10} = t\sqrt{\frac{g}{l}}.
$$

Hence,

$$\frac{\delta}{l} = \tilde{f}\left(\frac{M_s}{\rho\, l^3}, \ \frac{E_s}{\rho\, g\, l}, \ \frac{I_{ms}}{\rho\, l^5}, \ \frac{k_s}{\rho\, g\, l^2}, \ \frac{u_w}{\sqrt{g\, l}}, \ \frac{H_w}{l}, \ \frac{d_w}{l}, \ \frac{\mu}{\rho\, l\, \sqrt{g\, l}}, \ t\sqrt{\frac{g}{l}}\right).$$

$$\therefore \quad (5.79)$$

5.2   The Conditions of Similarity in Hydraulic Models

The great majority of hydraulic problems involve at most the following 9 variables:

$$l, \ t, \ V, \ p, \ \rho, \ \mu, \ g, \ \varepsilon, \ \sigma, \tag{5.80}$$

where l is a geometric dimension, t is time, V is velocity, p is pressure, ρ is density, μ is dynamic viscosity, g is gravity, ε is the bulk modulus and σ is the tension surface, thus excluding variables of electrical nature, temperature, etc. There are 3 fundamental quantities, and according to Buckingham's Theorem, it is possible to describe a physical process involving the 9 variables as a function of 6 dimensionless groups. The 6 dimensionless groups with a physical meaning that are usually selected are those already listed in Sect. 5.1. Based on the criteria of dimensional analysis, the complete similarity requires that the 6 dimensionless groups assume the same value in the model and prototype.

In practise, complete similarity is not feasible since some of the quantities involved are invariant or can be modified but with very expensive arrangements and high costs. For example, the acceleration of gravity cannot be modified (except for centrifuge models in geotechnical applications). In addition, the choice of the same fluid in the model and prototype (almost always the fluid is water) implies that the scale ratios of dynamic viscosity, surface tension, bulk modulus and density assume unit values. This reduces the number of degrees of freedom in the selection of scale ratios, with sometimes contradictory results. For example, in a physical model with the same fluid as the prototype, the scale ratios should satisfy a system of equations,

$$\begin{cases} r_g = r_\mu = r_\sigma = r_\varepsilon = r_\rho = 1, \\[2mm] \dfrac{r_\rho\, r_V\, \lambda}{r_\mu} = 1 & \text{(Reynolds)}, \\[4mm] \dfrac{r_V^2\, r_\rho}{r_\varepsilon} = 1 & \text{(Mach)}, \\[4mm] \dfrac{r_V^2\, r_\rho\, \lambda}{r_\sigma} = 1 & \text{(Weber)}, \\[4mm] \dfrac{r_V^2}{r_g\, \lambda} = 1 & \text{(Froude)}, \\[4mm] \dfrac{\lambda}{r_V\, r_t} = 1 & \text{(Strohual)}, \\[4mm] \dfrac{r_{\Delta p}}{r_\rho\, r_V^2} = 1 & \text{(Euler)}, \end{cases} \tag{5.81}$$

Table 5.1 The scale ratios for some derived quantities, calculated on the basis of the equality of the Froude number only and of the Reynolds number only

Variable	Froude	Reynolds
Length	λ	λ
Area	λ^2	λ^2
Volume	λ^3	λ^3
Time	$\sqrt{\lambda}$	λ^2
Velocity	$\sqrt{\lambda}$	λ^{-1}
Acceleration	1	λ^{-3}
Force	λ^3	1

that does not admit solutions: suffice it to say that the equality of Froude numbers requires $r_V = \sqrt{\lambda}$, while the equality of Reynolds numbers requires $r_V = 1/\lambda$. We can choose to respect the Froude number scales or the Reynolds number scales, but not both.

As an example, Table 5.1 lists the scale ratios calculated for some variables based on the equality of the Froude number only and of the Reynolds number only.

The lack of a solution in the system of equations (5.81) leads to an *approximate* or *partial similarity*, in which only some of the dimensionless groups are invariant. The choice of the dimensionless group to be left invariant in the model and in the prototype depends on the specific process: in free surface flows, the Froude number is selected; in confined flows, at low Reynolds numbers, the Reynolds number and the Euler number are selected. When the similarity is dominated with respect to one of the dimensionless groups, the similarity takes the name of the dominant dimensionless group and is commonly referred to as the similarity of Reynolds, similarity of Froude, similarity of Weber, etc.

5.2.1 Reynolds-Euler Similarity

Reynolds-Euler similarity applies to stationary flows confined by rigid boundaries, or extending to infinity, whenever the viscosity of the fluid is important. The variables of interest are reduced to 5, and the dimensionless groups that remain invariant are the Euler number and the Reynolds number. Using the criteria of similarity results in:

$$
\begin{cases}
\dfrac{r_{\Delta p}}{r_\rho\, r_V^2} = 1, \\[2ex]
\dfrac{r_\rho\, r_V\, \lambda}{r_\mu} = 1.
\end{cases}
\tag{5.82}
$$

If we use the same fluid in the model and prototype, we need to calculate 3 scale ratios constrained by 2 equations, and only one degree of freedom is left. Usually, we fix a geometrical scale and calculate $r_V = \lambda^{-1}$ and $r_{\Delta p} = \lambda^{-2}$. However, power scales such as $r_P = \lambda^{-1}$ and this creates quite a few problems in the realisation of Reynolds similar models with reduced geometric scale: both velocity and power in the model assume higher values than in the prototype. In a physical model with $\lambda = 1/10$, a current with $V_p = 2 \text{ m s}^{-1}$ in the prototype should be reproduced with $V_m = 20 \text{ m s}^{-1}$ in the model, which is excessive.

In fully turbulent flow regimes, the principle of asymptoticity of turbulence provides independence from the Reynolds number, and the Reynolds-Euler similarity is simplified to the similarity of Euler, which requires consistency with only the first equation of the system (5.82). With Euler similarity, velocity and pressure are not affected by the geometric scale, which can be fixed arbitrarily. This kind of partial similarity is adopted for hydraulic machines such as pumps and turbines.

5.2.2 Froude Similarity

Froude similarity applies to the study of flows in the presence of a free surface and when gravity plays an important role. It is customarily adopted to model natural rivers, weirs, dams, and sea gravity waves. In the general case of a hydraulic problem dependent on 8 variables, excluding only the bulk modulus from the 9 listed in Eq. (5.80), Buckingham's Theorem allows us to express the physical process as a function of 5 dimensionless groups:

$$f(\text{Re}, \text{We}, \text{Fr}, \text{St}, \text{Eu}) = 0, \tag{5.83}$$

and, for similarity, we can write 5 equations in the 8 unknown scale ratios:

$$
\begin{cases}
\dfrac{r_\rho \, r_V \, \lambda}{r_\mu} = 1 & \text{(Reynolds),} \\[2ex]
\dfrac{r_V^2 \, r_\rho \, \lambda}{r_\sigma} = 1 & \text{(Weber),} \\[2ex]
\dfrac{r_V^2}{r_g \, \lambda} = 1 & \text{(Froude),} \\[2ex]
\dfrac{\lambda}{r_V \, r_t} = 1 & \text{(Strohual),} \\[2ex]
\dfrac{r_{\Delta p}}{r_\rho \, r_V^2} = 1 & \text{(Euler).}
\end{cases}
\tag{5.84}
$$

If the fluid is the same in the model and prototype, with $r_g = r_\mu = r_\sigma = r_\varepsilon = r_\rho = 1$, we obtain the following system of equations:

$$\begin{cases} r_V = \dfrac{1}{\lambda} & \text{(Reynolds)}, \\[2mm] r_V = \sqrt{\dfrac{1}{\lambda}} & \text{(Weber)}, \\[2mm] r_V = \sqrt{\lambda} & \text{(Froude)}, \\[2mm] \dfrac{\lambda}{r_V\, r_t} = 1 & \text{(Strohual)}, \\[2mm] \dfrac{r_{\Delta p}}{r_V^2} = 1 & \text{(Euler)}, \end{cases} \qquad (5.85)$$

which, as already mentioned in Sect. 5.2, does not admit solutions. However, if the effects of surface tension are negligible because of the limited curvature of the interface and the flow is fully turbulent, rendering the effects of the Reynolds number negligible, the set of conditions reduces to

$$\begin{cases} r_V = \sqrt{\lambda}, \\[2mm] r_t = \sqrt{\lambda}, \\[2mm] r_{\Delta p} = \lambda, \end{cases} \qquad (5.86)$$

and the similarity is named *Froude similarity*. The flow must be turbulent, fully developed both in the model and in the prototype, on the basis of the criterion imposed by the Reynolds friction number and according to the indications given in Table 5.2.

Example 5.6 We analyse the physical model of a flood retention basin at the geometric scale $\lambda = 1/50$. The maximum flow rate in the prototype is $Q_p = 220\,\mathrm{m^3\,s^{-1}}$. We wish to calculate the maximum flow rate required in the model.

We adopt Froude similarity and the volumetric flow rate

$$Q = \Omega\, V \qquad (5.87)$$

scales like

$$r_Q = r_\Omega\, r_V = \lambda^{5/2}. \qquad (5.88)$$

Hence,

$$\frac{Q_m}{Q_p} = \left(\frac{1}{\lambda}\right)^{5/2} \rightarrow Q_m = 220\left(\frac{1}{50}\right)^{5/2} = 0.0125\,\mathrm{m^3\,s^{-1}}. \qquad \therefore \quad (5.89)$$

Figure 5.8 shows the physical model of the dam of the flood retention basin of the Parma torrent, reproduced in Froude similarity with $\lambda = 1/50$. The prototype is visible in Fig. 5.9. In a previous physical model, due to symmetry, only half of the model was reproduced, with a scale of 1/25 and with an evident economy. It should be noted that the symmetry of the structure does not necessarily mean symmetry of

Fig. 5.8 Physical model of the dam of the flood retention basin on the Parma torrent, realised on a geometric scale of $\lambda = 1/50$ (modified from Mignosa et al. 2008)

Fig. 5.9 Dam of the flood retention basin on the Parma torrent (courtesy by Paolo Mignosa)

Table 5.2 Flow regime as a function of the Reynolds friction number for open channels and ducts

Flow regime	Open channels	Ducts
Viscous	$Re_* < 4$	$Re_* < 5$
Turbulent in transition	$4 < Re_* < 100$	$5 < Re_* < 75$
Fully turbulent	$Re_* > 100$	$Re_* > 75$

the flow field, since phenomena of instability can give rise to flows that are non-symmetrical. An obvious case in which the instability of a flow in a symmetrical flow field is even exploited as a principle of flow measurement is the Coanda effect meter (Longo and Petti 2006).

5.2.3 Mach Similarity

The similarity of Mach applies to reproduce physical processes that take place in stationary and compressible flow fields; there are 6 variables involved, and they can be grouped into 3 dimensionless groups, with a physical process expressed as:

$$f(\text{Re, M, Eu}) = 0. \tag{5.90}$$

Physical processes where surface tension and gravity are important are excluded. Similarity criteria allow 3 equations in the 3 unknown scale ratios, namely:

$$\begin{cases} \dfrac{r_\rho\, r_V\, \lambda}{r_\mu} = 1 & \text{(Reynolds)}, \\[2mm] \dfrac{r_V^2\, r_\rho}{r_\varepsilon} = 1 & \text{(Mach)}, \\[2mm] \dfrac{r_{\Delta p}}{r_\rho\, r_V^2} = 1 & \text{(Euler)}. \end{cases} \tag{5.91}$$

If the fluid is the same in the model and prototype, the similarity is of little help since the set of equations reduces to

$$\begin{cases} r_V = \dfrac{1}{\lambda}, \\[2mm] r_V = 1, \\[2mm] r_{\Delta p} = r_V^2, \end{cases} \tag{5.92}$$

which admits only the trivial solution $\lambda = 1$. If the flow is turbulent with fully developed turbulence, we can neglect the Reynolds number, and we can realise a model with an invariant velocity scale compared to the prototype and with a pressure scale based on the Euler number. In many real cases, the Reynolds number cannot be overlooked (at least in some regions of the flow field), and the model must be realised on the same geometric scale as the prototype unless scale effects of some relevance are acceptable. Typically, models with Mach similarity are realised in wind tunnels of a size that can accommodate models at real or very close to real size.

If we change the characteristics of the fluid, for example, by lowering its temperature, a degree of freedom is recovered (see Sect. 9.2), sufficient to realise complete similarity even with a reduced geometric scale of the model.

5.2.4 Similarity in Filtration in the Darcy and Forchheimer Regimes

In filtration processes, there are some geometric variables that characterise the matrix through which the fluid filters and some physical properties of the fluid. We assume that the physical process is described by the following typical equation:

$$f(H,\ x,\ \rho,\ g,\ D,\ u,\ \mu,\ n) = 0, \tag{5.93}$$

where H is the total head, x is the length of the path of the fluid during filtration, ρ is the density of the fluid, g is the acceleration of gravity, D is a geometric scale of the pores, u is the velocity of the fluid, μ is the dynamic viscosity and n is the porosity of the medium.

The change in the specific total head along the path is expressed by the energy gradient:

$$J = -\frac{dH}{dx}, \tag{5.94}$$

which, neglecting the kinetic height (always modest in filtration processes), coincides with the piezometric gradient. Equation (5.93) becomes

$$J = f(\rho,\ g,\ D,\ u,\ \mu,\ n), \tag{5.95}$$

where the variable of interest, J, appears as a governed variable.

The dimensional matrix has rank 3, and applying Buckingham's Theorem, the typical equation becomes a function of 2 dimensionless groups and of the 2 dimensionless variables, J and n:

$$J = \tilde{f}\left(\frac{gD}{u^2},\ \frac{\rho D u}{\mu},\ n\right). \tag{5.96}$$

Experimental evidence suggests that Eq. (5.96) can be written as:

$$\frac{\gamma J D}{\rho u^2} \equiv \Pi_j = \Phi(\text{Re},\ n). \tag{5.97}$$

We observe the merging of the dimensionless groups that reduces the relevant groups from 4 to 3.

Based on the experimental evidence, it can be assumed that

$$\Phi(\text{Re},\ n) = \frac{\Phi_1(n)}{\text{Re}}, \tag{5.98}$$

and, by substitution into Eq. (5.97), we obtain the equation

$$u = \frac{1}{\Phi_1(n)}\frac{\gamma D^2}{\mu} J = kJ, \tag{5.99}$$

where k is the permeability coefficient. Equation (5.99) is known as *Darcy's equation*, valid for $\text{Re} \to 0$. A possible structure of the function Φ envisages independence from the Reynolds number for $\text{Re} \to \infty$, and for very large Re, we can assume

$$\Phi(\text{Re},\ n) = \Phi_2(n); \tag{5.100}$$

hence,

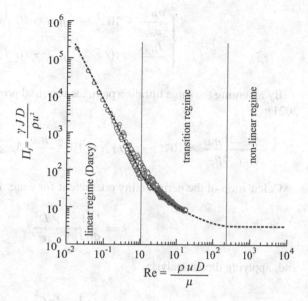

Fig. 5.10 Air permeability measurements in porous asphalt pavement (DICATeA Hydraulics Laboratory, 2006)

$$u = \sqrt{\frac{\gamma J D}{\rho \, \Phi_2(n)}}. \tag{5.101}$$

Figure 5.10 shows the results of some measurements of air permeability through a matrix of porous asphalt pavement. The linear regime zone is clearly identified, while at high Reynolds numbers, the curve is drawn by eye. The quadratic law is attributed to inertial effects, since turbulence is not developed under these conditions. The similarity requires that

$$\begin{cases} \dfrac{r_\gamma \, r_J \, \lambda}{r_\rho \, r_u^2} = 1, \\[2mm] \dfrac{r_\rho \, r_u \, \lambda}{r_\mu} = 1, \\[2mm] r_n = 1. \end{cases} \tag{5.102}$$

If we wish to extrapolate the results of permeability measurements performed with air instead of water (performed on exactly the same specimen), the imposed scale ratios are $r_n = 1$ and $\lambda = 1$, and the scales of the derived variables are:

$$\begin{cases} r_J = r_u^2, \\[2mm] r_u = \dfrac{r_\mu}{r_\rho}. \end{cases} \tag{5.103}$$

Given $r_\mu/r_\rho \equiv r_\nu = \nu_{H_2O}/\nu_{air} \approx 10^{-1}$, we obtain

$$\begin{cases} \dfrac{u_{H_2O}}{u_{air}} \approx 10^{-1} \rightarrow u_{H_2O} \approx 10^{-1}\,u_{air}, \\[2mm] \dfrac{J_{H_2O}}{J_{air}} \approx 10^{-2} \rightarrow J_{H_2O} \approx 10^{-2}\,J_{air}. \end{cases} \tag{5.104}$$

By assuming the same filtration path, the required pressure drop is (Giuliani et al. 2021):

$$\frac{\gamma_{air}\,\Delta p_{H_2O}}{\gamma_{H_2O}\,\Delta p_{air}} \approx 10^{-2} \rightarrow \Delta p_{H_2O} \approx 10^{-2}\frac{\gamma_{H_2O}}{\gamma_{air}}\Delta p_{air} \approx 10\Delta p_{air}. \tag{5.105}$$

Calculation of the permeability coefficient for water from Eq. (5.99) yields

$$k = \frac{1}{\Phi_1(n)}\frac{\gamma\,D^2}{\mu}, \tag{5.106}$$

and, applying direct analysis,

$$r_k = \frac{1}{r_{\Phi_1(n)}}\frac{r_g\,\lambda^2}{r_\nu}. \tag{5.107}$$

Since the same filtration matrix is used, we obtain $r_{\Phi_1(n)} = 1$, $\lambda = 1$ and $r_g = 1$. In summary:

$$\frac{k_{H_2O}}{k_{air}} = \frac{\nu_{air}}{\nu_{H_2O}} \rightarrow k_{H_2O} \approx 10\,k_{air}. \tag{5.108}$$

Forchheimer (1901) suggested a nonlinear relationship between the pressure drop and flow in porous media for a very large Reynolds number:

$$J = \frac{\mu}{\gamma\,k}u + \frac{C_1 d}{g\,k}u^2, \tag{5.109}$$

where C_1 is a constant coefficient and k is the intrinsic permeability, function of porosity n and of the representative size of the pores, d.

The similarity requires that

$$\begin{cases} r_J = \dfrac{r_\nu\,r_u}{r_k} = \dfrac{\lambda\,r_u^2}{r_k}, \\[2mm] r_n = 1, \end{cases} \tag{5.110}$$

or

$$r_u = \frac{r_\nu}{\lambda}, \quad r_J = \frac{r_\nu^2}{\lambda\,r_k}, \quad r_n = 1. \tag{5.111}$$

5.3 Geometrically Distorted Hydraulic Models

Physical models of very large environmental systems, such as rivers and lagoons, have the problem of inadequate reproduction of the vertical geometric scale once a reasonable geometric scale in the horizontal plane is selected: this last scale cannot be very large, and assuming the same scale in the vertical direction brings minuscule values, e.g., of the water depth in the model. To achieve good measurement accuracy and to avoid significant scale effects due, for example, to surface tension, it is necessary to check that the minimum water level in the model is a few centimetres. With this constraint, the adoption of a unique geometric scale requires very large models. In addition, for all shallow water models, if the geometrical scale is too small, it is not possible to measure with the necessary accuracy either the free surface level or the bottom, and it is also difficult to reproduce the roughness of the walls. However, even if the roughness is correctly reproduced, it may have characteristics that facilitate the establishment of laminar flow or transition flow in the model instead of turbulent flow, which is typical of most (if not all) environmental flows. To overcome these limitations, we realise the *distorted models*, with the adoption of a greater vertical geometric scale than the geometric scales in the horizontal plane.

In the more general case, it is possible to fix three distinct geometric scales, following the criteria of distorted similarity treated in Sect. 4.1.6. Indicating with x, y and z the coordinates in the average flow direction, in the vertical direction and in the direction orthogonal to the $x - y$ plane, the application of direct analysis to the equations of momentum balance and mass conservation

$$\begin{cases} \dfrac{\partial u}{\partial t} + u \dfrac{\partial u}{\partial x} + g \dfrac{\partial y}{\partial x} + g J = 0, \\ \dfrac{\partial Q}{\partial x} + \dfrac{\partial A}{\partial t} = 0, \end{cases} \tag{5.112}$$

leads to the following system of equations in the scales:

$$\begin{cases} \dfrac{r_u}{r_t} = \dfrac{r_u^2}{\lambda_x} = \dfrac{\lambda_y}{\lambda_x} = r_f \dfrac{r_u^2}{\lambda_R}, \\ \dfrac{r_u \lambda_y \lambda_z}{\lambda_x} = \dfrac{\lambda_y \lambda_z}{r_t}. \end{cases} \tag{5.113}$$

We have 4 independent equations with 7 unknowns. We can arbitrarily fix the three geometric scales and calculate all the other scales. It is possible to demonstrate that for two-dimensional shallow water processes (for instance, tidal flows in lagoons, two-dimensional flooding), the Saint-Venant equations require the use of the same geometric scale in all directions of the horizontal plane, and distortion can occur only with respect to the vertical geometric scale. With equal scales in the horizontal, $\lambda_x = \lambda_z$, the ratio λ_y/λ_x is called the *distortion ratio*.

5.4   Scale Effects in Hydraulic Models

Additionally, in physical hydraulic models, in most cases, it is necessary to neglect some dimensionless groups and to quantify the corresponding scale effect. Sometimes, scale effects originate from the difficulty of keeping unchanged some parameters that appear in the interaction processes between the bodies in the model, such as the friction between continuous contact solids and the *friction factor*.

What we define as friction is the contribution of adhesion and deformation phenomena in the contact surface between the two bodies. The deformation is partly due to the deformation of the roughness of the surfaces and partly to the deformation of the particles between the surfaces themselves. In theory, the coefficient of friction between continuous solids should be independent of the contact surface area; in reality, with the same nature and geometry of the materials, this coefficient is a function of the geometric scale (Bhushan and Nosonovsky 2004). Adhesion with single or multiple roughnesses in elastic contact increases as the geometric scale decreases, while in plastic contact, it increases or decreases depending on the material characteristics. The friction coefficient associated with deformation increases with the reduction of the geometric scale. Therefore, if the contact is elastic, the coefficient of friction is higher at smaller scales. The theoretical dependence of the friction coefficient on the geometric scale is often not monotonic; therefore, it is poorly related to the macroscale. For the deformation of surface roughness, we have

$$\mu = \mu_0 \left(\frac{L}{L_{1w}} \right)^{n-m}, \tag{5.114}$$

where L_{1w} is the asymptotic length of the contact area (the macroscale of the roughness calculated based on the autocorrelation function) and n and m are two empirical coefficients equal to 0.2 and 0.5, respectively. Hence,

$$r_\mu = \lambda^{-0.3}. \tag{5.115}$$

According to this expression, a reduction in geometric scale $\lambda = 1/10$ leads to doubling of the friction coefficient. This means that to achieve dynamic similarity, it is necessary to use materials with a lower friction coefficient in the model than in the prototype.

Example 5.7 We analyse a pier built with a series of independent floating caissons, vertically sliding along pairs of circular piles (see Fig. 5.11). We aim to realise a physical model to evaluate its dynamic behaviour under wave action and to estimate the efficiency of a pontoon, measured as the efficiency of shielding the area on the back from incident wave action.

To identify the variables involved in the physical process, we use the following simplified dynamic equation:

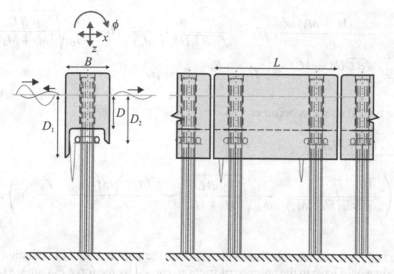

Fig. 5.11 Floating pontoon with floating caissons sliding vertically along circular piles

$$(M + M_a)\ddot{z} + \beta \dot{z} + \gamma\, BLz + \mu\, \frac{|\dot{z}|}{\dot{z}}\, F_x(t) = F_z(t), \qquad (5.116)$$

where M is the mass of the floating caisson, M_a is the added mass of water, β is the damping coefficient, B and L are the plan dimensions of the caisson, μ is the coefficient of friction with the piles, F_x is the horizontal thrust between the caisson and the pile, and $F_z = F_{z0} \sin \omega_0 t$ is the vertical thrust generated by the wave field. We neglect horizontal dynamics and rotations because the nature of the constraint greatly limits both horizontal displacement and roll, pitch, and yaw.

We are first interested in evaluating the maximum vertical displacement of the caisson z_{max}:

$$z_{max} = f((M + M_a),\ \beta,\ \gamma,\ BL,\ F_x,\ F_{z0},\ \omega_0,\ \mu). \qquad (5.117)$$

We observe that some variables are coupled since they do not intervene autonomously in the physical process; for example, the width B is coupled with the length L to define the cross-sectional area of the caisson at the waterline. The rank of the dimensional matrix is 3, and we can select 3 independent fundamental variables, for example, $(M + M_a)$, F_{z0} and ω_0, and express the physical process as a function of $(9 - 3) = 6$ dimensionless groups:

$$\Pi_1 = \frac{z_{max}\,(M + M_a)\,\omega_0^2}{F_{z0}}, \quad \Pi_2 = \frac{\beta}{\sqrt{\gamma\,BL\,(M + M_a)}}, \quad \Pi_3 = \frac{1}{\omega_0}\sqrt{\frac{\gamma\,BL}{(M + M_a)}},$$

$$\Pi_4 = \frac{\sqrt{BL}\,(M + M_a)\,\omega_0^2}{F_{z0}}, \quad \Pi_5 = \frac{F_x}{F_{z0}}, \quad \Pi_6 = \mu.$$

$$(5.118)$$

The typical equation becomes

$$\frac{z_{max}\,(M + M_a)\,\omega_0^2}{F_{z0}} =$$

$$\tilde{f}\left(\frac{\beta}{\sqrt{\gamma\,BL\,(M + M_a)}}, \frac{1}{\omega_0}\sqrt{\frac{\gamma\,BL}{(M + M_a)}}, \frac{\sqrt{BL}\,(M + M_a)\,\omega_0^2}{F_{z0}}, \frac{F_x}{F_{z0}}, \mu\right),$$

$$(5.119)$$

where z_{max} is non-dimensionalised with respect to the amplitude of the oscillation that the body would have in the absence of friction, caused by the force $F_{z0}\sin\omega_0 t$; the second dimensionless group is the damping coefficient, while the third dimensionless group is the ratio between the pulsation of the free body, in the absence of friction, and the pulsation of the forcing action.

To realise a physical model, we must calculate the scale ratios that render the dimensionless groups equal in the model and in the prototype; hence:

$$\begin{cases} \dfrac{r_z\,r_m}{r_{F_z}\,r_t^2} = 1, \\[2mm] \dfrac{r_\beta^2}{r_\gamma\,r_B\,r_L\,r_m} = 1, \\[2mm] \dfrac{r_\gamma\,r_B\,r_L\,r_t^2}{r_m} = 1, \\[2mm] \dfrac{r_B\,r_L\,r_m^2}{r_{F_z}^2\,r_t^4} = 1, \\[2mm] r_{F_x} = r_{F_z}, \\[2mm] r_\mu = 1. \end{cases} \qquad (5.120)$$

If we use the same fluid in the model and prototype, it results in $r_\gamma = 1$, and in an undistorted model, all geometric scales are equal. In addition, if the model is in Froude similarity, the result is $r_{F_z} = r_{F_x} = \lambda^3$ and $r_t = \sqrt{\lambda}$. Hence,

$$r_m = \lambda^3, \quad r_\beta = \lambda^{5/2}, \quad r_\mu = 1. \qquad \therefore \quad (5.121)$$

As already mentioned, the real problem of this physical model is represented by the friction coefficient, which is generally higher at reduced geometrical scales (see Sect. 5.4). We observe that the bodies are not continuously in contact due to the

Fig. 5.12 Geometry of the constriction between piles and floating caisson

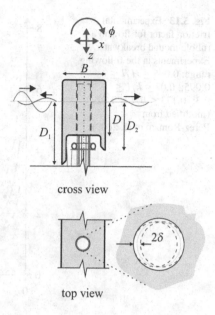

cross view

top view

presence of a constriction between the hole in the floating caisson and the circular pile (see Fig. 5.12); hence, the average horizontal thrust component is

$$\overline{F_x} = \frac{1}{T} \int_0^T F_x \, dt. \tag{5.122}$$

The reduction of the contact time between piles and caisson, controlled by the width of the constriction, could be a degree of freedom to reduce the scale effect of the friction: although during contact the instantaneous magnitude of thrust rises to ensure an invariant average thrust, the presence of a lubricant liquid in the constriction promotes a transfer of thrust from the caisson to the pile without direct contact, hence with a reduced friction. However, the calibration of the optimal width of the constriction could be problematic and could give discordant results for different incident wave characteristics.

Example 5.8 We wish to analyse a rubble mound breakwater under a wave attack. The incident waves are partially reflected and partially transmitted by virtue of the presence of connected paths in the porous structure of the breakwater, with the insurgence of a flow of filtration. The efficiency of the structure depends on the characteristics of the flow, which in turn is controlled by the friction factor f_c. The construction characteristics of the breakwater include an outer layer of material of sufficient dimensions to withstand wave attack and an inner layer of finer material, with the task of dissipating the transmitted energy. The most important geometric scales are the width of the breakwater B in the wave propagation direction, the diameter d representative of the loose material, the incident wavelength l and the

Fig. 5.13 Experimental friction factor for flow in a rubble mound breakwater. Experiments in the following range: $0.006 \leq H/l \leq 0.095$; $0.07 \leq h/l \leq 0.3$; $0.173 \leq B/l \leq 1.02$ (modified from Pérez-Romero et al. 2009)

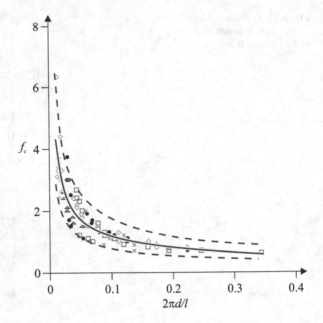

height of the incident wave H. The variable we wish to analyse is the friction factor of the flow generated in the porous structure of the breakwater. We can write the following typical equation:

$$f_c = f(B, H, l, d). \tag{5.123}$$

The dimensional matrix has rank 1 (its contents are all geometric variables), and we can render dimensionless the variables with respect to a geometric scale. The most significant groups lead to the following typical equation:

$$f_c = \tilde{f}\left(\frac{B}{l}, \frac{d}{l}, \frac{d}{H}\right). \tag{5.124}$$

Figure 5.13 shows the friction factor obtained experimentally as a function of d/l.

To estimate the scale ratio of the friction factor, we consider Bernoulli's equation in a porous medium:

$$s\frac{\partial \Phi}{\partial t} + \frac{p}{\rho_w} + gz + f_c\,\omega\,\Phi = 0, \tag{5.125}$$

where s is an inertial coefficient, Φ is the potential, ρ_w is the density of water, ω is the pulsation of the incident wave, and f_c is an equivalent friction coefficient. In dimensionless form, Eq. (5.125) can be written as

$$s\frac{\partial \tilde{\Phi}}{\partial \tilde{t}} + \tilde{p} + \frac{g}{\omega^2\,l}\tilde{z} + f_c\,\tilde{\Phi} = 0. \tag{5.126}$$

The condition of similarity requires that

$$r_{f_c} = \frac{r_g}{r_\omega^2 \lambda} = 1; \qquad\qquad \therefore \quad (5.127)$$

that is, $r_\omega \equiv r_t^{-1} = \lambda^{-1/2}$ (Froude similarity), and since $r_g = 1$, we must also have $r_{f_c} = 1$. However, the experiments suggest that by reproducing in Froude similarity all geometrical variables, including the diameter of the stones/pebbles/sediments, the friction factor assumes a different value in model and in the prototype. To correctly reproduce the filtration flow field in the porous medium, the wave reflection and the transmission, it is necessary to increase the diameter of the stone elements in the model with respect to the one calculated under Froude similarity.

Example 5.9 The action of sea waves breaking on a vertical wall of a pier can be classified, according to the duration of the impulse, as (i) quasi-static, if the duration of the impulse varies from 20% to 50% of the wave period, or (ii) impulsive, if the duration of the impulse is much shorter and up to 1% of the wave period. Froude similarity leads to the conclusion that the pressure peak scales with λ since $p \propto \rho u^2$ and usually $r_\rho = 1$. Actually, the (absolute) pressure peak varies experimentally according to the relationship (Takahashi et al. 1985)

$$\frac{\rho\, k_w\, u^2}{p_{atm}\, D} = 5 \left(\frac{p_{max}}{p_{atm}}\right)^{2/7} + 2 \left(\frac{p_{max}}{p_{atm}}\right)^{-5/7} - 7, \qquad (5.128)$$

where ρ is the density of the water, k_w is the thickness of the water layer considered active in the impulsive process, and D is the thickness of the air pocket initially interposed between the vertical wall and the breaking wave. The power function on the left is a dimensionless group defined as the *number of Bagnold* (we observe that this term is associated with another relationship that appears in the dynamics of granular mixtures; the meaning is completely different, but the titular author is the same). Under other conditions, a fraction of the air escapes with a reduction in the pressure peak. This results in the Bagnold number not being preserved in Froude similarity if the atmospheric pressure is the same in the model and in the prototype:

$$r_{Ba} \equiv \frac{Ba_m}{Ba_p} = \lambda, \qquad (5.129)$$

where λ is the geometric scale.

If the relationship between the pressure peak p_{max} and the Bagnold number were a power function, it would be straightforward to calculate the pressure scale ratio. Since the relationship is not power function, the scale factor depends also on the operating point, that is, the value of the measured peak (in the model or prototype). This limit can be overcome by interpolating Eq. (5.128) with a monotonic function through the origin (this requires the introduction of the peak pressure relative to atmospheric pressure). For example, for Ba $<$ 0.5 results, with fairly good approximation,

$$\text{Ba} = 0.31 \left(\frac{p^*_{max}}{p_{atm}}\right)^{3/2}, \tag{5.130}$$

where p^*_{max} is the maximum pressure relative to atmospheric pressure. Hence,

$$r_{p^*_{max}} \equiv \frac{p^*_{max,m}}{p^*_{max,p}} = r^{2/3}_{\text{Ba}} = \lambda^{2/3}. \qquad \therefore \quad (5.131)$$

The correction is quite relevant: in a physical model at scale $\lambda = 1/40$, we would expect a pressure in the prototype 40 times higher than the pressure in the model; instead, based on the above considerations, the peak pressure scale is $r_{p^*_{max}} = (1/40)^{2/3} \approx 1/11.7$, and the peak pressure in the prototype is only approximately 12 times higher than the peak pressure in the model.

Example 5.10 Some fuel cells operate with a direct electrochemical reaction between hydrogen and oxygen, which takes place in a polymeric membrane (PEM). The gas is at contact with the membrane and passes through a diffusion layer (GDL). The electrochemical reaction produces electric current and water, which must be removed to avoid degradation in membrane efficiency, especially because this degradation would disrupt the mass exchange that allows the reagents to come into contact with the CL catalyst layer and to avoid the damage caused by freezing water at low cell operating temperatures. An efficient method appears to be the introduction of a microporous layer (MPL) between the catalyst layer and the gas diffusion layer. The gas diffusion layer is hydrophobic and not wetted by water. The water tends to invade the gas diffusion layer, pushing the air away.

It is necessary to test with a physical model the efficiency of the microporous layer.

Having identified the important variables in the physical process, we assume that the process is described as

$$f\left(\mu_i, \ \mu_d, \ \sigma_{i-d}, \ \theta_{c,i}, \ V_i, \ \rho_i, \ \rho_d, \ g, \ l\right) = 0, \tag{5.132}$$

where μ_i and μ_d are the dynamic viscosity of the *invader* and *defender* fluid, respectively, σ_{i-d} is the tension at the interface between the two fluids, $\theta_{c,i}$ is the contact angle between the fluid and the material of the gas diffusion layer, V_i is the effective filtration velocity of the *invader* fluid, ρ_i and ρ_d are the densities of the two fluids, g is the acceleration of gravity and l is a scale length, for example, the diameter of the pores d_p. On the basis of experimental evidence, we observe that the tension at the interface and the angle of contact jointly participate in the form $\sigma_{i-d} \cos \theta_{c,i}$, abbreviated as σ_r. Similarly, the densities of the two fluids affect the system through their difference, $\Delta\rho = \rho_i - \rho_d$. In summary, the variables are reduced to 7, and the dimensional matrix is

$$
\begin{array}{c|ccccccc}
 & \mu_i & \mu_d & \sigma_r & V_i & \Delta\rho & g & l \\
\hline
M & 1 & 1 & 1 & 0 & 1 & 0 & 0 \\
L & -1 & -1 & 0 & 1 & -3 & 1 & 1 \\
T & -1 & -1 & -2 & -1 & 0 & -2 & 0
\end{array}
\tag{5.133}
$$

and has rank 3. By selecting as fundamental variables μ_i, σ_r and l and applying Rayleigh's method or the matrix method, the following 4 dimensionless groups are calculated:

$$
\Pi_1 = \frac{\mu_d}{\mu_i}, \quad \Pi_2 = \frac{\mu_i V_i}{\sigma_r}, \quad \Pi_3 = \frac{\sigma_r \Delta\rho\, l}{\mu_i^2}, \quad \Pi_4 = \frac{g\,\mu_i^2\, l}{\sigma_r^2}.
\tag{5.134}
$$

The dimensionless groups with a physical meaning are:

$$
\Pi_1 = \frac{\mu_d}{\mu_i}, \quad \mathrm{Ca} \equiv \Pi_2 = \frac{\mu_i V_i}{\sigma_r},
$$
$$
\mathrm{Bo} = \Pi_3 \cdot \Pi_4 = \frac{\Delta\rho\, g\, l^2}{\sigma_r}, \quad \mathrm{Re} = \Pi_3 \cdot \Pi_2 = \frac{\Delta\rho\, l\, V_i}{\mu_i},
\tag{5.135}
$$

where Ca is the *capillarity number*, or the ratio of viscous forces to interface forces, Bo is the *Bond number* (also known as the *Eötvös* number), or the ratio between the buoyancy forces and the forces at the interface, and Re is the Reynolds number.

To guarantee the similarity between the model and prototype, it is necessary to calculate the 7 scale ratios so that they satisfy the following 4 equations:

$$
\begin{cases}
r_{\mu_d} = r_{\mu_i}, \\
r_{\mu_i} r_{V_i} = r_{\sigma_r}, \\
r_{\Delta\rho}\, r_g\, \lambda^2 = r_{\sigma_r}, \\
r_{\Delta\rho}\, \lambda\, r_{V_i} = r_{\mu_i}.
\end{cases}
\tag{5.136}
$$

Having selected 3 scales, for example, λ, r_{μ_i} and r_g, the system of equations admits the following solution:

$$
\begin{cases}
r_{\mu_d} = r_{\mu_i}, \\
r_{V_i} = \sqrt{\lambda\, r_g}, \\
r_{\Delta\rho} = \dfrac{r_{\mu_i}}{\lambda^{3/2} \sqrt{r_g}}, \\
r_{\sigma_r} = r_{\mu_i} \sqrt{\lambda\, r_g}.
\end{cases}
\tag{5.137}
$$

The ratio r_g is unity, although it could be changed by performing the experiments in a centrifuge (see Sect. 8.2). Additionally, an enlarged geometric scale is suggested, with $\lambda \gg 1$. However, on the basis of experimental evidence that capillary effects are dominant with respect to viscosity and buoyancy and neglecting the Reynolds

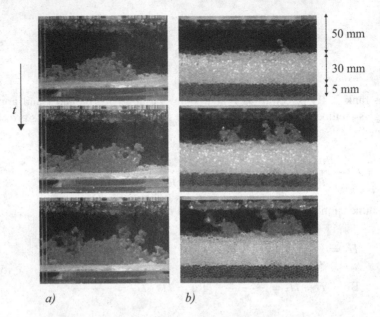

50 mm

30 mm

5 mm

t

a)   *b)*

Fig. 5.14 Evolution of the water percolation process in the gas diffusion layer: *a)* physical model without and *b)* with a microporous layer (modified from Kang et al. 2010)

and Bond numbers, the similarity equations become:

$$
\begin{cases}
r_{\mu_d} = r_{\mu_i}, \\
r_{\mu_i}\, r_{V_i} = r_{\sigma_r},
\end{cases}
\qquad \therefore \quad (5.138)
$$

constraining only some of the variables. In particular, the model can be realised on an arbitrary geometric scale (the geometric scale refers to the pore scale).

We observe that the negligibility of some dimensionless groups derives only from experimental evidence or from the structure of the equations that model the physical process. The fact that a dimensionless group assumes a very large or very small numerical value is irrelevant for the evaluation of its role in the dynamics of the physical process.

Figure 5.14 shows some photographs of a physical model in similarity, realised to verify the efficiency of the microporous layer between the catalyst layer and the gas diffusion layer (Kang et al. 2010).

5.5 Analogue Models

Alongside traditional physical models, it is sometimes possible and advantageous to use *analogue models*. Two physical processes of different natures are defined "in analogy" if they are described by formally identical mathematical equations and in which coefficients and parameters, although having different meanings, have the same role. A classic case of an analogue model is the one between two-dimensional filtration phenomena and the laminar flow of a fluid flowing between two very close flat and parallel plates, the *Hele-Shaw analogy*. In fact, the two components of velocity (u, v) of the two-dimensional filtration flow in a homogeneous and isotropic porous medium are given by the following equations:

$$\begin{cases} u = -k\, \dfrac{\partial h}{\partial x}, \\[2mm] v = -k\, \dfrac{\partial h}{\partial y}, \end{cases} \tag{5.139}$$

where k is the permeability coefficient and h is the piezometric head. The Navier–Stokes reduced equations for two-dimensional laminar flow between two flat plates at a relative distance d can be simplified as

$$\begin{cases} u = -\dfrac{g\,d^2}{12v}\, \dfrac{\partial h}{\partial x} = -k_r\, \dfrac{\partial h}{\partial x}, \\[2mm] v = -\dfrac{g\,d^2}{12v}\, \dfrac{\partial h}{\partial y} = -k_r\, \dfrac{\partial h}{\partial y}, \end{cases} \tag{5.140}$$

where v is the kinematic viscosity of the fluid. Equations (5.140) are valid as long as the flow is laminar ($\mathrm{Re} \equiv u\,(d/2)/v < 500$) and the inertial components and terms of the spatial derivatives $\partial/\partial x^2$ and $\partial/\partial y^2$ are negligible. The two models are analogous.

It is possible, for example, to reproduce two-dimensional filtration processes with an analogical model in a Hele-Shaw cell, realised with two parallel plates spaced from 0.1 mm to 2.0 mm, with one of the two transparent plates allowing visual observation. For larger gaps, it is necessary to use glycerol as a test fluid, with a much higher kinematic viscosity than water, to meet the condition $\mathrm{Re} < 500$. Hele-Shaw cells can be either vertical or horizontal. Vertical cells, see Fig. 5.15, are necessary to simulate a filtration process in the presence of a free surface at atmospheric pressure.

If the flow takes place in a stratified porous medium with zones of different permeabilities, it is possible to realise a Hele-Shaw model with two parallel plates that delimit a variable gap, larger in the zones of greater permeability. The relationship between the gap thickness d_1 and d_2 must satisfy

$$\frac{d_1^3}{d_2^3} = \frac{k_1}{k_2}. \tag{5.141}$$

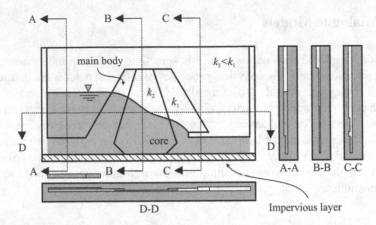

Fig. 5.15 Hele-Shaw cell to reproduce filtration in a stratified medium (modified from Bear 1972)

This condition seems to contradict the expression of hydraulic conductivity, which varies with the square of the distance between the plates. In fact, the cube dependence of the distance is because d has the double role of scale for hydraulic conductivity and scale for the width of the flow channel. To correctly reproduce the conductivity variation, it is necessary to correctly reproduce a hydraulic transmissivity parameter $T = k\,d \propto d^3$, both in the model and in the prototype (Bear 1972). The rest of the geometry must be realised through geometric similarity.

The analogy has been extended to simulate flows in porous media with vertically or horizontally varying porosity and permeability, possibly with non-Newtonian power-law fluids; see Ciriello et al. (2016).

There are numerous other examples of analogue models. For example, under certain conditions, there is an approximate analogy between filtration and the deformation of a membrane. The mass conservation can be written in cylindrical coordinates as

$$\frac{\partial^2 h}{\partial r^2} + \frac{1}{r}\frac{\partial h}{\partial r} + \frac{\partial^2 h}{\partial z^2} = 0, \tag{5.142}$$

while the deformation of a membrane can be expressed as

$$\frac{\partial^2 \delta}{\partial r^2} + \frac{1}{r}\frac{\partial \delta}{\partial r} + \frac{\gamma\,d}{\sigma} = 0, \tag{5.143}$$

where δ is the displacement, γ is the specific weight of the material of the membrane, d is the thickness of the plate and σ is the tension in the membrane. The only difference between the two equations is represented by the two terms $\partial^2 h / \partial z^2$ and $\gamma\,d/\sigma$, both almost always negligible. Figure 5.16 shows the schematic of a physical model for the simulation of the water table in the presence of drainage and recharge wells.

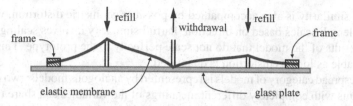

Fig. 5.16 Analogue model for the reproduction of filtration phenomena in a water table

Another analogy frequently adopted in the past is between filtration and electric circuits. An advantageous feature of electric analogue models stems from the fact that electrical phenomena are characterised by very fast transients (the steady-state condition is reached in a very short time), and the measurement instrumentation of electrical variables is easily available with characteristics of high accuracy and precision. This allows models on a very small geometric scale, guaranteeing cost savings and measurement accuracy. Among the phenomena that can be reproduced in electrical analogy, we remember the hydraulic pipeline networks, simulated by creating a circuit in which the pipelines are replaced by electrical conductors and the flow resistors by electrical light bulb resistors. Electric lamps have a nonlinear characteristic because the resistivity of the filament material increases with increasing temperature and, therefore, with increasing current passing through it. This is analogous to the resistance of a hydraulic pipe: the pressure drop (analogous to the voltage drop) increases according to the power of the volumetric flow (analogous to the electric current); that is, $\Delta H \propto Q^n$ with $n = 1$ in the laminar regime and $n = 1.75 - 2$ in the turbulent regime. Therefore, the differences in potential between the nodes of the electric network are analogous to the hydraulic head differences between the nodes of the hydraulic network; the currents crossing a branch of the electric network are analogous to the flow rates crossing a trunk of the hydraulic network.

Today, analogue models of hydraulic networks have been well replaced by mathematical models.

Summarising Concepts

- The similarity criteria applied to fluid mechanics and hydraulics are developed on the basis of the most important dimensionless groups, the result of possible combinations of 9 variables that control almost the entirety of the problems, with 3 of these variables independent.
- Complete similarity is practically never achieved because of numerous constraints on the nature of the fluid and because of the invariance of the acceleration of gravity. Almost always, the similarity is partial and is named after the most important dimensionless group that satisfies the invariance criterion.

218 5 Applications in Fluid Mechanics and Hydraulics

- Partial similarity is also accompanied by possible geometric distortion, with different length scales based on direction. Partial similarity involves scaling effects, with results of the model that do not scale perfectly to the prototype. This can be acceptable as long as distortion is quantified.
- A widespread category of models is represented by analogous models: two physical problems with completely different meanings of the variables can share the same structure of the differential (or algebraic) problem that approximates them. A well-known example is the Hele-Shaw analogy.

References

Bear, J. (1972). *Dynamics of fluids in porous media*. Dover.

Bhushan, B., & Nosonovsky, M. (2004). Scale effects in dry and wet friction, wear, and interface temperature. *Nanotechnology, 15*(7), 749.

Ciriello, V., Longo, S., Chiapponi, L., & Di Federico, V. (2016). Porous gravity currents: A survey to determine the joint influence of fluid rheology and variations of medium properties. *Advances in Water Resources, 92,* 105–115.

Forchheimer, P. (1901). Wasserbewegung durch boden. *Zeitschrift des Vereins deutscher Ingenieure, 45,* 1782–1788.

Giuliani, F., Petrolo, D., Chiapponi, L., Zanini, A., & Longo, S. (2021). Advancement in measuring the hydraulic conductivity of porous asphalt pavements. *Construction and Building Materials, 300,* 124110.

Goodridge, C. L., Shi, W. T., Hentschel, H. G. E., & Lathrop, D. P. (1997). Viscous effects in droplet-ejecting capillary waves. *Physical Review E, 56*(1), 472.

Hughes, S. A. (1993). *Physical models and laboratory techniques in coastal engineering* (vol. 7). World Scientific Publishing Co. Pte. Ltd.

Ivicsics, L. (1980). *Hydraulic models*. Water Resources Publications.

Kang, J. H., Lee, K.-J., Yu, S. H., Nam, J. H., & Kim, C.-J. (2010). Demonstration of water management role of microporous layer by similarity model experiments. *International Journal of Hydrogen Energy, 35*(9), 4264–4269.

Longo, S., & Petti, M. (2006). *Misure e controlli idraulici*. McGrawHill.

Maxwell, C. J. (1878). III. On stresses in rarefied gases arising from inequalities of temperature. *Proceedings of the Royal Society London, 27,* 304–308.

Mignosa, P., Giuffredi, F., Danese, D., La Rocca, M., Longo, S., Chiapponi, L., et al. (2008). *Prove su modello fisico del manufatto regolatore della cassa di espansione sul Torrente Parma (Physical model tests of the control systems for the detention basin of the Parma Torrent) (in Italian)*. DICATeA, University of Parma, and AIPo.

Novak, P., & Čábelka, J. (1981). *Models in hydraulic engineering: Physical principles and design applications* (Vol. 4). Pitman Publishing Ltd.

Pérez-Romero, D. M., Ortega-Sánchez, M., Moñino, A., & Losada, M. A. (2009). Characteristic friction coefficient and scale effects in oscillatory porous flow. *Coastal Engineering, 56*(9), 931–939.

Rouse, H., & Ince, S. (1963). *History of Hydraulics*. Dover.

Smeaton, J. (1759). XVIII. An experimental enquiry concerning the natural powers of water and wind to turn mills, and other machines, depending on a circular motion. *Philosophical Transactions of the Royal Society of London 51,* 100–174.

Takahashi, S., Tanimoto, K., & Miyanaga, S. (1985). Uplift wave forces due to compression of enclosed air layer and their similitude law. *Coastal Engineering in Japan, 28*(1), 191–206.

Yalin, M. S. (1971). *Theory of hydraulic models*. Macmillan Publishers Ltd.

Chapter 6
Applications to Heat Transfer Problems

The problems of heat exchange and heat transmission involve the temperature Θ in addition to mass, length and time. A greater number of fundamental variables almost always corresponds to a greater number of variables involved, suitable for characterising the behaviour of continuous solids or fluids in the presence of thermal fluxes, the latter with additional complication due to whether the flow regime is laminar or turbulent.

6.1 The Relevant Dimensionless Groups

In heat transfer problems the most frequent dimensionless groups, in addition to those listed in Sect. 5.1, are:

$$\text{Nu} = \frac{h\,l}{k} = \frac{heat\ flow\ by\ convection}{heat\ flow\ by\ conduction} \quad \text{(Nusselt)},$$

$$\text{Pr} = \frac{c_p\,\mu}{k} = \frac{diffusion\ of\ momentum}{thermal\ diffusion} \quad \text{(Prandtl)},$$

$$(6.1)$$

where c_p is the specific heat, h is the heat exchange coefficient per convection, and k is the thermal conductivity.

The relevant dimensionless groups can be partly computed using the heat diffusion equation,

$$\frac{\partial \theta}{\partial t} + \mathbf{v} \cdot \nabla \theta = k\,\nabla^2\theta. \quad (6.2)$$

© The Author(s), under exclusive license to Springer Nature Switzerland AG 2021
S. G. Longo, *Principles and Applications of Dimensional Analysis and Similarity*,
Mathematical Engineering, https://doi.org/10.1007/978-3-030-79217-6_6

Selecting the scales θ_0, t_0, u_0 and l_0 yields

$$\left(\frac{\theta_0}{t_0}\right)\frac{\partial\tilde{\theta}}{\partial\tilde{t}} + \left(\frac{u_0\theta_0}{l_0}\right)\tilde{\mathbf{v}}\cdot\tilde{\nabla}\tilde{\theta} = \left(\frac{k\theta_0}{l_0^2}\right)\tilde{\nabla}^2\tilde{\theta}, \tag{6.3}$$

where the symbol $\sim$ indicates a dimensionless variable. Dividing all the terms by $k\theta_0/l_0^2$ yields

$$\left(\frac{l_0^2}{k\,t_0}\right)\frac{\partial\tilde{\theta}}{\partial\tilde{t}} + \left(\frac{l_0 u_0}{k}\right)\tilde{\mathbf{v}}\cdot\tilde{\nabla}\tilde{\theta} = \tilde{\nabla}^2\tilde{\theta}, \tag{6.4}$$

where the first term in parentheses is the inverse of the Fourier number:

$$\text{Fo} = \frac{k\,t_0}{l_0^2}, \tag{6.5}$$

which is representative of the advancement of the thermal wavefront in the body; the second term in parentheses is the Péclet number, defined as the ratio of the flow of a physical quantity by convection and by conduction. If the physical quantity is heat, the Péclet number is also equal to the product of the Reynolds number and the Prandtl number, that is,

$$\text{Pe} = \frac{u_0 l_0}{k} \equiv \text{Re}\cdot\text{Pr} = \frac{convective\ heat\ transfer}{viscous\ diffusion\ of\ heat}. \tag{6.6}$$

If time is controlled by convective variables, then we obtain $t_0 = l_0/u_0$, the Fourier number is equal to the Péclet number, and we implicitly assume that convection and local inertia have equal intensity.

A modified version of the Péclet number is the Graetz number:

$$\text{Gz} = \frac{\dot{m}c_p}{k_f\,l_0} = \frac{thermal\ capacity\ of\ the\ fluid}{heat\ transferred\ by\ conduction}, \tag{6.7}$$

where $\dot{m}$ is the mass flow rate, and k_f is the thermal conductivity of the fluid.

Other alternatives are obtained by changing the meaning of some terms. For example, the Grashof number

$$G = \frac{\beta\theta g l_0^3 \rho^2}{\mu^2} = \frac{buoyancy\ for\ density\ variation}{viscous\ force}, \tag{6.8}$$

where β is the thermal expansion coefficient of volume, μ is the dynamic viscosity; the Grashof number is calculated from the Archimedes number, with density variation arising from temperature variation.

6.1.1 The Heat Exchanger

We wish to analyse the transfer in forced convection of thermal energy between the wall of a circular duct and a fluid in the turbulent flow regime.

Assume that V is the cross-section average velocity of the fluid, θ is the average temperature in the section, and $\theta + \Delta\theta$ is the temperature of the walls of the pipe; see Fig. 6.1. The heat transferred per unit time and unit surface area of the pipe wall is equal to $h\,\Delta\theta$, where h is the heat exchange coefficient. Near the wall, there is a viscous boundary layer where the flows (of momentum and of heat) are controlled by molecular diffusion, since the turbulent fluctuations are damped by the boundary condition. For this reason, although the flow regime of the fluid is turbulent, the coefficient h depends on the coefficient of thermal conductivity for conduction k. We observe that the thermal diffusivity and kinematic viscosity are representative of the same transfer mechanism: heat the former and momentum the latter. Usually, the thermal diffusivity in fluids is greater than the kinematic viscosity. Since the thickness of the viscous layer depends on the kinematic viscosity and the friction velocity, indirectly represented by the diameter of the pipe D and the average velocity V, these last quantities are also involved in the physical process. At steady state, the specific heat of the fluid is not expected to be relevant; however, if the transfer occurs by convection, the efficiency of a particle in transferring heat is a function of its thermal capacity.

On the basis of these considerations, we can write the following typical equation:

$$f\left(h,\ V,\ D,\ \mu,\ k,\ c_p,\ \rho,\ \Delta\theta\right) = 0. \tag{6.9}$$

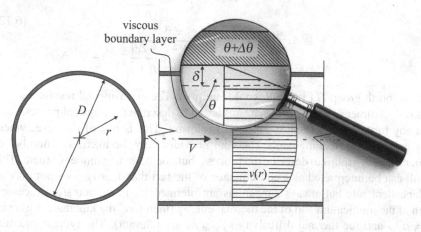

Fig. 6.1 Heat exchange between a fluid in the turbulent flow regime and the walls of a circular cross-section pipe

The dimensional matrix in an M, L, T, Θ system is

$$
\begin{array}{c|cccccccc}
 & h & V & D & \mu & k & c_p & \rho & \Delta\theta \\
\hline
M & 1 & 0 & 0 & 1 & 1 & 0 & 1 & 0 \\
L & 0 & 1 & 1 & -1 & 1 & 2 & -3 & 0 \\
T & -3 & -1 & 0 & -1 & -3 & -2 & 0 & 0 \\
\Theta & -1 & 0 & 0 & 0 & -1 & -1 & 0 & 1
\end{array}
\tag{6.10}
$$

and has rank 4. Applying Buckingham's Theorem, the dimensionless groups describing the physical process are $(8-4)=4$, and selecting V, D, μ and $\Delta\theta$ as the fundamental quantities (they are independent), we obtain the following dimensionless groups:

$$
\Pi_1 = \frac{hD\,\Delta\theta}{\mu\,V^2}, \quad \Pi_2 = \frac{k\,\Delta\theta}{\mu\,V^2}, \quad \Pi_3 = \frac{c_p\,\Delta\theta}{V^2}, \quad \Pi_4 = \frac{\rho\,V\,D}{\mu}. \tag{6.11}
$$

These groups are not uniquely defined, since their powers or power function combinations can be used as groups representing the phenomenon. In addition, our selection of the 4 fundamental quantities is not the only one possible. There is no useful indication for the choice of the most suitable dimensionless groups; the best ones have a physical meaning and are relevant for the experimental data interpretation. In the present analysis, the groups with a physical meaning are the Nusselt number, the Prandtl number and the Reynolds number and are obtained by combining the 4 groups listed in (6.11):

$$
\mathrm{Nu} = \frac{hD}{k} \equiv \frac{\Pi_1}{\Pi_2} = \frac{hD\,\Delta\theta}{\mu\,V^2}\,\frac{1}{\dfrac{k\,\Delta\theta}{\mu\,V^2}},
$$

$$
\mathrm{Pr} = \frac{c_p\,\mu}{k} \equiv \frac{\Pi_3}{\Pi_2} = \frac{c_p\,\Delta\theta}{V^2}\,\frac{1}{\dfrac{k\,\Delta\theta}{\mu\,V^2}}. \tag{6.12}
$$

The fourth group Π_4 is the Reynolds number. The experimental results indicate that only 3 dimensionless groups are sufficient to describe the physical process, and that any further dimensionless group is irrelevant. This is one of the cases where it is evident that Buckingham's Theorem provides only the maximum number of dimensionless groups to describe the process, but not the exact number of them. This result can be interpreted as a consequence of the fact that density does not play an independent role but always appears as an intermediary in the transfer processes, both of the momentum and of the thermal energy (therefore, the kinematic viscosity $v = \mu/\rho$ and the thermal diffusivity $k/(c_p\,\rho)$ are relevant). The typical equation (6.9) can be reduced to

$$
f\left(h,\ V,\ D,\ v,\ k,\ c_p\,\rho,\ \Delta\theta\right) = 0, \tag{6.13}
$$

Fig. 6.2 Interpretation of
some experimental results
for the calculation of the
exponent of the Reynolds
number in the heat exchange
process in a cylindrical
circular pipe (modified from
Dittus and Boelter 1985)

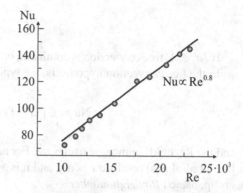

where $c_p \rho$ is the specific heat per unit volume. Having reduced the variables by
one unit, the physical process can be expressed as a function of the 3 dimensionless
groups only, namely,

$$\text{Nu} = \tilde{f}(\text{Re, Pr}) \rightarrow h = \frac{k}{D}\tilde{f}\left(\frac{VD}{\nu}, \frac{c_p \rho \nu}{k}\right). \tag{6.14}$$

For pipes with limited roughness and with flow in the range $\text{Re} > 10^4$ and $0.7 <
\text{Pr} < 170$, experimentally, the following relationship is derived (Dittus and Boelter
1985); see also Fig. 6.2:

$$h = 0.023 \frac{k}{D} \text{Re}^{0.8} \text{Pr}^n. \tag{6.15}$$

The exponent n of the Prandtl number assumes a value of 0.3 if the thermal flow
is from the fluid to the wall and 0.4 if the thermal flow is in the opposite direction.

We now extend the analysis to the more general case of heat transfer between the
fluid and the walls of a circular cross-section pipe by free and forced convection.
The set of variables already analysed for forced convection in Eq. (6.13) must be
supplemented by other variables describing free convection, such as the isobaric
volume thermal expansion coefficient β and the length of the pipe l. The physical
process is described by 5 dimensionless groups, for example:

$$\text{Nu} = \tilde{f}\left(\text{Re, Pr, G, }\frac{l}{D}\right) \rightarrow h = \frac{k}{D}\tilde{f}\left(\frac{VD}{\nu}, \frac{c_p \rho \nu}{k}, \frac{\beta \Delta\theta g D^3}{\nu^2}, \frac{l}{D}\right). \tag{6.16}$$

We observe that the Reynolds number is representative of forced convection and
that the Grashof number is representative of free convection. For forced convection,
the Grashof number disappears from the experimental relationships. For free convection, such as heating a room with a radiator, the Reynolds number is irrelevant.
In intermediate regimes, with free and forced convection of the same order, both
the Reynolds and Grashof numbers remain. A number representative of the relative
importance of free and forced convection is the Archimedes number:

$$\text{Ar} = \frac{G}{\text{Re}^2}. \tag{6.17}$$

If $\text{Ar} \ll 1$, free convection is dominant; otherwise, forced convection is dominant. In the free convection hypothesis, the typical equation becomes

$$\text{Nu} = C_1 \, G^{\alpha_1} \, \text{Pr}^{\alpha_2} f\left(\frac{l}{D}\right), \tag{6.18}$$

and the Reynolds number is irrelevant. For an ideal gas (air behaves, in many cases, like an ideal gas) results $\alpha_1 = \alpha_2$, and it is appropriate to define a new dimensional group, named *Rayleigh number*:

$$\text{Ra} = G \cdot \text{Pr}. \tag{6.19}$$

If the pipe is long enough, the edge effects parametrised by l/D are negligible, and Eq. (6.18) reduces to

$$\text{Nu} = C_1 \, \text{Ra}^{\alpha}. \tag{6.20}$$

The experiments indicate a relationship with a slightly different structure:

$$\text{Nu} = \left[\text{Nu}_0^{1/2} + \text{Ra}^{1/6}\left(\frac{f(\text{Pr})}{300}\right)^{1/6}\right]^2, \quad f(\text{Pr}) = \left[1 + \left(\frac{0.5}{\text{Pr}}\right)^{9/16}\right]^{-16/9}, \tag{6.21}$$

where Nu_0 takes on a different value depending on whether the pipe is horizontal, vertical or inclined.

In forced convection, the flow regime is almost always turbulent, while in free convection, the flow regime is almost always laminar. In natural free convection, the physical process is better interpreted by the Rayleigh number, which also defines the flow regime, which is laminar for $\text{Ra} < 10^9$.

The identification of the exact relation, with the calculation of numerical coefficients and exponents for equations, as well as the structure of the function in the more general case, requires experiments.

For some physical processes, it is possible to define the structure of the equations on a theoretical basis, as happens for the turbulence spectrum (see, e.g., Tennekes and Lumley 1972); the numerical coefficients can be estimated only with experiments.

6.1.2   Heat Transfer in Nanofluids

A class of artificial fluids is represented by *nanofluids*, with nanoparticles suspended in a liquid matrix that can increase the efficiency of heat exchange (Xuan and Li 2003). The possible mechanisms for increasing efficiency are more intense turbulence and a greater specific heat. Limiting the analysis to the forced convection regime, in addition to the variables involved in the case of an ordinary fluid, we have other variables related to the characteristics of the nanoparticles. The physical process can be described with the following typical equation:

$$f\left(h,\ V,\ D,\ \nu,\ k_f,\ k_p,\ c_f\,\rho_f,\ c_p\,\rho_p,\ \phi,\ d_p, \text{shape}_p,\ \Delta\theta\right) = 0, \qquad (6.22)$$

where ϕ is the volumetric concentration of nanoparticles and the subscript p/f refers to particles/fluid. There are 12 variables that can be reduced to $(12 - 4) = 8$ dimensionless groups:

$$\text{Nu} = \tilde{f}\left(\text{Re},\ \text{Pr},\ \text{Pe},\ \phi,\ \frac{c_p\,\rho_p}{c_f\,\rho_f},\ \text{shape}_p,\ \frac{d_p}{D}\right). \qquad (6.23)$$

The experiments indicate a relationship involving dimensionless groups calculated on the basis of the scales introduced by the nanoparticles, that is, on the basis of average scales of the properties of the pure fluid and the nanoparticles. The structure of the functional relationship is

$$\text{Nu}_{nf} = c_1\left(1.0 + c_2\,\phi^{m_1}\,\text{Pe}_d^{m_2}\right)\text{Re}_{nf}^{m_3}\,\text{Pr}_{nf}^{0.4}, \qquad (6.24)$$

where

$$\text{Pe}_d = \frac{V\,d_p\,c_{nf}\,\rho_{nf}}{k_{nf}},\quad \text{Re}_{nf} = \frac{V\,D}{\nu_{nf}},$$

$$\text{Pr}_{nf} = \frac{\nu_{nf}\,c_{nf}\,\rho_{nf}}{k_{nf}},\quad c_{nf}\,\rho_{nf} = (1 - \phi)\,c_f\,\rho_f + \phi\,c_p\,\rho_p. \qquad (6.25)$$

We observe that in the definition of some dimensionless groups, there is a function of the variables that govern the physical process. For example, the specific heat per unit volume of the nanofluid is the weight average of the fluid and particle values. Similarly, the kinematic viscosity of the nanofluid is a correction of the kinematic viscosity of the pure fluid due to the presence of nanoparticles. This means that some variables do not intervene independently, with consequent reduction in the number of dimensionless groups describing the process. The shape of the particles affects the coefficients c_1 and c_2. Some experimental results are reported in Fig. 6.3, with the interpolating curves having the following equation:

$$\text{Nu}_{nf} = 0.0059\left(1.0 + 7.6286\,\phi^{0.6886}\,\text{Pe}_d^{0.001}\right)\text{Re}_{nf}^{0.9238}\,\text{Pr}_{nf}^{0.4}. \qquad (6.26)$$

6.1.3 Heat Exchange in the Presence of Vapours

We consider a hydraulic circuit with steam at the saturation limit at the temperature θ, flowing in a smooth pipe inclined at an angle α to the horizontal, with wall temperature $\theta - \Delta\theta$. A layer of condensation grows, with thermal conductivity k that affects the

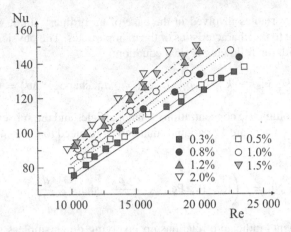

Fig. 6.3 Heat exchange between a fluid with nanoparticles in suspension, in the turbulent flow regime in a circular cross-section duct, and the walls of the duct. The results refer to experiments with different volumetric concentrations of nanoparticles (modified from Xuan and Li 2003)

Fig. 6.4 Condensation on the wall and formation of a thin film in motion due to gravity

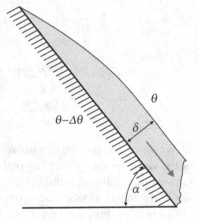

heat exchange (see Fig. 6.4). The most important geometric variable is the thickness of this layer, which is also a function of the latent heat of condensation per unit of mass λ; actually, it affects the latent heat of condensation per unit of volume, which can be expressed as $\lambda_v = \lambda \rho$. The film on the wall can be affected by the average speed of steam in the duct, unless that speed is very small. In addition, the film thickness varies along the pipe, and the length l of the pipe is a variable of the physical process; the diameter (or other relevant cross-section dimension for noncircular pipes) does not appear unless it is of the same order of magnitude as the thickness of the layer. The condensed fluid slides along the pipe in the laminar flow regime, controlled by the viscosity μ and the reduced specific weight $\gamma_r = \rho\, g\, \sin\alpha$. This is a typical gravity current in the viscous-buoyancy balance regime.

The physical process can be described as

$$f(h,\ \Delta\theta,\ l,\ \lambda_v,\ k,\ \gamma_r,\ \mu) = 0. \tag{6.27}$$

The rank of the dimensional matrix

$$
\begin{array}{c|ccccccc}
 & h & \Delta\theta & l & \lambda_v & k & \gamma_r & \mu \\
\hline
M & 1 & 0 & 0 & 1 & 1 & 1 & 1 \\
L & 0 & 0 & 1 & -1 & 1 & -2 & -1 \\
T & -3 & 0 & 0 & -2 & -3 & -2 & -1 \\
\Theta & -1 & 1 & 0 & 0 & -1 & 0 & 0
\end{array}
\tag{6.28}
$$

is 4. By selecting λ_v, k, γ_r and μ as fundamental quantities, the other 3 can be expressed as

$$
\begin{cases}
h = \dfrac{k\,\gamma_r}{\lambda_v} \equiv \dfrac{k\,g\,\sin\alpha}{\lambda}, \\[2mm]
\Delta\theta = \dfrac{\lambda_v^4}{k\,\gamma_r^2\,\mu} \equiv \dfrac{\rho^2\,\lambda^4}{k\,g^2\,\sin^2\alpha\,\mu}, \\[2mm]
l = \dfrac{\lambda}{g\,\sin\alpha},
\end{cases}
\tag{6.29}
$$

and the possible dimensionless groups are:

$$\Pi_1 = \frac{h\,\lambda}{k\,g\,\sin\alpha}, \quad \Pi_2 = \frac{\Delta\theta\,k\,g^2\,\sin^2\alpha\,\mu}{\rho^2\,\lambda^4}, \quad \Pi_3 = \frac{g\,l\,\sin\alpha}{\lambda}. \tag{6.30}$$

We can write

$$\frac{h\,\lambda}{k\,g\,\sin\alpha} = \tilde{f}\left(\frac{\Delta\theta\,k\,g^2\,\sin^2\alpha\,\mu}{\rho^2\,\lambda^4},\ \frac{g\,l\,\sin\alpha}{\lambda}\right). \tag{6.31}$$

To plan the experimental activity, the results can be parametric in Π_3 (which can be modified by changing the l length of the tube in the experimental apparatus) and can be plotted in a diagram with abscissa Π_1 and ordinate Π_2. The numerical value of Π_2 can be varied by changing the temperature difference $\Delta\theta$. Then, the corresponding experimental values of Π_1 are reported. The experimental values (Nusselt 1916) are interpolated by a power function:

$$\Pi_1 = \frac{0.943}{\sqrt[4]{\Pi_2\,\Pi_3}} \ \rightarrow\ h = 0.943\sqrt[4]{\frac{g\,\sin\alpha\,\rho^2\,\lambda\,k^3}{l\,\mu\,\Delta\theta}}. \tag{6.32}$$

If the condensate forms droplets, instead of slipping as a thin film on the pipe wall, the surface tension at the vapour-liquid condensate interface must also be introduced among the variables. In general, the condensation process is also influenced by the roughness of the surface and by grease or other substances deposited on the inner wall of the pipe.

6.1.4   The Heat Exchange of a Homogeneous Body

We analyse a body of thermal conductive material immersed in a fluid in a bath of high thermal capacity (ideally infinite), and we suppose that the fluid is stirred to ensure temperature uniformity. We wish to calculate the temperature variation of the body with respect to the initial temperature.

The variables involved in the transient heat exchange process are (i) the heat exchange coefficient h; (ii) the thermal conductivity coefficient of the fluid k; (iii) the specific heat per unit volume $c_p \, \rho$; (iv) a geometric scale of length l; (v) the time t; (vi) the initial temperature difference between fluid and body $\Delta\theta_{(1)} = \theta_f - \theta_0$; and (vii) the temperature change in the body, compared to the initial temperature, $\Delta\theta_{(2)} = \theta - \theta_0$. The temperature of the fluid does not change according to the hypothesis of infinite heat capacity. The typical equation of the physical process is

$$\Delta\theta_{(2)} = f\left(\Delta\theta_{(1)}, \ k, \ h, \ c_p \, \rho, \ l, \ t\right). \tag{6.33}$$

The dimensional matrix

	$\Delta\theta_{(2)}$	$\Delta\theta_{(1)}$	k	h	$c_p \rho$	l	t
M	0	0	1	1	1	0	0
L	0	0	1	0	-1	1	0
T	0	0	-3	-3	-2	0	1
Θ	1	1	-1	-1	-1	0	0

$$\tag{6.34}$$

has rank 4. After selecting the 4 fundamental quantities $\Delta\theta_{(1)}, \ k, \ h, \ c_p \, \rho$ (they are independent), we calculate the dimensions of the 3 residual quantities with respect to the fundamental quantities:

$$\begin{cases} \Delta\theta_{(2)} = \Delta\theta_{(1)}, \\[2mm] l = \dfrac{k}{h}, \\[3mm] t = \dfrac{k\, c_p \, \rho}{h^2}. \end{cases} \tag{6.35}$$

A set of 3 possible dimensionless groups are:

$$\Pi_1 = \frac{\theta - \theta_0}{\theta_f - \theta_0}, \quad \Pi_2 = \frac{h\,l}{k}, \quad \Pi_3 \equiv \frac{t\,h^2}{k\,c_p\,\rho}. \tag{6.36}$$

The second group is the Nusselt number. On an experimental basis, a third relevant group is a monomial function of Π_2 and Π_3,

$$\Pi_3' = \frac{\Pi_3}{\Pi_2^2} = \frac{t\,h^2}{k\,c_p\,\rho}\,\frac{k^2}{h^2\,l^2} = \frac{k\,t}{c_p\,\rho\,l^2}. \tag{6.37}$$

Hence, we obtain

$$\frac{\theta - \theta_0}{\theta_f - \theta_0} = \tilde{f}\left(\frac{k\,t}{c_p\,\rho\,l^2}, \frac{h\,l}{k}\right). \tag{6.38}$$

Equation (6.38) can also be re-written in another form, that is,

$$\frac{\Delta\theta}{\theta_f - \theta} = \tilde{f}\left(\frac{k\,\Delta t}{c_p\,\rho\,l^2}, \frac{h\,l}{k}\right), \tag{6.39}$$

where $\Delta\theta$ is the change in body temperature in the time interval Δt. Taylor series expansion for very small time increments results in

$$\frac{\Delta\theta}{\theta_f - \theta} = \tilde{f}\Big|_{(0,\,\frac{hl}{k})} + \frac{k\,\Delta t}{c_p\,\rho\,l^2}\,\tilde{f}'\Big|_{(0,\,\frac{hl}{k})} + O\left(\Delta t^2\right). \tag{6.40}$$

The prime indicates the derivation of the function with respect to the first argument. Since for $\Delta t = 0$ we also have $\Delta\theta = 0$, the first term in the series is null. Switching to differentials, we obtain

$$\frac{d\theta}{\theta_f - \theta} = \frac{A\,k\,d t}{c_p\,\rho\,l^2}, \tag{6.41}$$

where A is a function of $h\,l/k$. Integrating and imposing that at t_0 the temperature is θ_0 results in

$$\ln\frac{\theta_f - \theta}{\theta_f - \theta_0} = -\frac{A\,k\,(t - t_0)}{c_p\,\rho\,l^2}, \tag{6.42}$$

or

$$\frac{\theta - \theta_0}{\theta_f - \theta_0} = 1 - \exp\left[-\frac{A\,k\,(t - t_0)}{c_p\,\rho\,l^2}\right]. \tag{6.43}$$

The temperature adjustment of the body and the surrounding fluid shows an exponential decay over time.

6.2 Heat Transfer in Fractal Branching Networks

In many electronic devices, it is increasingly necessary to ensure an adequate cooling system, potentially based on networks of exchange ducts with a branched structure. We have already analysed in Sect. 4 how some biological networks, such as the cardiovascular circulatory network in mammals, have a *fractal* structure that allows us to identify for each rank the scale relationships between the characteristics of the network components.

We now analyse a cooling circuit that develops in only two dimensions, under the hypothesis that it is realised with a system of ducts in series and in parallel, with each duct bifurcating into two ducts (Chen and Cheng 2002). The relative diameter of the pipeline is identified by the rank, equal to (0, 1, ..., k), with 0 referring to the duct of maximum diameter and the subsequent values for the smallest ducts. If we assume that the ratio of geometric scale between the length of the pipeline at rank $(k + 1)$ and the length at rank k can be expressed as

$$\gamma_k = \frac{l_{k+1}}{l_k}, \tag{6.44}$$

the fractal network hypothesis requires that for each rank $\gamma_k = \gamma = \text{const}$. The D fractal dimension of channel length distribution satisfies the relationship (Mandelbrot 1982)

$$N_b = \gamma^{-D}, \tag{6.45}$$

where N_b is the number of branches into which each branch splits. If Δ is the fractal dimension of the hydraulic diameter distribution d of the circular ducts, we obtain

$$N_b = \left(\frac{d_{k+1}}{d_k}\right)^{-\Delta} \rightarrow \beta = \frac{d_{k+1}}{d_k} = N_b^{-1/\Delta}. \tag{6.46}$$

An example of two simple bifurcated fractal networks ($N_b = 2$), with the same maximum rank ($N = 7$) but different fractal dimensions, is shown in Fig. 6.5.

Fig. 6.5 Fractal network with $N_b = 2$ a) $D = 1.5$, $N = 7$; b) $D = 2, N = 7$

a)
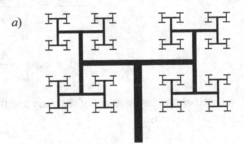

b)
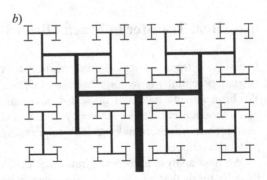

We now assume that the network reaches the maximum rank at the top of the circuit and connects with an identical network at the bottom through the ducts of maximum rank. The area of the heat exchange surface is equal to

$$
S = 2 \sum_{k=0}^{N} S_k = 2 \sum_{k=0}^{N} \pi \, d_k \, l_k \, N_b^k =
$$

$$
2 \sum_{k=0}^{N} \pi \, d_0 \, \beta^k \, l_0 \, \gamma^k \, N_b^k = 2 \pi \, d_0 \, l_0 \, \frac{1 - (N_b \, \beta \, \gamma)^{N+1}}{1 - N_b \, \beta \, \gamma}. \tag{6.47}
$$

The coefficient 2 appears because, between input and output, the network doubles, connecting at maximum rank. Assuming a laminar regime, both hydraulic and thermodynamic, the Nusselt number is invariant at each rank, and consequently, the heat exchange coefficient varies as follows:

$$
\frac{h_{k+1}}{h_k} = \frac{d_k}{d_{k+1}} = \beta^{-1}. \tag{6.48}
$$

Assuming the temperature jump $\Delta\theta$ is the same at each rank, the total heat flow is equal to

$$
Q_h = 2 \sum_{k=0}^{N} h_k \, S_k \, \Delta\theta = 2 \pi \, d_0 \, l_0 \, h_0 \, \frac{1 - (N_b \, \gamma)^{N+1}}{1 - N_b \, \gamma} \, \Delta\theta. \tag{6.49}
$$

For a single duct with diameter d_0 having the same exchange surface area of the network and the same Nusselt number and temperature jump, we obtain

$$
Q_{hpl} = h_0 \, S \, \Delta\theta = 2 \pi \, d_0 \, l_0 \, h_0 \, \frac{1 - (N_b \, \beta \, \gamma)^{N+1}}{1 - N_b \, \beta \, \gamma} \, \Delta\theta. \tag{6.50}
$$

The ratio between the heat flow in the fractal network and the heat flow in the equivalent duct is equal to

$$
\frac{Q_h}{Q_{hpl}} = \frac{\left[1 - (N_b \, \gamma)^{N+1}\right] (1 - N_b \, \beta \, \gamma)}{\left[1 - (N_b \, \beta \, \gamma)^{N+1}\right] (1 - N_b \, \gamma)}. \tag{6.51}
$$

Figure 6.6 shows the efficiency curves calculated for $N_b = 2$, $\Delta = 3$ for various maximum rank and fractal dimension D values. The fractal network is always more efficient than a duct with equivalent exchange surface area.

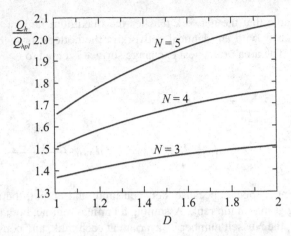

Fig. 6.6 Relative efficiency of a fractal network exchanger compared to a duct having the same area of the exchange surface. D is the fractal dimension, and N is the maximum rank. The number of bifurcations is $N_b = 2$ and the fractal dimension of the hydraulic diameter distribution is $\Delta = 3$

Summarising Concepts

- The approach for heat transfer problems is very similar to that adopted for fluid mechanics: identification of the most significant dimensionless groups and modelling in almost always partial similarity. The independent variables are generally 4, including the temperature.
- The problems can present a greater complexity in the case of two-phase systems, with the presence, for example, of water and steam, and in the presence of reactive fluids.

References

Chen, Y., & Cheng, P. (2002). Heat transfer and pressure drop in fractal tree-like microchannel nets. *International Journal of Heat and Mass Transfer, 45*(13), 2643–2648.

Dittus, F. W., & Boelter, L. M. K. (1985). Heat transfer in automobile radiators of the tubular type. *International Communications in Heat and Mass Transfer, 12*(1), 3–22.

Mandelbrot, B. B. (1982). *The Fractal Geometry of Nature* (Vol. 2). Freeman & Co. Ltd.

Nusselt, W. (1916). Die Oberflachenkondensation des Wasserdampfes (in German). *Zeitschrift Vereines Deutscher Ingenieure, 60*, 541.

Tennekes, H., & Lumley, J. L. (1972). *A first course in turbulence*. The MIT Press.

Xuan, Y., & Li, Q. (2003). Investigation on convective heat transfer and flow features of nanofluids. *Journal of Heat Transfer, 125*(1), 151–155.

Chapter 7
Applications to Problems of Forces and Deformations

Structural science and technology have greatly benefited from the application of dimensional analysis and physical modelling. Complex structures can be studied through physical models, and the building practise itself, codified in regulations and codes, is mostly based on experimental results obtained from physical structural models. Dimensional analysis and similarity criteria have been adopted by engineers since the early times of building construction.

7.1 Classification of Structural Models

A physical structural model is a representation of a structure, or of a part of it, almost always with a reduced geometric scale, with reproduction of the shape and the static, dynamic, thermal or wind loads.

A general classification distinguishes the structural models into (i) elastic, (ii) direct, (iii) indirect, (iv) ultimate strength models, and (v) models for studying the effects of wind.

The *elastic models* are geometrically similar to the prototype but are made of homogeneous elastic material, not necessarily in full respect of similarity criteria of the material in the prototype. These models are used to investigate the behaviour in the elastic regime and cannot reproduce a different regime, such as the inelastic post-fracturing behaviour of reinforced concrete or the post-yield strength of steel.

The *direct models* are geometrically similar to the prototype, with loads corresponding to the loads applied to the prototype. The forces and deformations in the model reproduce in scale the behaviour in the prototype.

The *indirect models* are a particular type of elastic models used to calculate lines of influence, in which the loads in the model are not related to the real loads (the effects on the prototype are calculated by superposition of effects); even the geometry may not be reproduced accurately, while it is sufficient to reproduce the component

S. G. Longo, *Principles and Applications of Dimensional Analysis and Similarity*,
Mathematical Engineering, https://doi.org/10.1007/978-3-030-79217-6_7

that determines the performance of interest. For example, if we wish to study the torsional behaviour of a structural element, it is sufficient to reproduce only the torsional stiffness. This model class is currently out of date because the analysis can benefit from numerical simulations.

Ultimate strength models are direct models made with materials that reproduce inelastic behaviour until the structure collapses. The realisation of such models is technically difficult for structures made of composite material since reproduction of the similarity of the components (e.g., steel bars, aggregates) is required. The difficulties also arise for structures made of homogeneous material (steel, wood) because it is difficult to find materials that reproduce in scale the post-elastic and fracture behaviour of the materials of the prototype.

Models for the study of wind effects can be used to identify only wind forces (in this case it is sufficient the reproduction of the shape of the structure) and to analyse the behaviour of the structure in the presence of wind forces. In the latter case, these are named *aeroelastic models* and require the reproduction of the shape and of the structural characteristics of the prototype.

Other, different models are available for the study of thermal actions, normally direct and elastic models. They include photomechanical models in which the photoelastic effect of the materials is exploited to estimate the stresses. Additionally, other optical principles (for example, interference phenomena or image analysis) are used to study the displacements of plane elements.

For a long time, many codes have allowed the use of physical models for the design of structures, although with some limitations. This is perfectly acceptable because many analytical relationships and numerous numerical models currently applied are the result of experiments on physical models.

Physical structural models are almost always on a reduced geometric scale. Sabnis et al. (1983) suggest the limits of geometric scale, in relation to the type of structure, reported in Table 7.1. The models for the study of wind effects (typically only elastic models) suggest a geometric scale ratio between 1/300 and 1/50. The models for ultimate resistance analysis have a limit of reduction in scale dictated by the difficulty of achieving the similarity of materials with excessively small construction components.

7.2 Similarity in Structural Models

Structural models can be realised (i) in *complete similarity*, (ii) in *partial similarity*, and (iii) in *distorted similarity*.

The optimal choice would be complete similarity, but several reasons lead to models with partial similarity or distorted models.

Full similarity models require that all dimensionless groups describing the processes assume the same value in the model and in the prototype. Any dimensionless parameters, such as Poisson's ratio, must have the same value in the model and in the prototype. The number of scale equations equals the number of dimensionless groups

Table 7.1 Recommended geometric scale depending on the type of structure (modified from Sabnis et al. 1983)

Type of structure	Elastic model	Ultimate strength model
Membranal roofings	$\dfrac{1}{200} - \dfrac{1}{50}$	$\dfrac{1}{30} - \dfrac{1}{10}$
Bridges	$\dfrac{1}{25}$	$\dfrac{1}{20} - \dfrac{1}{40}$
Reactors	$\dfrac{1}{100} - \dfrac{1}{50}$	$\dfrac{1}{20} - \dfrac{1}{4}$
Plates	$\dfrac{1}{25}$	$\dfrac{1}{10} - \dfrac{1}{4}$
Dams	$\dfrac{1}{400}$	$\dfrac{1}{75}$

and is always lower than the number of unknowns. In theory, we should have several degrees of freedom, but in practise, there are further constraints deriving from the selection of materials for the model, in most cases the same material of the prototype. As a consequence, some scales are forced to unity, and we can have more equations than unknowns, with a system of equations admitting only the trivial-solution, unitary value for all the scales. The acceleration of gravity is always on a unitary scale, except for experiments in microgravity conditions, dedicated to specific applications well different from structural physical models, or in centrifuges (see Sect. 8.2), for models with specific characteristics and of limited size.

There are other sources of deviation from complete similarity, such as the use of concentrated loads in the model instead of the distributed loads in the prototype or the lack of similarity, for instance, in adhesion for steel bars or the use of concrete with different behaviour at failure, in the model and in the prototype.

In summary, the similarity is partial if, for various reasons, it is not possible to meet the strict criteria resulting from dimensional analysis or direct analysis. A partial similarity can be acceptable in which only some dimensionless groups, those most representative for the physical process of interest, assume the same numerical value in the model and in the prototype. In partial similarity, it is necessary to quantify the scale effects, which should be monitored and minimised in any case.

If the partial similarity is also inadequate, a possible solution is represented by distorted models. In the realisation of distorted models, we assume that, in principle, any deviation from perfectly similar models is possible, as long as the deviation of the results from the case of complete similarity can be quantified. Distortion can result from a lack of similarity in the initial and boundary conditions, geometry or material properties. The latter type of distortion is the most commonly used structural model.

For example, in a uniaxial stress system, the complete similarity also requires constitutive similarity, with the material in the model and in the prototype following a stress-strain relationship according to the diagram in Fig. 7.1a: the two curves $\sigma/E = f(\varepsilon)$ must be geometrically similar (see Sect. 4.1.1). If the material in the

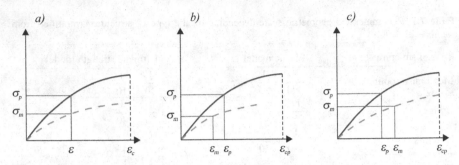

Fig. 7.1 Behaviour *a)* for a material in constitutive similarity; *b)* and *c)* for a material in distorted constitutive similarity. The continuous curves refer to the prototype, and the hatched curves refer to the model. The subscripts *p*, *m* refer to prototype and model, respectively, *c* refers to the critical condition (modified from Sabnis et al. 1983)

model follows a stress-strain curve such as that shown in Fig. 7.1b and c, the specific deformations in the model are defective (or in excess) compared to those obtained for an undistorted model. In cases where the structural behaviour of the elements is only marginally affected by deformation, distortion is acceptable.

The same reasoning applies if the distortion relates to Poisson's ratio: if the structural behaviour is characterised by plane stress states, the distortion introduced by Poisson's ratio being different in the model and in the prototype is perfectly acceptable; however, if the stress state is unknown, a distortion in Poisson's ratio can give severely incorrect results.

7.3   Statically Loaded Structures

In physical problems involving forces and moments statically applied to structures, the rank of the dimensional matrix is usually 2, although mass, length and time appear. In fact, mass always appears in groups in which a term of the type $M\,LT^{-2}$, having the dimensions of a force, can be identified, and the adoption of a system $F\,L$ is the best choice. In general, the reduction in the number of fundamental quantities is not advantageous (an increase is desirable) since it increases by one unit the maximum number of dimensionless groups necessary to describe the physical process. As we shall see, in dynamically loaded systems, mass also appears as an inertial term and is not simply embedded in forces. In this case, mass must be considered as a fundamental quantity.

The analysis is simpler if we consider that, in some conditions, the effects of moments of forces can be considered equivalent to the effects of forces, and we can analyse actions referred to forces only.

7.3.1 Scale Ratios in Undistorted Structural Similarity for Static Elastic Models

We analyse the case of structures with large deformations, a geometrically nonlinear problem, in which the load-deformation relationship is generally nonlinear.

We consider an elastic body loaded by a force P and a moment M_f. The generic component of the stress σ acting on an infinitesimal area around a point can be expressed with the typical equation

$$\sigma = f\left(P,\ M_f,\ l,\ E,\ \nu\right),\tag{7.1}$$

where l is a geometric scale, E is Young's modulus and ν is Poisson's ratio of the material. Choosing as the fundamental quantities F and L, the dimensional matrix

$$\begin{array}{c|cccccc} & \sigma & P & M_f & l & E & \nu \\ \hline F & 1 & 1 & 1 & 0 & 1 & 0 \\ L & -2 & 0 & 1 & 1 & -2 & 0 \end{array}\tag{7.2}$$

has rank 2, and we can express the typical equation with a new function of 3 dimensionless groups and ν (already dimensionless), for example

$$\frac{\sigma l^2}{P} = \tilde{f}\left(\frac{P}{El^2},\ \frac{M_f}{Pl},\ \nu\right).\tag{7.3}$$

In a similar way, if we are interested in the deflection δ, it results in

$$\delta = f\left(P,\ M_f,\ l,\ E,\ \nu\right),\tag{7.4}$$

equivalent to

$$\frac{\delta l^2}{EP} = \tilde{f}\left(\frac{P}{El^2},\ \frac{M_f}{Pl},\ \nu\right).\tag{7.5}$$

To ensure a complete similarity between the model and prototype, it is necessary that:

$$\begin{cases} \dfrac{r_\sigma \lambda^2}{r_P} = 1, \\[2mm] \dfrac{r_P}{r_E \lambda^2} = 1, \\[2mm] \dfrac{r_{M_f}}{r_P \lambda} = 1, \\[2mm] r_\nu = 1. \end{cases}\tag{7.6}$$

This is a system of 4 equations in 6 unknowns with 2 degrees of freedom. By selecting the geometric scale λ and the elastic modulus scale, we calculate

$$\begin{cases} r_\nu = 1, \\ r_P = r_E \, \lambda^2, \\ r_{M_f} = r_P \, \lambda \equiv r_E \, \lambda^3, \\ r_\sigma = \dfrac{r_P}{\lambda^2} = r_E. \end{cases} \qquad (7.7)$$

In many situations, Poisson's ratio is not relevant, and the condition $r_\nu = 1$ is unnecessary.

Example 7.1 Suppose we have made a model of an arrow arch on a geometric scale $\lambda = 3/4$ made of wood with Young's modulus $E = 1.45 \cdot 10^{10}$ Pa. The prototype is aluminium, with Young's modulus $E = 7.23 \cdot 10^{10}$ Pa. The force required to stretch the model arc by a length equal to the length of an arrow is $P_m = 19.8$ N. We wish to calculate the force required in the prototype.

Using the similarity condition $r_P = r_E \, \lambda^2$, the result is

$$P_p = \frac{P_m}{\dfrac{E_m}{E_p} \lambda^2} = \frac{19.8}{\dfrac{1.45 \cdot 10^{10}}{7.23 \cdot 10^{10}} \times \left(\dfrac{3}{4}\right)^2} = 175.5 \text{ N}. \qquad \therefore \quad (7.8)$$

Example 7.2 A model of an arched dam is made of resin, which has the same Poisson's ratio as concrete. The pressure of the water is simulated by the pressure of mercury, and the model is on a geometric scale $\lambda = 1/50$. One of the aims is measurements of deflection.

The pressure at each point is equal to the product of the specific gravity by the fluid level, and the scale ratio is equal to

$$\frac{p_m}{p_p} \equiv r_p = \frac{\gamma_m}{\gamma_p} \frac{\zeta_m}{\zeta_p} = 13.6 \times \frac{1}{50} = 0.272, \qquad (7.9)$$

where γ is the specific gravity of the fluid and ζ is the depth relative to the free surface. The same scale ratio also applies to the internal stresses of the dam. The scale of forces is equal to

$$F = p L^2 \rightarrow r_F = r_p \lambda^2 = 0.272 \times \left(\frac{1}{50}\right)^2 = 1.09 \cdot 10^{-4}. \qquad \therefore \quad (7.10)$$

Table 7.2 lists the scales for the other variables of interest.

The main advantages of a reduced scale model are the very small scale ratios for loads: for example, if the model is made of plastic material, the elastic modulus scale ratio varies from 1/8, for concrete structures, to 1/75, for steel structures. With a geometric scale of 1/50, we calculate a corresponding scale of concentrated loads

Table 7.2 Scale ratios in the similarity of static elastic models

Variable	Dimensions	Scale ratio
Stress	$M\,L^{-1}\,T^{-2}$	r_E
Young's modulus	$M\,L^{-1}\,T^{-2}$	r_E
Poisson's ratio	–	1
Density	$M\,L^{-3}$	r_E/λ
Specific deformation	–	1
Linear dimension	L	λ
Angular rotation	–	1
Surface area	L^2	λ^2
Moment of inertia	L^4	λ^4
Concentrated load	$M\,L\,T^{-2}$	$r_E\,\lambda^2$
Linear load	$M\,T^{-2}$	$r_E\,\lambda$
Pressure or distributed load	$M\,L^{-1}\,T^{-2}$	r_E
Bending and torsional moment	$M\,L^2\,T^{-2}$	$r_E\,\lambda^3$
Force	$M\,L\,T^{-2}$	$r_E\,\lambda^2$

equal to 1/20 000 and 1/187 500. To increase the deformation in the model to gain accurate measurements, the loads applied in the model are distorted in excess with respect to the loads in the prototype, and the results must be corrected to include this distortion.

7.3.2 The Plastic Behaviour

When the stress in the material exceeds the limit of elastic behaviour, part of the deformation becomes permanent, and there is no biunivocal correspondence between stress and strain. However, it is possible to limit the analysis to the case in which this regime is reached for a progressive and monotonic increase in the stresses in the absence of fractures, eliminating any effect of hysteresis.

In the von Mises plasticity criterion plasticity model, the stress-strain relation in the plastic regime is defined by the same relation used in the elastic regime and by Poisson's ratio. Two materials share the same stress-strain function if the dimensionless curves $\sigma/E = f(\varepsilon)$ are geometrically similar, where ε is the specific deformation. Under these conditions, from the point of view of dimensional analysis, the distinction between materials with elastic behaviour and materials with plastic behaviour is irrelevant, and we can use the scales reported in (7.7). The geometric similarity of the $\sigma/E = f(\varepsilon)$ curves implies that the forces scale with λ^2, the moments scale with λ^3, the displacements scale with λ and the stresses are the same in the model and in the prototype. As usual, λ is the geometric scale.

The analysis carried out so far assumes that the weight of the structures is negligible. In many cases, the stresses that cause the collapse of a structural element do not depend on the size of the element itself. For example, it has been experimentally verified that the rivet joints between aluminium plates collapse at a constant stress value, regardless of the size of the joints. In other cases, however, (for instance, for samples of magnesium alloy under tensile stress), the sizes are important in determining the collapse.

7.3.3 Models of Reinforced or Pre-compressed Concrete Structures

The modelling of reinforced concrete or pre-compressed reinforced concrete structures is complicated by the rheological behaviour of the concrete, both in compression and in traction, which is generally inelastic, and by the surface characteristics of the reinforcement elements. The details of the anchorages of the precompression cables, for example, must be carefully modelled. Since these structures are usually modelled up to their collapse, the behaviour under multi-axial stresses should be the same in both the model and the prototype. However, the lack of a uniquely defined failure criterion alleviates requirements in the modelling phase, and what is normally required is that the uniaxial stress-strain curves in the model and prototype must be similar (see Sect. 4.1.7), with a constant ratio of Young's modulus, at the same specific strain, both under traction and under compression. In addition, the deformation at failure (traction and compression) must be the same in the model and in the prototype. These criteria must be valid for both concrete and steel bars (or strands); see the diagrams in Fig. 7.2. The scales for many variables of interest are listed in Table 7.3.

Since the scale ratio of stresses coincides with the scale ratio of Young's modulus, it is necessary that $r_\sigma \neq 1$ and $r_{\sigma'} \neq 1$ for concrete and steel, respectively (in the following, the apex refers the variable to the steel). However, a series of constraints

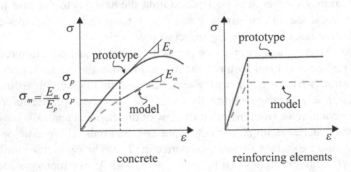

concrete   reinforcing elements

Fig. 7.2 Stress-strain diagrams for the model and the prototype required for the similarity in reinforced or precompressed concrete structures (modified from Sabnis et al. 1983)

Table 7.3 Scales for models of reinforced and pre-compressed concrete

Variable	Dimensions	Model	Practical model
Stress in concrete	$M\,L^{-1}\,T^{-2}$	r_σ	1
Specific deformation in the concrete	–	1	1
Young's modulus of the concrete	$M\,L^{-1}\,T^{-2}$	r_σ	1
Poisson's ratio of the concrete	–	1	1
Density of the concrete	$M\,L^{-3}$	r_σ/λ	$1/\lambda$
Stress in the steel	$M\,L^{-1}\,T^{-2}$	r_σ	1
Specific deformation in the steel	–	1	1
Young's modulus in the steel	$M\,L^{-1}\,T^{-2}$	r_σ	1
Tangential stress of adhesion	$M\,L^{-1}\,T^{-2}$	r_σ	1
Linear dimension	L	λ	λ
Linear displacement	L	λ	λ
Angle of rotation	–	1	1
Steel cross-section area	L^2	λ^2	λ^2
Concentrated load	$M\,L\,T^{-2}$	$r_\sigma\,\lambda^2$	λ^2
Linear load	$M\,T^{-2}$	$r_\sigma\,\lambda$	λ
Pressure or distributed load	$M\,L^{-1}\,T^{-2}$	r_σ	1
Bending and torsional moment	$M\,L^2\,T^{-2}$	$r_\sigma\,\lambda^3$	λ^3

also force the use of steel in the model, with a model called the *practical model* where $r_E = r_\sigma = r_{\sigma'} = 1$. The last column in Table 7.3 refers to the scales for a practical model.

In general, we are not able to satisfy the similarity of the stress-strain curve for concrete, and we cannot satisfy the similarity of the density unless we realise a distorted model. In the next section, we briefly analyse two kinds of distorted models with $r_{E'} = 1$, with the use of steel in the prototype and in the model.

In the first distorted model, indicated with *a)*, the results are $r_\varepsilon \neq 1$, $r_\sigma = r_\varepsilon$, $r_E = 1$ for concrete and $r_{\varepsilon'} = r_{\sigma'} = r_\varepsilon$, $r_{E'} = 1$ for steel. The behaviour of the materials that justifies these relationships is shown in the diagrams in Fig. 7.3.

In a second distorted model, indicated with *b)*, the results are $r_{E'} = 1$ and $r_E \neq 1$. To ensure the same distortion for specific deformations in both steel and concrete, that is, $r_{\varepsilon'} = r_\varepsilon$, since the Young's modulus of the steel has a ratio of unity, we must have $r_{\sigma'} = r_{\varepsilon'}$, and the yield strength of the steel must satisfy the relationship $\sigma'_{r,p} = \sigma'_{r,m}/r_{\varepsilon'}$, as displayed in the diagrams in Fig. 7.4.

The ratio of forces for the steel is

$$\frac{F_m}{F_p} \equiv r_F = r_{\sigma'}\,\lambda^2. \qquad (7.11)$$

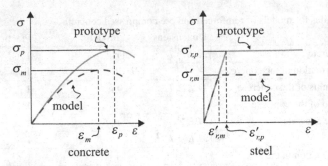

Fig. 7.3 Stress-strain diagrams for the model and prototype required to achieve the similarity in reinforced or precompressed concrete structures in a distorted model with $r_\varepsilon \neq 1$, $r_\sigma = r_\varepsilon$, $r_E = 1$ for concrete and $r_{\varepsilon'} = r_{\sigma'} = r_\varepsilon$, $r_{E'} = 1$ for steel (modified from Sabnis et al. 1983)

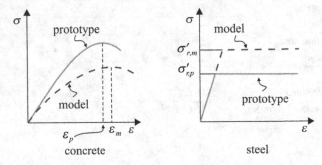

Fig. 7.4 Stress-strain diagrams for the model and prototype needed to achieve the similarity in reinforced or precompressed concrete structures in a distorted model, with $r_{E'} = 1$ and $r_E \neq 1$ (modified from Sabnis et al. 1983)

The force in the steel in the prototype is equal to

$$F_p = \sigma'_{r,p} A_p \equiv \frac{\sigma'_{r,m}}{r_{\varepsilon'}} A_p, \qquad (7.12)$$

and the force in the steel in the model is equal to $F_m = \sigma'_{r,m} A_m$. Substituting, we calculate the scale ratio of the steel area:

$$r_A \equiv \frac{A_m}{A_p} = \frac{r_\sigma \lambda^2}{r_\varepsilon}. \qquad (7.13)$$

Table 7.4 lists some scales of interest for these two distorted models.

Example 7.3 We wish to analyse the action of wind on large glass surfaces.

The pressure of the wind induces a deflection that can be even greater than the thickness of the sheets of glass: these are structures with large deformations. Normally, the perimetral constraint does not limit the rotations and cannot apply moments

Table 7.4 Scales for two distorted models of reinforced and pre-compressed concrete

Variable	Dimensions	Model a)	Model b)
Stress in concrete	$M L^{-1} T^{-2}$	r_σ	r_σ
Specific deformation in the concrete	–	r_ε	r_ε
Young's modulus of the concrete	$M L^{-1} T^{-2}$	r_σ/r_ε	r_σ/r_ε
Poisson's ratio of the concrete	–	1	1
Density of the concrete	$M L^{-3}$	r_σ/λ	r_σ/λ
Stress in the steel	$M L^{-1} T^{-2}$	r_σ	r_σ
Specific deformation in the steel	–	r_ε	r_ε
Young's modulus of the steel	$M L^{-1} T^{-2}$	1	1
Tangential stress of adhesion	$M L^{-1} T^{-2}$	r_σ	*
Linear dimension	L	λ	λ
Linear displacement	L	$r_\varepsilon \lambda$	$r_\varepsilon \lambda$
Angle of rotation	–	r_ε	r_ε
Steel cross-section area	L^2	λ^2	$r_\sigma \lambda^2/r_\varepsilon$
Concentrated load	$M L T^{-2}$	$r_\sigma \lambda^2$	$r_\sigma \lambda^2$
Linear load	$M T^{-2}$	$r_\sigma \lambda$	$r_\sigma \lambda$
Pressure or distributed load	$M L^{-1} T^{-2}$	r_σ	r_σ
Bending and torsional moment	$M L^2 T^{-2}$	$r_\sigma \lambda^3$	$r_\sigma \lambda^3$

*Depending on the cross-section area of the steel

to the slab. A physical model can be realised with a sheet of glass with the same type of constraint as the actual sheet and loaded to the breaking point.

We assume that the physical process can be described by the typical equation

$$\sigma = f(p, E, B, \delta, l, \nu), \tag{7.14}$$

where σ is the stress in the generic section of the slab, p is the pressure exerted by the wind, E is Young's modulus of the glass, $B = (E_t I_t)$ is the bending stiffness of the plate elements, δ is the thickness of the slab, l is a scale length representative of the slab sizes, and ν is Poisson's ratio.

The dimensional matrix is

$$\begin{array}{c|ccccccc} & \sigma & p & E & B & \delta & l & \nu \\ \hline M & 1 & 1 & 1 & 1 & 0 & 0 & 0 \\ L & -1 & -1 & -1 & 3 & 1 & 1 & 0 \\ T & -2 & -2 & -2 & -2 & 0 & 0 & 0 \end{array} \tag{7.15}$$

and has rank 2. We can fix a pair of fundamental quantities, for example, E and l, and check that they are independent. By applying Buckingham's Theorem, we can re-write equation (7.14) as a function of $(7 - 2) = 5$ dimensionless groups including Poisson's ratio (already dimensionless).

The dimensionless groups with physical meaning can be those included in the following typical equation:

$$\frac{\sigma}{p} = \tilde{f}\left(\frac{p}{E}, \frac{B}{E\,l^4}, \frac{\delta}{l}, \nu\right). \tag{7.16}$$

By imposing the equality of the dimensionless groups in the model and in the prototype, we obtain the following system of 5 equations in 7 unknowns scales:

$$\begin{cases} r_\sigma = r_p, \\ r_p = r_E, \\ r_B = r_E\,\lambda^4, \\ r_\delta = \lambda, \\ r_\nu = 1. \end{cases} \tag{7.17}$$

If the model is made of the same material as the prototype, the condition $r_\nu = 1$ is met, and since $r_E = 1$, an additional degree of freedom is lost. For example, if we set the geometric scale λ, we calculate the scales:

$$\begin{cases} r_p = r_\sigma = 1, \\ r_B = \lambda^4, \\ r_\delta = \lambda. \end{cases} \qquad \therefore \quad (7.18)$$

We observe that the slab thickness scales with λ. Since it can be difficult to find commercial glass slabs with the desired thickness, λ is selected on the basis of the available glass slabs.

If the weight of the glass plate in the model cannot be neglected, it can be considered as part of the applied load (in the prototype, the plate is vertical; in the model, it is almost always horizontal).

7.3.4 The Bending of a Beam Made of Ductile Material

We consider a cylindrical beam with a prismatic cross-section that is symmetrical with respect to the vertical and horizontal axes and stressed by a bending moment. Here, h is the distance of the most stressed fibre with respect to the beam axis. The deformation, represented by the local curvature of the neutral axis $1/r$, is a function of the moment M_f, of h and of Young's modulus E, and the typical equation is

$$r = f(M_f, E, h). \tag{7.19}$$

The rank of the dimensional matrix is 2, and the physical process becomes a function of $(4-2)=2$ dimensionless groups. By setting fundamental quantities r and E, we obtain

$$\frac{h}{r} = \tilde{f}\left(\frac{M_f}{E r^3}\right). \tag{7.20}$$

Instead of the $M_f/(E r^3)$ dimensionless group, the $M_f h/(E I)$ dimensionless group is usually employed, where I is the moment of inertia of the section with respect to the neutral axis. Equation (7.20) can be re-written as

$$\frac{h}{r} = \tilde{f}\left(\frac{M_f h}{E I}\right), \tag{7.21}$$

and introducing the strain at the yield point, ε_y, results in

$$\frac{h}{r \varepsilon_y} = \tilde{f}\left(\frac{M_f h}{E I \varepsilon_y}\right). \tag{7.22}$$

In a diagram with abscissa $M_f h/(E I \varepsilon_y)$ and ordinate $h/(r \varepsilon_y)$, this function is a $45°$ straight line in the elastic regime, which grows asymptotically after yielding. The moment can be expressed as a function of the normal stress distribution in the cross-section,

$$M_f = 2 \int_0^h b(z)\, \sigma(z)\, z \, dz, \tag{7.23}$$

where z is the vertical coordinate with origin on the neutral axis and b and σ are respectively the width and normal stress at the coordinate z. Introducing the dimensionless variables $\zeta = z/h$ and $\beta = b/h$ yields

$$M_f = 2 h^3 \int_0^1 \beta(\zeta)\, \sigma(\zeta)\, \zeta \, d\zeta. \tag{7.24}$$

The strain is $\varepsilon = z/r = h\zeta/r$. Assuming that

$$\frac{\sigma}{E} = \mathscr{G}\left(\frac{\varepsilon}{\varepsilon_y}\right) \equiv \mathscr{G}\left(\frac{h\zeta}{r \varepsilon_y}\right), \tag{7.25}$$

we can write

$$\frac{M_f h}{E I \varepsilon_y} = \frac{2 h^4}{I \varepsilon_y} \int_0^1 \beta(\zeta)\, \zeta\, \mathscr{G}\left(\frac{h\zeta}{r \varepsilon_y}\right) d\zeta. \tag{7.26}$$

Fig. 7.5 Schematic
stress-strain function

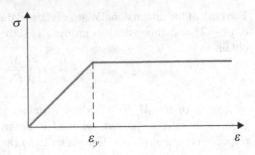

The function argument of the integral identifies the relation between dimensionless stress and deformation (relative to the deformation at the yield point), while the width $\beta(\zeta)$ depends only on the shape of the cross-section. It follows that for a given section, the integral depends only on $h/(r\,\varepsilon_y)$. Additionally, since h^4/I depends only on the shape of the cross-section, the value of $M_f\,h/(E\,I\,\varepsilon_y)$ is calculated as a function of $h/(r\,\varepsilon_y)$. The corresponding curve does not depend on the width of the cross-section since a variation of b of a constant factor determines a variation in β that is cancelled out by the variation in I.

If we assume the schematic representation for the $\mathscr{G}$ function shown in Fig. 7.5, analytically described as

$$\frac{\sigma}{E} = \mathscr{G}\left(\frac{h\zeta}{r\,\varepsilon_y}\right) = \begin{cases} \left(\dfrac{h\zeta}{r\,\varepsilon_y}\right)\varepsilon_y, & \zeta < \zeta_1, \\[2ex] \varepsilon_y, & \zeta > \zeta_1, \end{cases} \tag{7.27}$$

with $\zeta_1 = r\,\varepsilon_y/h$, Eq. (7.26) becomes

$$\frac{M_f\,h}{E\,I\,\varepsilon_y} = \frac{2\,h^4}{I}\left[\frac{1}{\zeta_1}\int_0^{\zeta_1}\beta(\zeta)\,\zeta^2\,\mathrm{d}\zeta + \int_{\zeta_1}^{1}\beta(\zeta)\,\zeta\,\mathrm{d}\zeta\right]. \tag{7.28}$$

For a rectangular cross-section,

$$\frac{M_f\,h}{E\,I\,\varepsilon_y} = \begin{cases} \dfrac{3}{2} - \dfrac{1}{2}\left(\dfrac{r\,\varepsilon_y}{h}\right), & \dfrac{h}{r\,\varepsilon_y} \geq 1, \\[3ex] \dfrac{h}{r\,\varepsilon_y}, & \dfrac{h}{r\,\varepsilon_y} < 1, \end{cases} \tag{7.29}$$

also shown in Fig. 7.6. For a cross-section of a generic shape, numerical integration is required.

Fig. 7.6 Curvature $1/r$ as a function of bending moment for a rectangular cross-section of height $2\,h$, with material characterised by the stress-strain diagram shown in Fig. 7.5, where ε_y is the strain at the yield point (modified from Langhaar 1951)

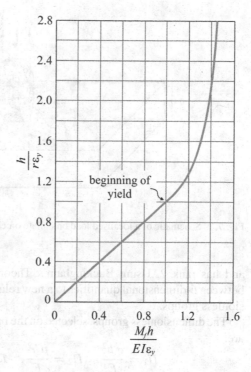

Example 7.4 We consider an airport runway pavement supported by a plate with the foundation behaving like elastic Winkler soil.

In the presence of a load acting on the pavement, such as the wheel of an aircraft or of a service truck, the lower fibres of the plate are stretched. The maximum tensile stress depends on the characteristics of the plate and the foundation, as well as the nature of the load. If the load is transferred from a tire, in addition to the F vertical component, the footprint area is also important; see Fig. 7.7. The footprint is a function of the inflation pressure and of the tire casing structure.

The tensile stress σ on the stretched (lower) fibres of the plate is

$$\sigma = f(F, p, h, k, E), \tag{7.30}$$

where $p = F/(\pi a^2)$ is the pressure, a is the equivalent radius of the footprint area, h is the thickness of the plate, k is the elastic constant of the foundation, and E is Young's modulus of the plate material. The dimensional matrix in terms of M, L and T is

	σ	F	p	h	k	E
M	1	1	1	0	1	1
L	−1	1	−1	1	−2	−1
T	−2	−2	−2	0	−2	−2

$$\tag{7.31}$$

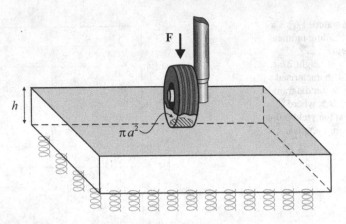

Fig. 7.7 Schematic of a localised load on a plate on elastic soil

and has rank 2. Using Buckingham's Theorem, we can reduce the relationship between 6-dimensional quantities to a new relationship between $(6 - 2) = 4$ dimensionless groups.

The dimensionless groups, selected on the basis of intuition and physical meaning, are:

$$\Pi_1 = \frac{\sigma\, h^2}{F}, \quad \Pi_2 = \frac{p\, h^2}{F}, \quad \Pi_3 = \frac{E}{k\, h}, \quad \Pi_4 = \frac{p}{E}, \tag{7.32}$$

and Eq. (7.30) can be re-written as

$$\Pi_1 = \tilde{f}(\Pi_2,\ \Pi_3,\ \Pi_4). \tag{7.33}$$

More information on the structure of the function $\tilde{f}$ can be gained from the analysis of the phenomenon. If the load F increases, the area of the footprint increases and σ is reduced since a larger part of the plate is active in transferring the load to the foundation. Therefore, the $\tilde{f}$ function monotonically increases with respect to Π_2. If the stiffness of the foundation increases, σ is reduced because the load is locally balanced, and $\tilde{f}$ is monotonically increasing with respect to Π_3. Finally, $\tilde{f}$ also increases with increasing tire inflation pressure p since a rise in p reduces the footprint area and results in more concentrated loads.

Using concepts from the theory of elasticity, Westergaard (1926) suggested the equation

$$\Pi_1 = \tilde{f}_1\left(\Pi_2^2\, \Pi_3\right) \ \rightarrow \ \sigma = \frac{F}{h^2}\tilde{f}_1\left(\frac{E\, p^2\, h^3}{k\, F^2}\right). \tag{7.34}$$

The variable

$$l = \sqrt[4]{\frac{E\, h^3}{12\, (1 - \nu^2)\, k}} \tag{7.35}$$

has the dimensions of a length and is defined as the *equivalent radius of the relative stiffness of the ground*. The formal relationship can be re-written as

$$\sigma = \frac{F}{h^2} \tilde{f_2}\left(\frac{l}{a}\right). \tag{7.36}$$

For a plate of finite size, there are two other geometric scales, namely, the length l_1 and the width l_2 of the plate. Equation (7.36) becomes

$$\sigma = \frac{F}{h^2} \tilde{f_3}\left(\frac{l}{a}, \frac{l}{l_1}, \frac{l}{l_2}\right). \qquad \therefore \quad (7.37)$$

For an infinite plate Westergaard (1926) suggested the relation

$$\sigma = \frac{3\,(1+\nu)}{2\,\pi} \frac{F}{h^2} \left[\ln\left(\frac{2}{a} \sqrt[4]{\frac{E\,h^3}{12\,(1-\nu^2)\,k}}\right) + 0.5 - \gamma \right], \tag{7.38}$$

where $\gamma = 0.577\ldots$ is Euler's constant. Poisson's ratio has been inserted, but it does not modify the previous dimensional analysis, being dimensionless.

In a similar way, the deformation and the stress in the ground can also be analysed.

7.3.5 The Phenomenon of Instability

We analyse an axially loaded column, which is stable but with a threshold load triggering the instability, with the column undergoing large transversal deformations. This phenomenon can be described with the following typical equation:

$$P_{crit} = f\,(E,\ l,\ \nu,\ \text{b.c., geometry}), \tag{7.39}$$

where P_{crit} is the value of the load that triggers the instability (critical load), E is Young's modulus, l is the length of the column, ν is Poisson's ratio, and b.c. indicates the boundary conditions. For a given geometry and boundary condition, the typical equation reduces to

$$P_{crit} = f\,(E,\ l,\ \nu). \tag{7.40}$$

Selecting a fundamental system F, L, the dimensional matrix is

$$
\begin{array}{c|cccc}
 & P_{crit} & E & l & \nu \\
\hline
F & 1 & 1 & 0 & 0 \\
L & 0 & -2 & 1 & 0 \\
\end{array}
\tag{7.41}
$$

and has rank 2. Poisson's ratio is dimensionless, and we can express the functional
relation between the single dimensionless group and the Poisson coefficient:

$$\frac{P_{crit}}{E\,l^2} = \tilde{f}(\nu). \tag{7.42}$$

In many cases, there is a weak dependence on Poisson's ratio, and $\tilde{f}$ is almost
constant. There is no dependence at all if the deflection of the beam is only flexural.
    Using the criteria of dimensional analysis, the conditions of similarity read

$$\begin{cases} \dfrac{P_{crit,m}}{P_{crit,p}} = \dfrac{E_m}{E_p}\left(\dfrac{l_m}{l_p}\right)^2 \to r_{P_{crit}} = r_E\,\lambda^2, \\[2mm] \dfrac{\nu_m}{\nu_p} \equiv r_\nu = 1. \end{cases} \tag{7.43}$$

Using the same material in the model and prototype results in $r_E = 1$, and the
critical load varies according to λ^2, being smaller in the model than in the prototype
if $\lambda < 1$. Given the weak dependence on Poisson's ratio, the condition $r_\nu = 1$ can
be overlooked.
    The typical equation can also be formulated in terms of critical stress σ_{crit},

$$\sigma_{crit} = f(E,\ l,\ \nu), \tag{7.44}$$

obtaining the following condition of similarity:

$$\frac{\sigma_{crit,m}}{\sigma_{crit,p}} = \frac{E_m}{E_p} \to r_{\sigma_{crit}} = r_E. \tag{7.45}$$

If the model and prototype are of the same material, the normal critical stress is
unvaried.
    By neglecting the dependence on Poisson's ratio, we can write

$$\frac{\sigma_{crit}}{E} = \text{const} \longrightarrow \frac{\sigma_{crit}}{\sigma_y}\frac{\sigma_y}{E} = \text{const}, \tag{7.46}$$

where σ_y is the yield strength. In similarity conditions, results

$$\left(\frac{\sigma_{crit}}{\sigma_y}\right)_m = \left(\frac{\sigma_{crit}}{\sigma_y}\right)_p \frac{\left(\dfrac{\sigma_y}{E}\right)_p}{\left(\dfrac{\sigma_y}{E}\right)_m}. \tag{7.47}$$

By selecting the material in the model so that $(\sigma_y/E)_m > (\sigma_y/E)_p$, we can ensure
that instability in the model develops in an elastic regime, even if this instability
appears in a plastic regime in the prototype. This simplifies the analysis of instability

behaviour because the model continues to deform elastically even after the critical condition of instability is attained. For example, if the structure to be modelled is of steel, with $\sigma_y/E = 1.3 \cdot 10^{-3}$, we build the model in titanium ($\sigma_y/E = 6 \cdot 10^{-3}$), yielding $(\sigma_{crit}/E)_m \approx 5\,(\sigma_{crit}/E)_p$. The ratio is even more advantageous for the testing of shells and thin plates, using materials in the model such as polyester Mylar, which has a ratio of $\sigma_y/E = 20 \cdot 10^{-3}$, much higher than that of most building materials.

For instability in the plastic regime, the analysis becomes more complex.

Suppose that the material has perfectly plastic behaviour after reaching the limit of elasticity; in this case, we can assume a dependence of the following type:

$$\sigma_{max} = f(\sigma_y, E, l), \tag{7.48}$$

where σ_{max} is the maximum normal compressive stress that causes the strut or plate to collapse, σ_y is the yield strength, E is Young's modulus and l is a characteristic dimension of the structure. The dimensional matrix of the 4 variables has rank 2, and 2 dimensionless groups can be calculated; for instance, we can write

$$\frac{\sigma_{max}}{\sigma_y} = \tilde{f}\left(\frac{\sigma_y}{E}\right). \tag{7.49}$$

This means that if the model and prototype are made of the same material, with $r_{\sigma_y} = r_E = 1$, it also results in $r_{\sigma_{max}} = 1$, and the maximum stress that causes the collapse in the model equals the stress that causes the collapse in the prototype.

7.3.6 The Plastic Rotation of a Reinforced Section

We analyse the behaviour of a section of reinforced concrete to verify its plastic rotation capacity.

The physical process can be described with a functional relation of the type

$$M_f = f(\sigma_u, \sigma_c, \mathcal{G}_f, \mathcal{G}_c, E_c, \sigma_y, r_t, h, b, l, \theta), \tag{7.50}$$

where M_f is the bending resistant moment, σ_u, σ_c, $\mathcal{G}_f$, $\mathcal{G}_c$ and E_c are the tensile strength and compressive strength, fracture energy and crushing energy, and Young's modulus of concrete, respectively, σ_y is the yield strength of the reinforcement material, r_t is the percentage of reinforcement, h, b, l are the geometric characteristics of the beam, and θ is the local rotation. The dimensional matrix in an M, L, T system is

$$
\begin{array}{c|ccccccccccc}
 & M_f & \sigma_u & \sigma_c & \mathcal{G}_f & \mathcal{G}_c & E_c & \sigma_y & h & b & l & r_t & \theta \\
\hline
M & 1 & 1 & 1 & 1 & 1 & 1 & 1 & 0 & 0 & 0 & 0 & 0 \\
L & 2 & -1 & -1 & 0 & 0 & -1 & -1 & 1 & 1 & 1 & 0 & 0 \\
T & -2 & -2 & -2 & -2 & -2 & -2 & -2 & 0 & 0 & 0 & 0 & 0
\end{array}
\tag{7.51}
$$

and has rank 2. The first and last rows are a linear combination, and we proceed by replacing M, L and T with a pair of quantities that allow the expression of all the other quantities, such as $F = M L T^{-2}$ and L, obtaining the following dimensional matrix:

$$
\begin{array}{c|cccccccccccc}
 & M_f & \sigma_u & \sigma_c & \mathscr{G}_f & \mathscr{G}_c & E_c & \sigma_y & h & b & l & r_t & \theta \\
\hline
F & 1 & 1 & 1 & 1 & 1 & 1 & 1 & 0 & 0 & 0 & 0 & 0 \\
L & 1 & -2 & -2 & -1 & -1 & -2 & -2 & 1 & 1 & 1 & 0 & 0
\end{array}
\qquad (7.52)
$$

which still has rank 2. By virtue of Buckingham's Theorem, we can express the functional relationship using $(12 - 2) = 10$ dimensionless groups, including the already dimensionless parameters r_t and θ. Once we have selected 2 fundamental dimensions, we can proceed using the Rayleigh method or the matrix method.

In the particular case where the analysis is limited to a cross-section that has already undergone a rotation such that the concrete is cracked, the tensile strength of the concrete σ_u and the fracture energy $\mathscr{G}_f$ are physically irrelevant. The number of variables is reduced to 10, and 2 of them are independent. The dimensional matrix becomes

$$
\begin{array}{c|cccccccccc}
 & M_f & \sigma_c & \mathscr{G}_c & E_c & \sigma_y & h & b & l & r_t & \theta \\
\hline
F & 1 & 1 & 1 & 1 & 1 & 0 & 0 & 0 & 0 & 0 \\
L & 1 & -2 & -1 & 2 & -2 & 1 & 1 & 1 & 0 & 0
\end{array}
\qquad (7.53)
$$

and the rank is still 2. Upon selecting $(\mathscr{G}_c E_c)$ and h as fundamentals, we have

$$
\begin{array}{c|cccccccc}
 & M_f & \sigma_c & E_c & \sigma_y & b & l & r_t & \theta \\
\hline
\mathscr{G}_c E_c & 1/2 & 1/2 & 1/2 & 1/2 & 0 & 0 & 0 & 0 \\
h & 5/2 & 1/2 & -1/2 & -1/2 & 1 & 1 & 0 & 0
\end{array}
\qquad (7.54)
$$

The possible 8 dimensionless groups are:

$$
\Pi_1 = \frac{M_f}{h^{5/2}\sqrt{\mathscr{G}_c E_c}}, \quad \Pi_2 = \frac{\sigma_c \sqrt{h}}{\sqrt{\mathscr{G}_c E_c}}, \quad \Pi_3 = \frac{E_c \sqrt{h}}{\sqrt{\mathscr{G}_c E_c}},
$$

$$
\Pi_4 = \frac{\sigma_y \sqrt{h}}{\sqrt{\mathscr{G}_c E_c}}, \quad \Pi_5 = \frac{b}{h}, \quad \Pi_6 = \frac{l}{h}, \quad \Pi_7 = r_t, \quad \Pi_8 = \theta.
\qquad (7.55)
$$

If we wish to investigate the case with invariant b, l and h, we can calculate the number of dimensionless groups on the basis of the extension of Buckingham's Theorem (see Sect. 1.4.7). Of the $n_f = 3$ invariant variables, only $k_f = 1$ are independent since their dimensional matrix

$$
\begin{array}{c|ccc}
 & b & l & h \\
\hline
F & 0 & 0 & 0 \\
L & 1 & 1 & 1
\end{array}
\qquad (7.56)
$$

has rank 1. Thus, the number of dimensionless groups is reduced to $(n - k) - (n_f - k_f) = (10 - 2) - (3 - 1) = 6$.

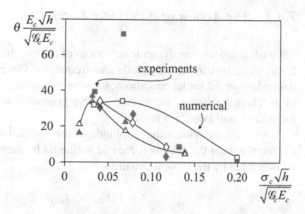

Fig. 7.8 Comparison between the results of a numerical model (curves) and experimental results (symbols) for the plastic rotation of a reinforced concrete section (modified from Carpinteri and Corrado 2010)

This result is also evident since the two groups Π_5 and Π_6 in Eq. (7.55) become constants if b, l and h are invariant and, therefore, are unnecessary, and the typical equation becomes

$$\frac{M_f}{h^{5/2}\sqrt{\mathscr{G}_c E_c}} = \tilde{f}\left(\frac{\sigma_c\sqrt{h}}{\sqrt{\mathscr{G}_c E_c}}, \frac{E_c\sqrt{h}}{\sqrt{\mathscr{G}_c E_c}}, \frac{\sigma_y\sqrt{h}}{\sqrt{\mathscr{G}_c E_c}}, r_t, \theta\right). \qquad (7.57)$$

Some analytical models of this physical process include only 4 dimensionless groups, with the following typical equation (Carpinteri and Corrado 2010):

$$\frac{M_f}{h^{5/2}\sqrt{\mathscr{G}_c E_c}} = \Phi\left(\frac{\sigma_c\sqrt{h}}{\sqrt{\mathscr{G}_c E_c}}, \frac{r_t\sigma_y\sqrt{h}}{\sqrt{\mathscr{G}_c E_c}}, \frac{\theta E_c\sqrt{h}}{\sqrt{\mathscr{G}_c E_c}}\right), \qquad (7.58)$$

where r_t and θ are combined with other groups. We recall that Buckingham's Theorem indicates the maximum number of dimensionless groups, but experimental relations or analytical models can involve a smaller number of groups.

Figure 7.8 shows the comparison between a numerical model (Carpinteri and Corrado 2010) and some experimental results.

7.4 Dynamically Loaded Structures

As already mentioned in the Introduction, the dimensional analysis of dynamically loaded structures involves 3 fundamental quantities, with mass also appearing in inertial terms and not only as a dimension of forces or moments. The typical dynamic loads are periodic forces and impulsive forces.

7.4.1 The Action of a Periodic Force

We wish to analyse the dynamic behaviour of a structure in the presence of a periodic force, described by its amplitude and frequency. Once the transient is exhausted, depending on the initial conditions, the system oscillates with an amplitude function of the characteristics of the driving force, a geometric scale l, the density ρ, Young's modulus E and Poisson's ratio ν.

If we indicate with A the amplitude of the oscillations of the structure, with F_0 the amplitude of the driving force and with n its frequency, the physical process can be described by the typical equation

$$A = f(F_0, \ n, \ l, \ \rho, \ E, \ \nu). \tag{7.59}$$

Since the problem is dynamic, we expect 3 fundamental quantities involved, namely, M, L and T. The dimensional matrix is

$$\begin{array}{c|ccccccc}
 & A & F_0 & n & l & \rho & E & \nu \\
\hline
M & 0 & 1 & 0 & 0 & 1 & 1 & 0 \\
L & 1 & 1 & 0 & 1 & -3 & -1 & 0 \\
T & 0 & -2 & -1 & 0 & 0 & -2 & 0
\end{array} \tag{7.60}$$

and has rank 3. Applying Buckingham's Theorem, we calculate $(7 - 3) = 4$ dimensionless groups including ν, which is already dimensionless. Based on the physical knowledge of the process, the possible dimensionless groups are those listed in the following typical equation:

$$\frac{A E l}{F_0} = \tilde{f}\left(\frac{F_0}{E l^2}, \ n l \sqrt{\frac{\rho}{E}}, \ \nu\right). \tag{7.61}$$

The 4 equations expressing the similarity condition in the 7 unknowns scales are:

$$\begin{cases}
r_A \, r_E \, \lambda = r_{F_0}, \\
r_{F_0} = r_E \, \lambda^2, \\
r_n^2 \, \lambda^2 \, r_\rho = r_E, \\
r_\nu = 1.
\end{cases} \tag{7.62}$$

If the model and prototype are of the same material, with $r_E = r_\rho = r_\nu = 1$, the similarity equations reduce to

$$\begin{cases}
r_A = \lambda, \\
r_{F_0} = \lambda^2, \\
r_n = \dfrac{1}{\lambda}.
\end{cases} \tag{7.63}$$

We have only one degree of freedom, and once the geometric scale has been set, we can calculate the scales of all the other variables. If the oscillations are of small amplitude, they depend linearly on the applied load. Observing Eq. (7.61), we can deduce that a possible nonlinear dependence of the amplitude on the driving force can derive only from the group $F_0/(E\,l^2)$. This suggests that for $F_0/(E\,l^2) \to 0$, Eq. (7.61) reduces to

$$\frac{A\,E\,l}{F_0} = \tilde{f}\left(n\,l\sqrt{\frac{\rho}{E}},\ \nu\right),$$
(7.64)

and the similarity conditions are:

$$\begin{cases} r_A\,r_E\,\lambda = r_{F_0}, \\ r_n^2\,\lambda^2\,r_\rho = r_E, \\ r_\nu = 1; \end{cases}$$
(7.65)

for $r_E = r_\rho = r_\nu = 1$,

$$\begin{cases} r_A = \dfrac{r_{F_0}}{\lambda}, \\ r_n = \dfrac{1}{\lambda}. \end{cases}$$
(7.66)

There are 2 degrees of freedom left, and we can arbitrarily fix the geometric scale and the scale of forces and derive the scale of amplitude of oscillations and of frequency. However, in the hypothesis of small oscillations, the scale of the amplitude is of negligible interest, except for the identification of possible phenomena of resonance when the amplitude of the oscillations grows.

7.4.2 The Action of an Impulsive Force: Impact Phenomena

When a rigid body hits a structure, the local effect depends on the mass, speed and size of the body, while at a distance from the impact point, the effects do not depend on the size. For this reason, the geometric scale of the body in a physical model is not constrained by the geometric scale of the structure. The stress induced by the impact depends on the mass m and velocity V of the impacting body, on a characteristic length l, the density ρ of the structure, Young's modulus E and Poisson's ratio ν. It can be demonstrated that E and ν are sufficient to characterise the behaviour of the material even beyond the yield point. In a pragmatic approach, we neglect the velocity of deformation of the material, which in reality plays a very important role, especially in the yield strength. The problem involves inertia and, consequently, 3 fundamental quantities since mass is autonomously present, not only as a dimension of force. The typical equation in the 7 variables is

$$\sigma = f(m, \; V, \; l, \; \rho, \; E, \; \nu). \tag{7.67}$$

The dimensional matrix has rank 3, and on the basis of Buckingham's Theorem, we can select 3 dimensionless groups and Poisson's ratio. The groups with a physical meaning are:

$$\Pi_1 = \frac{\sigma \, l^3}{m \, V^2}, \quad \Pi_2 = \frac{E \, l^3}{m \, V^2}, \quad \Pi_3 = \frac{m}{\rho \, l^3}, \quad \Pi_4 = \nu, \tag{7.68}$$

and the typical equation becomes

$$\Pi_1 = \tilde{f}(\Pi_2, \; \Pi_3, \; \Pi_4) \rightarrow \frac{\sigma \, l^3}{m \, V^2} = \tilde{f}\left(\frac{E \, l^3}{m \, V^2}, \; \frac{m}{\rho \, l^3}, \; \nu\right). \tag{7.69}$$

The first dimensionless group is similar to the damage number of Johnson (1983), defined as

$$\mathrm{Dn} = \frac{\rho_c \, V_0^2}{\sigma_y}, \tag{7.70}$$

where ρ_c is the density of the impacting body, V_0 is its speed, and σ_y is the yield strength of the material. This dimensionless group naturally derives from the nondimensionalization of the momentum balance equation in the simplified one-dimensional form and without body forces:

$$\rho_c \frac{\partial V}{\partial t} = \frac{\partial \sigma}{\partial x}. \tag{7.71}$$

Once we have selected the scales σ_y, t_0 and V_0, we can write

$$\frac{\rho_c \, V_0^2}{\sigma_y} \frac{\partial \tilde{V}}{\partial \tilde{t}} = \frac{\partial \tilde{\sigma}}{\partial \tilde{x}}, \tag{7.72}$$

with $\tilde{V} = V/V_0, \tilde{t} = t/t_0, \tilde{\sigma} = \sigma/\sigma_y, \tilde{x} = x/(V_0 \, t_0)$. The damage number has a structure equal to the Reynolds times the Strohual numbers in fluid mechanics, representing the ratio of local inertia forces to resistance forces associated with the rheological characteristics of the material. Similarity can be achieved by requiring that the following system of 4 equations in the 7 unknown scales is satisfied:

$$\begin{cases} r_\sigma \, \lambda^3 = r_m \, r_V^2, \\ r_E \, \lambda^3 = r_m \, r_V^2, \\ r_m = r_\rho \, \lambda^3, \\ r_\nu = 1. \end{cases} \tag{7.73}$$

Fig. 7.9 Normalised
displacement as a function of
the geometric scale ratio
(modified from Booth et al.
1983)

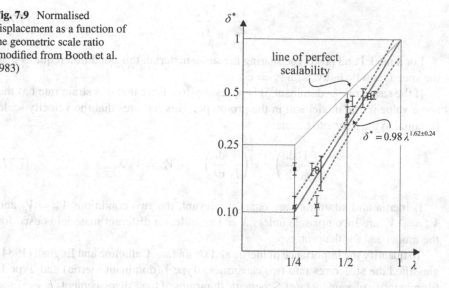

There are 3 degrees of freedom, but using the same materials in the model and
prototype, with $r_E = r_\nu = r_\rho = 1$, we lose 2 more degrees of freedom. Fixing the
geometric scale, we obtain:

$$\begin{cases} r_\sigma = 1, \\ r_V = 1, \\ r_m = \lambda^3. \end{cases} \qquad (7.74)$$

Such a similarity rule can be adopted, for example, in a physical model for the
analysis of the effects of a collision between two ships.

We consider, however, that the strain rate of the material is not reproduced at all,
which brings some relevant scale effects (see Fig. 7.9).

To ensure that the deformed configurations in the model and prototype are similar,
it is necessary that the energy dissipated per unit volume of the structure is invariant
in the model and in the prototype. In the absence of other dissipative phenomena,
this requires that the energy of the impact per unit volume of the structure is the same
in the model and in the prototype:

$$\frac{E_{kin,m}}{l_m^3} = \frac{E_{kin,p}}{l_p^3}. \qquad (7.75)$$

Ultimately, the similarity rules depend on the mechanism of shock absorption.

If the inertia of the structure is dominant, the impact generates waves that propagate depending only on the characteristics of the continuous medium. In this case,
the similarity requires that the ratio between the velocity of the impacting body and
the speed c of propagation of the waves (of various nature) be the same for the model
and for the prototype (this ratio is equivalent to the Mach number for fluids), and

$$\frac{V_m}{c_m} = \frac{V_p}{c_p}. \tag{7.76}$$

For a model and prototype sharing the same material, this condition requires that the speed ratio be unity, with $c_m = c_p$.

If the strain rate is dominant, it is necessary to ensure that this strain rate has the same value in the model and in the prototype. This requires that the velocity scale be equal to the geometric scale:

$$\left(\frac{1}{l_0}\frac{dl}{dt}\right)_m = \left(\frac{1}{l_0}\frac{dl}{dt}\right)_p \rightarrow V_m = V_p\,\lambda. \tag{7.77}$$

If inertia and strain rate are equally relevant, the two conditions $V_m = V_p$ and $V_m = \lambda V_p$ are incompatible unless $\lambda = 1$ or unless a different material is used for the model and for the prototype.

To quantify the importance of inertia and strain rate, Calladine and English (1984) classified the structures into two categories: Type I (dominant inertia) and Type II (dominant displacement rate). Schematic diagrams of load-displacement $(F - \delta)$ and energy dissipated-displacement $(U - \delta)$ conditions are shown in Fig. 7.10.

Beams, plates and membranes generally behave like Type I, with $U \propto \delta$, while axially loaded columns or panels loaded into their plane behave like Type II, with $U \propto \delta^{1/2}$. Load-strain diagrams are calculated directly since $F = dU/d\delta$. Presumably, the structures studied by Booth et al. (1983), with results shown in Fig. 7.9, belong to Type II, with a fast-decreasing load-strain curve. In such structures, the inertia is attributed to the strong transversal accelerations induced by the impact and to the fast rotation of the plastic hinges that are formed, and the impact implies the dissipation of a significant fraction of the kinetic energy of the impacting body, no longer available for the deformation of the structure.

In Fig. 7.11, diagrams of the deflection as a function of the velocity of the impacting body with the same kinetic energy are shown. The linear relationship of perfect

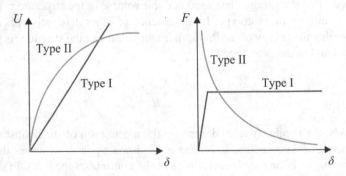

Fig. 7.10 Scheme for the classification of structures according to their inertia: energy-absorbed-displacement and load-displacement curves (modified from Calladine and English 1984)

Fig. 7.11 Results of some experiments performed with Type I and Type II structures. The energy of the impacting body is constant and equal to 122 J (modified from Calladines and English 1984)

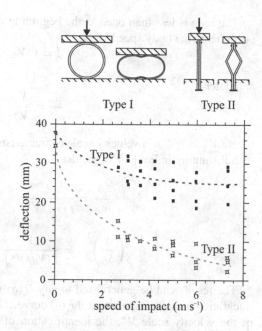

scalability is not satisfied because of the exclusion from the analysis of some important mechanisms underlying the phenomenon. From this point of view, the analysis by Calladine and English (1984), shortly resumed here, is very instructive.

We assume that:

a) the impact velocity is not so high as to affect the quasi-static collapse pattern of the structure;
b) the material is deformed in a viscoplastic regime;
c) the deformation process in the first stage is controlled by the strain rate.

An approximation of the constitutive relationship in a viscoplastic regime may be

$$\frac{\sigma_y}{\sigma_{y0}} = f\left(\frac{\dot{\varepsilon}}{\dot{\varepsilon}^*}\right), \qquad (7.78)$$

where σ_y is the yield strength under dynamic conditions, σ_{y0} is the corresponding stress under static conditions, $\dot{\varepsilon}$ is the strain rate and $\dot{\varepsilon}^*$ is a material property. Condition c) allows us to calculate the energy U_r dissipated under static conditions (with a reference yield strength equal to σ_{y0}) in relation to the energy actually dissipated (with a yield strength value equal to σ_y):

$$\frac{U_r}{U_0} = \frac{\sigma_{y0}}{\sigma_y} = \frac{1}{f(\dot{\varepsilon}/\dot{\varepsilon}^*)}. \qquad (7.79)$$

The ratio is less than one. At the beginning of the impact, the strain rate is proportional to the body speed V_0,

$$\dot{\varepsilon} = C\, V_0, \tag{7.80}$$

and, later

$$\frac{\dot{\varepsilon}}{\dot{\varepsilon}^*} = \frac{V_0}{V^*}, \tag{7.81}$$

where $V^* = \dot{\varepsilon}/C$ is a velocity scale characteristic of the structure.

Substituting in Eq. (7.79), yields:

$$\frac{\delta}{\delta_0} = \frac{1}{f(V_0/V^*)} \qquad \text{Type I,} \tag{7.82}$$

$$\left(\frac{\delta}{\delta_0}\right)^{1/2} = \frac{1}{f(V_0/V^*)} \qquad \text{Type II.} \tag{7.83}$$

The results can be generalised to any $U(\delta)$ function, which can be obtained by integration from the experimental $F(\delta)$ curve. The procedure requires the estimation of the velocity scale V^*, the identification of the relationship between the yield strength and the quasi-static yield strength, the calculation of the dissipated energy, and the calculation of the deformation. The conversion of the experimental data obtained in the model to the real case is immediately applicable.

We observe that this analysis does not include the effect of inertia.

Example 7.5 We analyse the shear collapse of a beam clamped at the ends, made of a material that deforms plastically and with an impulsive load due to an impacting body. The deformation at the joints is a function of the impulsive load through the density ρ_c and the impact speed V_0 of the body, the yield strength σ_y and the beam height H. We assume unity width of the beam so that the typical equation is

$$\delta = f\left(\rho_c,\ V_0,\ \sigma_y,\ H\right). \tag{7.84}$$

The 5 variables involved have the dimensional matrix

$$
\begin{array}{c|ccccc}
 & \delta & \rho_c & V_0 & \sigma_y & H \\
\hline
M & 0 & 1 & 0 & 1 & 0 \\
L & 1 & -3 & 1 & -1 & 1 \\
T & 0 & 0 & -1 & -2 & 0
\end{array}
\tag{7.85}
$$

of rank 3. It is possible to express the relationship as a function of only 2 dimensionless groups, for example,

$$\frac{\delta}{H} = \tilde{f}\left(\frac{\rho_c\, V_0^2}{\sigma_y}\right) \equiv \tilde{f}\,(\text{Dn}). \qquad \qquad \therefore \tag{7.86}$$

The relative deflection of the joint depends on the number of damage Dn. If we assume that the collapse occurs at a given value of the relative deflection (for example, $\delta/H = 1$), this results in Dn = const. The value of the collapse damage number in Zhao (1998) is Dn = 8/9.

7.5 Structures Subject to Thermal Loads

Modelling the effects of thermal loads has successfully been carried out in numerous structures, such as nuclear reactors, arches, spacecrafts, and civil buildings in the presence of fire. The process involves at least 10 variables: stress σ, strain ε, Young's modulus E, Poisson's ratio ν, linear thermal expansion coefficient α, thermal diffusivity $D = k/(c\gamma)$ (where k is the thermal conductivity, c is the specific heat, and γ is the specific gravity), geometric scale l, scale of displacements δ, temperature θ and time t. The typical equation is

$$\sigma = f(\varepsilon,\ E,\ \nu,\ \alpha,\ D,\ l,\ \delta,\ \theta,\ t), \tag{7.87}$$

and the dimensional matrix

$$\begin{array}{c|cccccccccc}
 & \sigma & \varepsilon & E & \nu & \alpha & D & l & \delta & \theta & t \\
\hline
M & 1 & 0 & 1 & 0 & 0 & 0 & 0 & 0 & 0 & 0 \\
L & -1 & 0 & -1 & 0 & 0 & 2 & 1 & 1 & 0 & 0 \\
T & -2 & 0 & -2 & 0 & 0 & -1 & 0 & 0 & 0 & 1 \\
\Theta & 0 & 0 & 0 & 0 & -1 & 0 & 0 & 0 & 1 & 0
\end{array} \tag{7.88}$$

has rank 4.

Four possible fundamental quantities are E, α, l and t (since the corresponding minor extract from the matrix is not singular). Following the procedure in Sect. 2.3.2, we calculate the matrix of the exponents $\mathbf{E}$ and impose a matrix of the known terms $\mathbf{H}$ that coincides with the identity matrix 6×6. The matrix $\mathbf{P} = \mathbf{E} \cdot \mathbf{H}$, transposed and composed of the dimensional matrix, gives

$$\begin{array}{c|cccccc|cccc}
 & \varepsilon & \sigma & \nu & D & \delta & \theta & E & \alpha & l & t \\
\hline
M & 0 & 1 & 0 & 0 & 0 & 0 & 1 & 0 & 0 & 0 \\
L & 0 & -1 & 0 & 2 & 1 & 0 & 0 & 0 & 1 & 0 \\
T & 0 & -2 & 0 & -1 & 0 & 0 & -2 & 0 & 0 & 1 \\
\Theta & 0 & 0 & 0 & 0 & 0 & 1 & 0 & -1 & 0 & 0 \\
\hline
\Pi_1 & 1 & 0 & 0 & 0 & 0 & 0 & 0 & 0 & 0 & 0 \\
\Pi_2 & 0 & 1 & 0 & 0 & 0 & 0 & -1 & 0 & 0 & 0 \\
\Pi_3 & 0 & 0 & 1 & 0 & 0 & 0 & 0 & 0 & 0 & 0 \\
\Pi_4 & 0 & 0 & 0 & 1 & 0 & 0 & 0 & 0 & -2 & 1 \\
\Pi_5 & 0 & 0 & 0 & 0 & 1 & 0 & 0 & 0 & -1 & 0 \\
\Pi_6 & 0 & 0 & 0 & 0 & 0 & 1 & 0 & 1 & 0 & 0
\end{array} \tag{7.89}$$

The 6 possible dimensionless groups are:

$$\Pi_1 = \varepsilon, \quad \Pi_2 = \frac{\sigma}{E}, \quad \Pi_3 = \nu, \quad \Pi_4 = \frac{Dt}{l^2}, \quad \Pi_5 = \frac{\delta}{l}, \quad \Pi_6 = \alpha\,\theta. \quad (7.90)$$

Under similarity conditions, the 6 dimensionless groups must assume the same value in the model and in the prototype. This requires that the following 6 equations in the 10 scales are satisfied:

$$\begin{cases} r_\varepsilon = 1, \\ r_\sigma = r_E, \\ r_\nu = 1, \\ r_t = \lambda^2/r_D, \\ r_\delta = \lambda, \\ r_\theta = 1/r_\alpha. \end{cases} \quad (7.91)$$

Four degrees of freedom remain, with the constraint that Poisson's ratio of the material is identical in the model and in the prototype. If we use the same material in the model and in the prototype, it results in $r_E = r_D = r_\alpha = 1$, and the only remaining choice is the geometric scale λ, with the constraint that the temperature is the same in the model and in the prototype.

The scale ratios for some variables are given in the third and fourth columns of Table 7.5.

However, it is not possible to satisfy the similarity of the boundary condition of the superficial thermal flow, which can be expressed as

$$q = \frac{h}{s}\,\Delta\theta, \quad (7.92)$$

Table 7.5 Scale ratios in the similarity of models with thermal load

Variable	Dimensions	Model standard	With the same material	Distorted and the same temperature
Stress	$M\,L^{-1}\,T^{-2}$	r_E	1	$r_\alpha\,r_\theta\,r_E$
Specific deformation	–	1	1	$r_\alpha\,r_\theta$
Young's modulus	$M\,L^{-1}\,T^{-2}$	r_E	1	r_E
Poisson's ratio	–	1	1	1
Coefficient of thermal linear expansion	Θ^{-1}	r_α	1	r_α
Thermal diffusivity	$L^2\,T^{-1}$	r_D	1	r_D
Linear dimension	L	λ	λ	λ
Linear displacement	L	λ	λ	$r_\alpha\,r_\theta\,\lambda$
Temperature	Θ	$1/r_\alpha$	1	r_θ
Time	T	λ^2/r_D	λ^2	λ^2/r_D

where $h = k/l$ is equal to the ratio between the thermal conductivity and a length scale, s is the thickness, and $\Delta\theta$ is the temperature difference between the ambient environment and the model. The ratio hl/k is the Nusselt number. Since adopting the same material in the model and prototype results in $r_h = r_k = 1$, similarity is possible only for $\lambda = 1$, while for $\lambda \neq 1$, scale effects are expected.

A thermal model can be distorted on the temperature scale. If we set $r_\theta \neq 1/r_\alpha$, we define the distortion ratio as the value d_θ, such that $r_\theta = d_\theta (1/r_\alpha) \rightarrow d_\theta = r_\alpha r_\theta$. The scales of strain, stress and displacement are calculated by multiplying their values in the undistorted model by the distortion ratio. The results are summarised in the last column of Table 7.5.

7.6 The Vibrations of the Elastic Structures

Elastic vibrations are typical of many civil structures, including buildings. Their analysis requires the inclusion of the resonance frequency and the inertial mass.

The physical process can be expressed as

$$n = f(l, F, E, \nu, \rho, \delta, \sigma, g), \tag{7.93}$$

where n is the resonance frequency, l is a geometric scale of the structure, F is the load (a force, for example), E is Young's modulus, ν is Poisson's ratio, ρ is the density, δ is the deflection, σ is the stress and g is the acceleration of gravity. The dimensional matrix has rank 3, and it is possible to express the typical equation as a function of 5 dimensionless groups and of Poisson's ratio,

$$\frac{n^2 l}{g} = \tilde{f}\left(\frac{\delta}{l}, \frac{\sigma}{E}, \frac{\rho g l}{E}, \frac{F}{E l^2}, \nu\right). \tag{7.94}$$

The dimensionless group σ/E is usually replaced by $\sigma l^2/F$. The condition of similarity requires that the dimensionless groups assume the same value in the model and in the prototype, and this results in a system of 6 equations in the 9 scales:

$$\begin{cases} r_n^2 \lambda = r_g, \\ r_\delta = \lambda, \\ r_\sigma = r_F/\lambda^2, \\ r_\rho r_g \lambda = r_E, \\ r_F = r_E \lambda^2, \\ r_\nu = 1. \end{cases} \tag{7.95}$$

Since the acceleration of gravity has a unity scale ratio, a further constraint is added, and 2 degrees of freedom are left. If we arbitrarily select the geometric scale λ and the scale of Young's modulus r_E, we obtain:

$$
\begin{cases}
r_n = 1/\sqrt{\lambda} \rightarrow r_t = \sqrt{\lambda}, \\
r_F = r_E \lambda^2, \\
r_g = 1, \\
r_\delta = \lambda, \\
r_\sigma = r_E, \\
r_\nu = 1, \\
r_\rho = r_E/\lambda,
\end{cases}
\tag{7.96}
$$

where r_t is the ratio of the time scale.

If we neglect the own weight of the structure, we are in the same condition regarding the effects of a periodic driving force (see Sect. 7.4.1), obtaining the following scale ratios:

$$
\begin{cases}
r_F = r_E \lambda^2, \\
r_g = 1, \\
r_t = \lambda, \\
r_\delta = \lambda, \\
r_n = 1/\lambda, \\
r_\sigma = r_E, \\
r_\nu = 1.
\end{cases}
\tag{7.97}
$$

We observe that the frequency scale ratio $r_n > 1$ if $\lambda < 1$, and the measuring instruments in the model (accelerometers, strain gauges) must have a frequency response equal to the maximum frequency expected in the prototype multiplied by the scale ratio of the frequency.

Example 7.6 We wish to measure in a physical model with $\lambda = 1/50$ the pulsating actions of a water current in the stilling basin downstream of a dam. A rigid plate supported by three load cells is installed, schematically reproduced in Fig. 7.12 (Mignosa et al. 2008). The cells are connected to the plate with spherical joints, resulting in a statically determinate system for the vertical component of the loads; this immediately allows calculation of the point of application of the resulting force P.

The plate reproduces one of the bottom plates of the stilling basin, made of concrete in the prototype. Since we are not interested in the dynamic response of the plate but only the action of the current and the fluctuations of pressure due to the macrovortices, the plate in the model is designed in order to have the minimum mass compatible with a high stiffness; it is made with a wafer of two aluminium plates and interposed honeycomb for avionic applications, also in aluminium.

Fig. 7.12 Schematic
diagram of the plate
connected to three load cells
(modified from Mignosa
et al. 2008)

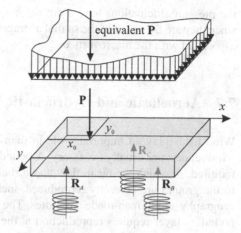

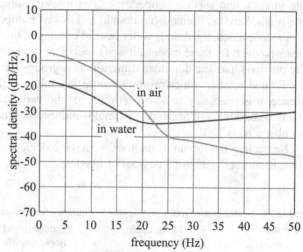

Fig. 7.13 Dynamic response
of the plate in the model
(modified from Mignosa
et al. 2008)

It is necessary to check that the frequency response of the load cell-plate system
is high enough to correctly reproduce the expected spectrum of forces exerted by the
water current.

The water flow exhibits Froude similarity, with geometric scale $\lambda = 1/50$, time
scale $r_t = \lambda^{1/2} = 0.14$ and frequency scale $r_n = \lambda^{-1/2} = 7.07$. The response spectrum of the plate is estimated on the basis of experimental measurements obtained
by exciting the plate with an electromagnetic shaker driven by a white noise signal.
Model scale results are shown in Fig. 7.13 for both dry plate and submerged plate
conditions. The presence of a cushion of water increases inertia and thus reduces the
frequency response.

From the analysis of the response spectrum, it can be inferred that the dynamic
response in water is flatter than that in air, with a gain of approximately -30 dB
up to 50 Hz. This frequency corresponds to 7 Hz in the prototype. It is assumed that

the pressure fluctuations have a cut-off frequency of a few hertz in the prototype, since they are the result of the spatial average over the whole plate of the fluctuations associated with the macrovortices.

7.7 Aeroelastic and Hydroelastic Models

Wind loads play an important role in many civil buildings, and models tested in wind tunnels are usually necessary. A boundary layer wind tunnel is almost always required, in which the mean flow and turbulence of the wind in the region closest to the ground are correctly reproduced, including the effects of roughness due to orography and to man-made structures. The complete similarity of the atmospheric boundary layer requires reproduction at the scale of the topography, roughness of the ground, and surface temperature. The relevant dimensionless groups are the Reynolds, Rossby, Richardson, Prandtl and Eckert numbers.

In some cases, the fluid is water instead of air, and the models are called *hydroelastic models*; for these models, it is also necessary to include the Froude number in the previous list. The dominant dimensionless groups are the Reynolds number and the Froude number, which result in contrasting values of the scales for the variables. Hence, it is necessary to select only one of the two. Table 7.6 lists the scales for the two conditions of negligible Reynolds number effects and of negligible Froude number effects.

Neglecting the Reynolds number, we have 2 degrees of freedom; neglecting the Froude number, we have 3 degrees of freedom.

Table 7.6 Scales in the Froude and Reynolds similarity for aero-elastic models

Variable	Dimensions	Froude model neglecting Re	Reynold model neglecting Fr
Force	$M\,L\,T^{-2}$	$r_\rho \lambda^3$	$r_\rho \lambda^3$
Pressure	$M\,L^{-1}\,T^{-2}$	$r_\rho \lambda$	$r_\rho \lambda$
Acceleration of gravity	$L\,T^{-2}$	1	1
Velocity	$L\,T^{-1}$	$\sqrt{\lambda}$	r_V
Time	T	$\sqrt{\lambda}$	λ/r_V
Linear dimension	L	λ	λ
Linear displacement	L	λ	λ
Frequency	T^{-1}	$1/\sqrt{\lambda}$	r_V/λ
Young's modulus	$M\,L^{-1}\,T^{-2}$	$r_\rho \lambda$	$r_\rho \lambda$
Stress	$M\,L^{-1}\,T^{-2}$	$r_\rho \lambda$	$r_\rho \lambda$
Poisson's ratio	−	1	1
Density	$M\,L^{-3}$	r_ρ	r_ρ

Of particular interest is the study of the interaction between liquids and containers in dynamic conditions, with physical models realised on shaking tables. These models are applied to the study of the behaviour of tanks for the storage of water and other liquids in the presence of earthquakes and of sloshing in tanks on trucks, for instance. We use the direct analysis method to identify the dimensionless groups that control the process. While the Navier-Stokes equation applies to the liquid, the Navier equation applies to the continuous medium of the structure:

$$\frac{\partial^2 \delta}{\partial t^2} - \frac{\mu' + \mu}{\rho_s} \nabla \operatorname{div} \delta - \frac{\mu}{\rho_s} \nabla^2 \delta - \mathbf{f} = 0, \tag{7.98}$$

where δ is the displacement vector, ρ_s is the density of the walls of the container, μ' and μ are the constants of Lamé, expressed in terms of Young's modulus and Poisson's ratio, and $\mathbf{f}$ are the body forces. The equation can be written in dimensionless form:

$$\frac{\partial^2 \tilde{\delta}}{\partial \tilde{t}^2} - \left[\frac{(\mu' + \mu) \, t_0^2}{\rho_s \, l_0^2} \right] \tilde{\nabla} \, \widetilde{\operatorname{div}} \, \tilde{\delta} - \left(\frac{\mu \, t_0^2}{\rho_s \, l_0^2} \right) \tilde{\nabla}^2 \tilde{\delta} - \left(\frac{f_0 \, t_0^2}{\delta_0} \right) \tilde{\mathbf{f}} = 0, \tag{7.99}$$

where the symbol $\tilde{\ }$ indicates a dimensionless variable, f_0 is the body force scale, δ_0 is the displacement scale, l_0 is the length scale and t_0 is the time scale. The dimensionless groups are:

$$\Pi_1 = \frac{E \, t_0^2}{\rho_s \, l_0^2}, \quad \Pi_2 = \frac{f_0 \, t_0^2}{\delta_0}, \tag{7.100}$$

where Π_1 is representative of the first two dimensionless coefficients involving $\mu' + \mu$ and μ separately, through the relationship between Lamé's constants, Poisson's ratio and Young's modulus. The conditions at the interface between the walls of the container and the liquid are:

$$\begin{cases} p = -\sigma_n, \\ \mathbf{v} \cdot \mathbf{n} = -\dfrac{\partial \delta}{\partial t} \cdot \mathbf{n}, \end{cases} \tag{7.101}$$

where p is the pressure in the liquid, σ_n is a component of the stress in the walls of the container along the local normal $\mathbf{n}$, and $\mathbf{v}$ is the velocity of the liquid. Equation (7.101) can be expressed in nondimensional form as:

$$\begin{cases} \tilde{p} = -\left(\dfrac{E \, \delta_0}{p_0 \, l_0} \right) \tilde{\sigma}_n, \\ \tilde{\mathbf{v}} \cdot \mathbf{n} = -\left(\dfrac{\delta_0}{u_0 \, t_0} \right) \dfrac{\partial \tilde{\delta}}{\partial \tilde{t}} \cdot \mathbf{n}, \end{cases} \tag{7.102}$$

providing the two dimensionless groups:

$$\Pi_3 = \frac{E \, \delta_0}{p_0 \, l_0}, \quad \Pi_4 = \frac{\delta_0}{u_0 \, t_0}, \tag{7.103}$$

where u_0 is the velocity scale for the liquid and p_0 is the pressure scale. The other dimensionless groups derived from the equation of Navier-Stokes (see Sect. 5) are:

$$\Pi_5 = \frac{u_0 \, t_0}{l_0}, \quad \Pi_6 = \frac{u_0^2}{g \, l_0}, \quad \Pi_7 = \frac{p_0}{\rho \, u_0^2}, \quad \Pi_8 = \frac{u_0 \, l_0}{\nu}, \tag{7.104}$$

representing the Strohual number, the Froude number, the Euler number and the Reynolds number.

The process is described by 10 variables, with a dimensional matrix of rank 3, leaving 7 degrees of freedom. Thus, it is possible to eliminate one of the groups in the subset Π_4, Π_5, Π_6 to obtain a set of 7 independent groups. In practise, only the following 4 groups, obtained by recombining the groups derived from the direct analysis, are significant:

$$\Pi_1' = \frac{u_0^2}{g \, l_0}, \quad \Pi_2' = \frac{\rho_s \, \delta_0}{\rho \, l_0}, \quad \Pi_3' = \frac{E}{\rho_s \, u_0^2}, \quad \Pi_4' = \frac{u_0 \, l_0}{\nu}. \tag{7.105}$$

The conditions of similarity are:

$$\begin{cases} r_V = \lambda^{1/2}, \\ r_{\rho_s} \, r_\delta = r_\rho \, \lambda, \\ r_E = r_{\rho_s} \, r_V^2, \\ r_V \, \lambda = r_\nu. \end{cases} \tag{7.106}$$

The first (Froude similarity) and the last (Reynolds similarity) equations are in contrast unless a fluid with a given kinematic viscosity is available. Thus, if we do not respect the Reynolds similarity, the system (7.106) reduces to

$$r_\delta = \frac{\lambda^2 \, r_\rho}{r_E}, \quad r_{\rho_s} = \frac{r_E}{\lambda}, \tag{7.107}$$

which allows us to calculate the thickness and density of the container in the model once we have selected the scales λ, r_E and r_ρ.

7.8 Models with Explosive Loads External to the Structure

In some cases, it is necessary to simulate the actions and the effects of an explosion. The difficulty encountered in numerically modelling these actions requires the creation of physical models in which the structural and geometric characteristics and the explosive load are reproduced.

Fig. 7.14 Scale law of Hopkinson-Cranz

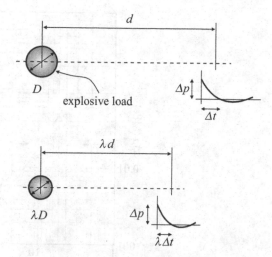

The explosion dynamics follow the law of Hopkinson-Cranz (modified from Hopkinson 1915; Cranz 1910): if an explosive load with the shape of a sphere of diameter D generates an overpressure Δp at a distance d, with a duration of the positive phase equal to Δt, then a load of the same explosive of diameter λD produces the same overpressure at a distance λd, with a positive phase duration equal to $\lambda \Delta t$, see Fig. 7.14. The overpressure phase is followed by the phase of reduced pressure, which is much less dangerous and less important. Since the mass m of the explosive varies with the cube of the diameter, we can define the scaled distance equal to

$$\tilde{d} = \frac{d}{m^{1/3}}. \tag{7.108}$$

An example of a scaling distance for trinitrotoluene (TNT) is shown in Fig. 7.15.

Once the scale of the explosive action is established, we can calculate the scales for the variables involved, bearing in mind that explosions usually stress the structures in an elastic-plastic regime, possibly up to failure, and that the characteristics of the material are a function of the deformation rate. The main scales are listed in Table 7.7.

Explosive loads can be reproduced either in suitably constructed chambers using TNT loads in quantities calculated on the basis of Hopkinson-Cranz's law or in tunnels that allow the generation of shock waves. In these tunnels, there is a high-pressure section separated by a low-pressure section where the physical model is placed. A diaphragm separating the two sections during failure enerates the shock wave.

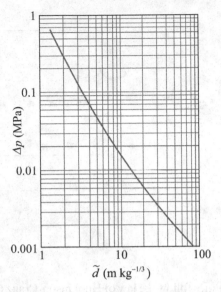

Fig. 7.15 Scaling distance for a TNT explosion (modified from Grossel 1996)

Table 7.7 Scale ratios in the similarity of models with loads due to an explosion

Variable	Dimensions	Full model	Neglected gravity forces, same material
Force	$M\,L\,T^{-2}$	$r_E\,\lambda^2$	λ^2
Pressure	$M\,L^{-1}\,T^{-2}$	r_E	1
Acceleration of gravity	$L\,T^{-2}$	1	–
Velocity	$L\,T^{-1}$	1	1
Time	T	λ	λ
Linear dimension	L	λ	λ
Linear displacement	L	λ	λ
Specific deformation	–	1	1
Young's modulus	$M\,L^{-1}\,T^{-2}$	r_E	1
Stress	$M\,L^{-1}\,T^{-2}$	r_E	1
Poisson's ratio	–	1	1
Density	$M\,L^{-3}$	r_ρ	1

7.9 Dynamic Models with Earthquake Action

The action of earthquakes on structures must be adequately considered at the design stage. For obvious economic reasons, it is almost always necessary to activate the inelastic behaviour of the structures to counteract the dynamic loads during earth-

Table 7.8 Scale ratios in the similarity of dynamic models with earthquake action

Variable	Dimensions	Full model	Neglected gravity forces, same material
Force	$M L T^{-2}$	$r_E \lambda^2$	λ^2
Pressure	$M L^{-1} T^{-2}$	r_E	1
Acceleration	$L T^{-2}$	1	$1/\sqrt{\lambda}$
Acceleration of gravity	$L T^{-2}$	1	–
Velocity	$L T^{-1}$	$\sqrt{\lambda}$	1
Time	T	$\sqrt{\lambda}$	λ
Linear dimension	L	λ	λ
Linear displacement	L	λ	λ
Frequency	T^{-1}	$1/\sqrt{\lambda}$	$1/\lambda$
Young's modulus	$M L^{-1} T^{-2}$	r_E	1
Stress	$M L^{-1} T^{-2}$	r_E	1
Specific deformation	–	1	1
Poisson's ratio	–	1	1
Density	$M L^{-3}$	r_E/λ	1
Energy	$M L^2 T^{-2}$	$r_E \lambda^3$	λ^3

quakes, which makes the realisation of physical models difficult, especially because of the difficulty of adequately reproducing the rheological characteristics of the materials.

Table 7.8 lists the scales calculated according to the criteria of dimensional analysis.

With a full model, there are still 2 degrees of freedom, and we can arbitrarily set, for example, the geometric scale λ and the scale ratio of Young's modulus r_E. However, the selection of a material with density following the scale $r_\rho = r_E/\lambda$ is impractical. For this reason, the inertial effect of the structural mass is replaced by equivalent masses, disregarding the scale for density. If gravity forces are negligible and the same material is used in the model and in the prototype, we obtain the scales listed in the last column in Table 7.8.

For earthquake simulation, the *shaking tables* are used; see Sect. 8.1.

7.10 Scale Effects in Structural Models

In the physical modelling of structural elements, scale effects are particularly important because they almost always overestimate the performances of materials and structures.

Fig. 7.16 Stress-strain curve for concrete as a function of deformation speed (modified from US Department of the Army 1990)

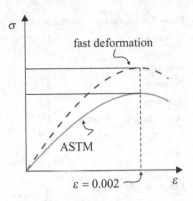

We can estimate these effects, at least those inherent to the properties of the materials, by carrying out measurements of these properties on very small specimens. Reducing the size of steel and concrete samples leads to an increase in strength. In addition, the detailed characteristics of the structures can be significantly influenced by the construction of the model. If the same material is used in the model and in the prototype, deviations from the expected behaviour inevitably occur on a reduced geometric scale. These deviations are due to many factors; for example, the rheological behaviour of materials is a function of the loading rate, and the strength of concrete and many metals is higher if loads are applied quickly (see Fig. 7.16).

In metals, the yield strength and yield strain increase with increasing speed of application of the load, while Young's modulus remains constant. Modest increments are also recorded for the ultimate failure stress.

Thus, the theoretically calculated stress concentration factors are too high when applied to small test specimens under fatigue stress. However, if samples are large enough to have an adequate number of crystals in the cross-section and if theoretically calculated stress amplification coefficients are adopted, the resistance limits tend to become the same as those recorded for standard specimens. This is not surprising since crystals do not scale geometrically as the specimens.

The same considerations apply to the case of composite framework models.

While the elastic behaviour of the samples at different geometric scales is independent of the size of the sample, the resistance is significantly influenced, mainly because micro-cracks and imperfections are more numerous in the larger samples (which show a lower resistance than the smaller ones, see Jackson et al. 1992).

When performing impact tests, the results are also influenced by the fact that the specimens are stressed at a different frequency than the frequency of the stress in the real structures.

Summarising Concepts

- Structural physical models have long been a tool used for writing codes. They are classified according to the regime of behaviour and the nature of the loads.
- A broad classification lists models for statically and dynamically loaded structures. Special attention is paid to models for composite structures. Similarity criteria are almost always approximate; sometimes the correct reproduction of only one of the structural characteristics (e.g., bending strain) is sufficient, without the need to fully reproduce the static or dynamic behaviour of the entire structure.
- The complexity of the constitutive equations requires a strong simplification, abandoning full constitutive similarity. Other categories of models refer to the case of thermally stressed structures and structures with explosion loads.

References

Booth, E., Collier, D., & Miles, J. (1983). Impact scalability of plated steel structures. In N. Jones & T. Wierzbicki (Eds.), *Structural crashworthiness* (136–174).

Calladine, C. R., & English, R. W. (1984). Strain-rate and inertia effects in the collapse of two types of energy-absorbing structure. *International Journal of Mechanical Sciences, 26*(11–12), 689–701.

Carpinteri, A., & Corrado, M. (2010). Dimensional analysis approach to the plastic rotation capacity of over-reinforced concrete beams. *Engineering Fracture Mechanics, 77*(7), 1091–1100.

Cranz, K. J. (1910). *Lehrbuch der ballistik* (Vol. 1). B. G. Teubner, Leipzig.

Grossel, S. S. (1996). Guidelines for evaluating the characteristics of vapour cloud explosions, flash fires and BLEVEs. *Journal of Loss Prevention in the Process Industries, 3*(9), 247.

Hopkinson, B. (1915). UK Ordnance Board Minutes 13565. *Public records office, London Vol Sup6/187.*

Jackson, K. E., Kellas, S., & Morton, J. (1992). Scale effects in the response and failure of fiber reinforced composite laminates loaded in tension and in flexure. *Journal of Composite Materials, 26*(18), 2674–2705.

Johnson, W. (1983). *Impact strength of materials*. Edward Arnold.

Langhaar, H. L. (1951). *Dimensional Analysis and Theory of Models*. Wiley.

Mignosa, P., Giuffredi, F., Danese, D., La Rocca, M., Longo, S., Chiapponi, L., et al. (2008). *Prove su modello fisico del manufatto regolatore della cassa di espansione sul Torrente Parma (Physical model tests of the control systems for the detention basin of the Parma Torrent) (in Italian)*. DICATeA, University of Parma, and AIPo.

Sabnis, G. M., Harris, H. G., White, R. N., & Mirza, R. N. (1983). *Structural Modeling and Experimental Techniques*. Prentice-Hall Inc.

US Department of the Army. (1990). Structures to resist the effects of accidental explosions. *Technical manual TM5-1300*. Picatinny Arsenal.

Westergaard, H. M. (1926). Stresses in concrete pavements computed by theoretical analysis. *Public Roads, 7*, 25–35.

Zhao, Y.-P. (1998). Prediction of structural dynamic plastic shear failure by Johnson's damage number. *Forschung im Ingenieurwesen, 63*(11–12), 349–352.

Chapter 8
Applications in Geotechnics

Many complex problems typical of geotechnics can be adequately addressed with the tools of dimensional analysis and physical modelling. Geotechnical models in which body forces are important can be classified into two main categories: (i) models in the presence of acceleration of gravity, with the adoption of the *shaking table*, and (ii) models with an increased acceleration in a *centrifuge*.

8.1 The Shaking Table

Shaking tables, see Fig. 8.1, are rigid plates with controlled movements for the 6 degrees of freedom, or only for some of them for cheaper devices. They are widely used for the study of soil liquefaction phenomena, soil behaviour in the aftermath of an earthquake, behaviour of foundations and analysis of soil lateral pressure. They require sophisticated instrumentation both for the actuation of movements (hydraulic actuators are almost always used) and for the measurement of displacements and accelerations. Physical geotechnical models require special attention to ensure an adequate reproduction of boundary conditions.

8.1.1 Conditions of Similarity for a Model on a Shaking Table

We wish to reproduce in a physical model the time evolution of the stress state in a continuum as a consequence of the acceleration due to an earthquake. The typical equation of the process is

© The Author(s), under exclusive license to Springer Nature Switzerland AG 2021
S. G. Longo, *Principles and Applications of Dimensional Analysis and Similarity*,
Mathematical Engineering, https://doi.org/10.1007/978-3-030-79217-6_8

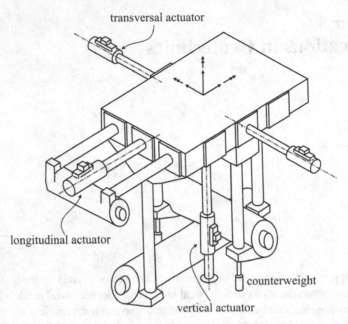

transversal actuator

longitudinal actuator

counterweight

vertical actuator

Fig. 8.1 Overview of a 6-degree freedom shaking table installed at LNEC (modified from Bairrao and Vaz 2000)

$$\sigma = f(x, \ y, \ z, \ t, \ \rho, \ E, \ a, \ g, \ l, \ \sigma_0, \ x_0, \ y_0, \ z_0), \qquad (8.1)$$

where σ is representative of the stress state, x, y, z are the coordinates of the generic point, ρ is the density, E is Young's modulus, a is the imposed acceleration, g is the gravity acceleration, and l is a characteristic length. Variables with the '0' subscript indicate the initial conditions. The characteristics of the material are synthesised by Young's modulus E, implicitly assuming that the similarity of the residual characteristic properties is respected. The dimensional matrix has rank 3, and by applying Buckingham's Theorem, it is possible to re-write the typical equation as

$$\frac{\sigma}{E} = \tilde{f} \left(\frac{x}{l}, \ \frac{t}{l} \sqrt{\frac{E}{\rho}}, \ \frac{a}{g}, \ \frac{g\,l\,\rho}{E}, \ \frac{\sigma_0}{E}, \ \frac{x_0}{l} \right). \qquad (8.2)$$

For the sake of simplicity, we have considered a single spatial coordinate. The conditions of similarity are:

$$\begin{cases} r_\sigma = r_E, \\ r_x = \lambda, \\ r_t\, r_E^{1/2} = \lambda\, r_\rho^{1/2}, \\ r_a = r_g, \\ r_g\, \lambda\, r_\rho = r_E, \\ r_{\sigma_0} = r_E, \\ r_{x_0} = \lambda. \end{cases} \tag{8.3}$$

If gravity cannot be ignored, $r_g = 1$ also results in $r_a = 1$. The corresponding similarity condition is the *Froude similarity*, with an imposed acceleration comparable to the acceleration of gravity.

If the elastic forces are dominant, from the third equation in (8.3),

$$\frac{\lambda}{r_t} \equiv r_V = \frac{r_E^{1/2}}{r_\rho^{1/2}}. \tag{8.4}$$

This condition of similarity is called *similarity of Cauchy* because it is based on the invariance of the Cauchy number in the model and in the prototype. The scale ratios for some of the most interesting variables in Cauchy and in Froude similarity are listed in Table 8.1.

If $\lambda = r_E\, r_\rho^{-1}$, both Froude and Cauchy similarity are satisfied. In addition, since $r_\rho = r_E\, \lambda^{-1} > 1$ if $r_E = 1$ and $\lambda < 1$, it is necessary to add auxiliary masses to the model. These masses must be connected to the structure in the model to ensure their inertial action without affecting the control system of the shaking table.

Table 8.1 The scale ratios for similarity of Cauchy and of Froude

Variable	Cauchy similarity	Froude similarity
Length	λ	λ
Young's modulus	r_E	r_E
Poisson's ratio	1	1
Density	r_ρ	r_ρ
Velocity	$r_E^{1/2}\, r_\rho^{-1/2}$	$\lambda^{1/2}$
Acceleration	$r_E\, r_\rho^{-1}\, \lambda^{-1}$	1
Force	$r_E\, \lambda^2$	$r_\rho\, \lambda^3$
Stress	r_E	$r_\rho\, \lambda$
Strain	1	$\lambda\, r_E^{-1}\, r_\rho$
Time	$\lambda\, r_E^{-1/2}\, r_\rho^{1/2}$	$\lambda^{1/2}$
Frequency	$\lambda^{-1}\, r_E^{1/2}\, r_\rho^{-1/2}$	$\lambda^{-1/2}$

Shaking tables are frequently used to simulate dynamic earthquake actions in civil buildings and constructions.

8.2 The Centrifuge

We wish to reproduce with a physical model the collapse of a mine or a tunnel in compact rock. A necessary but not sufficient condition for the similarity is that the ratio between the resistance of the material in the model and in the prototype is equal to the ratio between the stress induced by the weight in the material in the model and in the prototype. All other conditions being equal, it follows that

$$
\begin{cases}
\dfrac{\sigma_{0-max,m}}{\sigma_{0-max,p}} = \dfrac{\sigma_{0,m}}{\sigma_{0,p}} \rightarrow r_{\sigma_{0-max}} = r_{\sigma_0}, \\[4mm]
\dfrac{\sigma_{0,m}}{\sigma_{0,p}} = \dfrac{\rho_m\, g\, h_m}{\rho_p\, g\, h_p} \rightarrow r_{\sigma_0} = r_\rho\, \lambda,
\end{cases}
\tag{8.5}
$$

where σ_0 is the stress, σ_{0-max} is the maximum stress at failure, ρ is the density, g is the acceleration of gravity and h is the thickness of the overlying material.

When selecting the geometrical scale, 2 degrees of freedom remain, and in theory, similarity is possible by selecting the material in the model with characteristics of density and stiffness such as to satisfy Eq. (8.5). A material denser but more compliant than rock should be used; however, despite the rapid evolution of *intelligent* materials with controllable rheological or structural characteristics at the production stage, at present, there are no materials that can meet these needs. Resistance is difficult to program, and the variations that can be obtained are usually less than one order of magnitude. Moreover, even if such materials existed, collapse could occur during the realisation of the model, under its own weight.

In some models, elastic or plastic materials and viscous fluids, such as high-density oil or soft clay, are used, but with several difficulties during preparation and during tests.

From all these practical limitations comes the idea to increase gravity acceleration.

A device used to increase gravity is the centrifuge (see Fig. 8.2), in which body forces are generated that are almost indistinguishable from gravity, recreating a high level of stress, which otherwise would not be reproducible on a reduced geometric scale.

The centrifuge, first applied by Bucky (1931), exploits the idea of using materials similar or identical to natural materials, increasing body forces.

In these conditions, the scale of the stress reads

$$
r_{\sigma_0} = \frac{\sigma_{0,m}}{\sigma_{0,p}} \equiv \frac{\rho_m\,(g+a)\,h_m}{\rho_p\, g\, h_p} \approx r_\rho\, r_a\, \lambda,
\tag{8.6}
$$

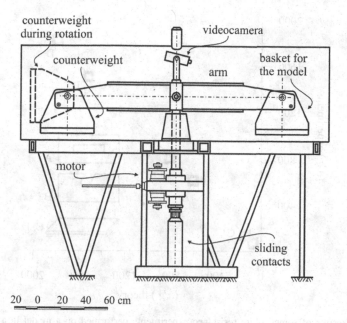

Fig. 8.2 Schematic diagram of a hinged basket centrifuge

where r_a is the ratio of model acceleration to gravity acceleration. A simple calculation indicates that if the centrifuge induces an acceleration equal to $200\,g$, it is possible to realise the model using material with resistance 200 times greater than the needed resistance in standard gravity conditions. Moreover, the possibility of increasing the acceleration at will over a wide range provides an additional degree of freedom and no longer necessitates a material with specific mechanical characteristics. The advantage becomes even more evident by considering the possibility of carrying out experiments by increasing centrifugal acceleration to collapse, equivalent to a progressive load increment.

Example 8.1 We wish to estimate in a centrifuge the ultimate tensile strength of a simply supported plate of granitic material, reproduced with a material model with a maximum tensile strength of $\sigma_{maxt} = 1.11$ MPa, on a geometric scale $\lambda = 4.66 \cdot 10^{-4}$. The density of the material in the model is 1850 kg m^{-3}, and the density of the granite is 2500 kg m^{-3}. Failure is reached with an acceleration of the centrifuge $a = 140\,g$.

If the plate is schematically represented as a simply supported beam, the maximum tensile stress is equal to

$$\sigma_{max} = \frac{3}{4}\,\rho\,a\,\frac{l^2}{h}. \tag{8.7}$$

Using the criteria of direct analysis, we obtain:

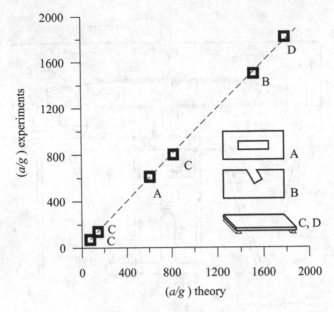

Fig. 8.3 Results of some failure resistance experiments performed on a model in a centrifuge compared to the theoretical solution (modified from Ramberg and Stephansson 1965)

$$\frac{\sigma_{0,m}}{\sigma_{0,p}} = \frac{\rho_m}{\rho_p}\frac{a}{g}\lambda \rightarrow \sigma_{0,p} = \frac{\sigma_{0,m}}{r_\rho\, r_a\, \lambda} \rightarrow$$

$$\sigma_{0,p} = \frac{11.1 \cdot 10^5}{0.74 \times 140 \times 4.66 \cdot 10^{-4}} = 22.9 \text{ MPa.} \qquad (8.8)$$

Figure 8.3 shows the results of some experiments where the result can be analytically calculated, with a fairly good agreement between theory and experiments in scaled variables. The advantage of the physical model is the reproduction of phenomena for which there is no analytical solution, or so complex as to make difficult even a numerical solution.

8.2.1 Scales in Centrifuge Models

The fundamental scales in centrifuge models derive from the need to reproduce in the model the same stress occurring in the prototype. This requires that, for stresses due to body forces,

$$\frac{\sigma_m}{\sigma_p} \equiv r_\sigma = 1 \rightarrow \frac{\rho_m}{\rho_p}\frac{a_m}{g}\frac{l_m}{l_p} \equiv r_\rho\, r_a\, \lambda = 1. \qquad (8.9)$$

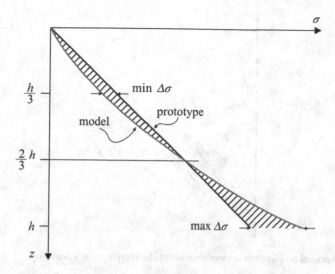

Fig. 8.4 Deviation between normal stress in the prototype and in the model. Here, h is the height of the model, or the distance between the highest and lowest point of the model (modified from Taylor 1995)

If the material in the model is the same as in the prototype,

$$r_a = \frac{1}{\lambda},$$
(8.10)

and the centrifuge should generate an acceleration equal to g/λ, where g is the acceleration of gravity.

However, there are some differences between the gravity field and the acceleration field reproduced in a centrifuge. A first difference is that in the centrifuge, the acceleration increases linearly with increasing radius. A typical trend of normal stress in the vertical direction (with a distance measured from the centrifuge axis) is shown in Fig. 8.4. The result is a distortion of the stress state compared to that in the prototype, although if h/R (h is the height of the model, and R is the radius of the centrifuge) is less than 0.2, the difference between stresses is less than 3%.

In addition, the gravity field admits potentials with spherical equipotential surfaces (locally flat), while the field of centrifugal acceleration admits potentials only if the axis of the centrifuge is parallel to the axis of gravity, and the equipotential surfaces are paraboloids with vertex on the axis, with equation

$$z = \frac{\omega^2 r^2}{2g} + \text{const.}$$
(8.11)

Centrifuges usually have a load basket that automatically orients itself according to the acceleration. Distortion of the acceleration occurs in the two directions of

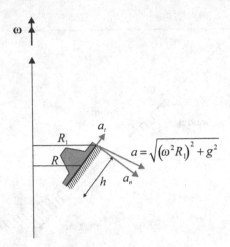

Fig. 8.5 Tangential acceleration components along the meridians in a centrifuge

the basket plane. If l is the dimension of the plane along the parallels and h is the dimension along the meridians, the relative difference of the acceleration, in the direction of the parallels, is equal to

$$\left.\frac{\Delta a}{a}\right|_{par} = \sqrt{1 + \frac{l^2}{4R^2}} - 1 \approx \frac{l^2}{8R^2},\tag{8.12}$$

while the relative difference of the acceleration, in the direction of the meridians, is equal to

$$\left.\frac{\Delta a}{a}\right|_{mer} \approx 2\frac{h}{R}\lambda.\tag{8.13}$$

For $l = h = 400$ mm and $R = 2000$ mm, with a scale model of $\lambda = 1/100$, we have $\Delta a/a|_{par} \approx 0.5\%$ and $\Delta a/a|_{mer} \approx 0.4\%$.

Much more important is the presence of acceleration components parallel to the plane on which the model rests.

Following the schematic in Fig. 8.5, along the meridians, we calculate a maximum ratio (at the edges) between tangential acceleration and normal acceleration equal to

$$\left.\frac{a_t}{a_n}\right|_{mer} = \tan(\alpha_0 - \alpha_1) \approx \frac{h}{R}\frac{\lambda^2}{2}.\tag{8.14}$$

For $h = 400$ mm and $R = 2000$ mm, with a scale model of $\lambda = 1/100$, we have $a_t/a_n|_{mer} = 10^{-5}$, which is a negligible value.

Along the parallels, see Fig. 8.6, we calculate a maximum ratio (at the edges) between tangential acceleration and normal acceleration equal to

Fig. 8.6 Tangential acceleration components along the parallels in a centrifuge

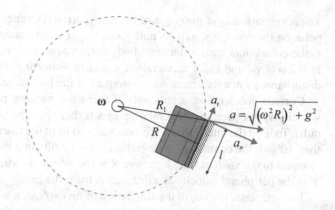

$$\left.\frac{a_t}{a_n}\right|_{par} = \frac{l}{2R}.$$ (8.15)

For $l = 400$ mm and $R = 2000$ mm, we have $a_t/a_n|_{par} = 0.1$. To avoid slip, this tangential acceleration must be adequately considered when realising the physical constraints of the model. A solution to eliminate a_t is to bend the whole model, building it on a curved surface with the same radius of curvature as the centrifuge arm.

8.2.2 Scale Effects and Anomalies in Centrifuges

The rotating reference system attached to the model is non-inertial, which causes fictitious accelerations that may generate interference in the physical process under study.

We consider the transformation of the acceleration from an inertial reference system to a non-inertial one:

$$\mathbf{a}_a = \mathbf{a}_r + \mathbf{a}_0 + \boldsymbol{\omega} \times (\boldsymbol{\omega} \times (\mathbf{x} - \mathbf{x}_0)) + 2\,\boldsymbol{\omega} \times \mathbf{v}_r + \frac{d\boldsymbol{\omega}}{dt} \times (\mathbf{x} - \mathbf{x}_0),$$ (8.16)

where the symbol $\times$ indicates a vector product, $\mathbf{a}_a$ is the absolute acceleration, $\mathbf{a}_r$ is the acceleration in the relative reference system, $\mathbf{a}_0$ is the acceleration of the origin of the relative reference system, $(\mathbf{x} - \mathbf{x}_0)$ is the position vector with respect to the origin of the relative reference system, and $\mathbf{v}_r$ is the relative velocity. In the rotating (non-inertial) reference frame, the acceleration is

$$\mathbf{a}_r = \mathbf{a}_a - \mathbf{a}_0 + \boldsymbol{\omega} \times (\boldsymbol{\omega} \times (\mathbf{x} - \mathbf{x}_0)) - 2\,\boldsymbol{\omega} \times \mathbf{v}_r - \frac{d\boldsymbol{\omega}}{dt} \times (\mathbf{x} - \mathbf{x}_0).$$ (8.17)

The acceleration $\mathbf{a}_0$ of the origin of the non-inertial reference system (for example, a point on the centrifuge axis) is null. Centrifugal acceleration $-\boldsymbol{\omega} \times (\boldsymbol{\omega} \times (\mathbf{x} - \mathbf{x}_0))$ is the component used to increase body forces at will. Coriolis acceleration, equal to $-2\,\boldsymbol{\omega} \times \mathbf{v}_r$, and Euler acceleration, equal to $-d\boldsymbol{\omega}/dt \times (\mathbf{x} - \mathbf{x}_0)$, are considered disturbances since they have no counterpart in the prototype.

Coriolis acceleration occurs only if there are moving parts in the model with velocity $\mathbf{v}_r$ not parallel to the rotation axis (otherwise, the vector product $\boldsymbol{\omega} \times \mathbf{v}_r$ is null). To limit the interference, it is necessary to limit the Coriolis acceleration to less than 10% of the centrifugal acceleration. This results in a maximum speed value of v_r equal to $0.05\,\omega R \equiv 0.05\,V$, where R is the nominal radius of the centrifuge and V is the peripheral velocity at a distance R from the axis.

In some tests, such as in the simulation of an explosion with expulsion of ground and launch of projectiles, the speed of the moving parts in the model is very high; in this case, the limits of velocity are calculated on the basis of the trajectories of the projectiles, according to the following scheme (see Fig. 8.7).

The equations of motion for a point mass are

$$\begin{cases} \dfrac{d^2x}{dt^2} - \omega^2 x - 2\omega \dfrac{dy}{dt} = 0, \\ \dfrac{d^2y}{dt^2} - \omega^2 y + 2\omega \dfrac{dx}{dt} = 0, \end{cases} \tag{8.18}$$

where $dx/dt = u_r$ and $dy/dt = v_r$ are the components of relative velocity. Equation (8.18) can be reduced to a system of ordinary linear differential equations,

$$\begin{cases} \dot{x} = u_r, \\ \dot{y} = v_r, \\ \dot{u}_r = \omega^2 x + 2\omega v_r, \\ \dot{v}_r = \omega^2 y - 2\omega u_r. \end{cases} \tag{8.19}$$

The dots on the symbols indicate the time derivative operation. With the initial conditions $(x, y, u_r, v_r)|_{t=0} = (x_0, y_0, u_0, v_0)$, we calculate the following solution:

$$\begin{cases} x(t) = \left[x_0 + (u_0 - y_0\,\omega)\,t\right]\cos\omega t + \left[y_0 + (v_0 + x_0\,\omega)\,t\right]\sin\omega t, \\ y(t) = \left[y_0 + (v_0 + x_0\,\omega)\,t\right]\cos\omega t - \left[x_0 + (u_0 - y_0\,\omega)\,t\right]\sin\omega t. \end{cases} \tag{8.20}$$

Assuming, for simplicity, that the projectile is initially in $x_0 = R$, $y_0 = 0$ and that the initial relative velocity is only radial, with the component equal to a multiple of the drag velocity at a distance R from the axis, $u_0 = -\beta\,\omega R$ and $v_0 = 0$, the trajectories in the relative reference system are almost parabolic, as shown in Fig. 8.7, depending on the parameter β.

The trajectory has a vertex at $x = 0$ in the time interval $[0, \pi/(2\omega)]$ if $\beta = 2.26$. For nonzero velocity in the tangential direction, this condition is fulfilled for $\beta >$

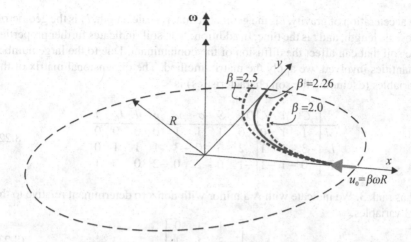

Fig. 8.7 Trajectories of a projectile in the centrifuge rotating reference system for different initial velocities u_0

2.26, depending on the angle of inclination and the modulus of the initial velocity. To minimise the effects of the acceleration of Coriolis, it is necessary that $u_0 > 2.26\,V$; otherwise, the curvature of the trajectories is excessive. We observe that the lower the minimum initial velocity of the projectile is, the greater the radius of the centrifuge. With the same centrifugal acceleration, the angular velocity is lower, and all disturbance effects are minimised. Hence, the indication is to build large-radius centrifuges.

The last source of disturbance is represented by the acceleration of Euler related to angular acceleration. The angular acceleration is a vector parallel to ω that opposes the variation in peripheral speed and that can be limited at will during transient by reducing the magnitude of the angular accelerations of the centrifuge.

8.2.3 Contaminant Transport Models in Centrifuges

We wish to study the transport of a contaminant in a porous medium with a physical model in a centrifuge.

If the fluid and sediments are incompressible, the typical equation for the concentration C of the contaminant is

$$C = f\left(\mu,\ V_s,\ D_m,\ S,\ \sigma,\ \rho_f,\ g,\ l,\ l_\mu,\ t,\ \text{car. soil}\right), \qquad (8.21)$$

where μ is the dynamic viscosity of the fluid, V_s is the interstitial velocity of the fluid, D_m is the molecular diffusion coefficient, S is the mass of contaminant absorbed per unit volume, σ is the fluid-sediment interface tension, ρ_f is the density of the fluid, g

is the acceleration of gravity, l is the geometric macroscale length, l_μ is the geometric microscale length, and t is the time. In addition, 'car. soil' indicates further properties of the soil that can affect the diffusion of the contaminant. Due to the large number of quantities involved, we apply the matrix method. The dimensional matrix of the 11 variables (excluding the soil characteristics) is

$$
\begin{array}{c|ccc|cccccccc}
 & C & \mu & V_s & D_m & S & \sigma & \rho_f & g & l & l_\mu & t \\
\hline
M & 1 & 1 & 0 & 0 & 1 & 1 & 1 & 0 & 0 & 0 & 0 \\
L & -3 & -1 & 1 & 2 & -3 & 0 & -3 & 1 & 1 & 1 & 0 \\
T & 0 & -1 & -1 & -1 & 0 & -2 & 0 & -2 & 0 & 0 & 1
\end{array}
\tag{8.22}
$$

and has rank 3. We indicate with $\mathbf{A}$ a minor with nonzero determinant relative to the first 3 variables,

$$
\mathbf{A} = \begin{bmatrix} 1 & 1 & 0 \\ -3 & -1 & 1 \\ 0 & -1 & -1 \end{bmatrix},
\tag{8.23}
$$

and $\mathbf{B}$ is the residual matrix (see Sect. 2.3),

$$
\mathbf{B} = \begin{bmatrix} 0 & 1 & 1 & 1 & 0 & 0 & 0 & 0 \\ 2 & -3 & 0 & -3 & 1 & 1 & 1 & 0 \\ -1 & 0 & -2 & 0 & -2 & 0 & 0 & 1 \end{bmatrix}.
\tag{8.24}
$$

Then, we calculate the matrix

$$
\mathbf{C} = \mathbf{A}^{-1}\mathbf{B} \equiv \begin{bmatrix} -1 & 1 & 0 & 1 & 1 & -1 & -1 & -1 \\ 1 & 0 & 1 & 0 & -1 & 1 & 1 & 1 \\ 0 & 0 & 1 & 0 & 3 & -1 & -1 & -2 \end{bmatrix},
\tag{8.25}
$$

which is equivalent to the dimensional matrix of the variables D_m, S, σ, ρ_f, g, l, l_μ, t, as a function of C, μ, V_s:

$$
\begin{array}{c|cccccccc}
 & D_m & S & \sigma & \rho_f & g & l & l_\mu & t \\
\hline
C & -1 & 1 & 0 & 1 & 1 & -1 & -1 & -1 \\
\mu & 1 & 0 & 1 & 0 & -1 & 1 & 1 & 1 \\
V_s & 0 & 0 & 1 & 0 & 3 & -1 & -1 & -2
\end{array}
\tag{8.26}
$$

The most straightforward dimensionless groups are:

$$
\Pi_1 = \frac{D_m C}{\mu}, \quad \Pi_2 = \frac{S}{C}, \quad \Pi_3 = \frac{\sigma}{\mu V_s},
$$

$$
\Pi_4 = \frac{\rho_f}{C}, \quad \Pi_5 = \frac{g \mu}{C V_s^3}, \quad \Pi_6 = \frac{l C V_s}{\mu},
\tag{8.27}
$$

$$
\Pi_7 = \frac{l_\mu C V_s}{\mu}, \quad \Pi_8 = \frac{t C V_s^2}{\mu},
$$

Table 8.2 Summary of significant dimensionless numbers in the physical process of transport of a contaminant in a porous medium

Group	Expression
Concentration	$\Pi_1' = \dfrac{C}{\rho_f} = \dfrac{1}{\Pi_4}$
Advection	$\Pi_2' = \dfrac{V_s t}{l} = \dfrac{\Pi_8}{\Pi_6}$
Diffusion	$\Pi_3' = \dfrac{D_m t}{l^2} = \dfrac{\Pi_1 \Pi_8}{\Pi_6^2}$
Capillarity	$\Pi_4' = \dfrac{l_\mu \, g \, l \, \rho_f}{\sigma} = \dfrac{\Pi_4 \Pi_5 \Pi_6 \Pi_7}{\Pi_3}$
Absorption	$\Pi_5' = \dfrac{S}{\rho_f} = \dfrac{\Pi_2}{\Pi_4}$
Reynolds	$\Pi_6' = \dfrac{l_\mu V_s \rho_f}{\mu} = \Pi_4 \Pi_7$
Péclet	$\Pi_7' = \dfrac{V_s l_\mu}{D_m} = \dfrac{\Pi_7}{\Pi_1}$
Dynamic	$\Pi_8' = \dfrac{g t^2}{l} = \dfrac{\Pi_5 \Pi_8^2}{\Pi_6}$

although the dimensionless groups with a physical meaning are those reported in Table 8.2. If the material in the model is equal to the material in the prototype, to guarantee the equality of the stresses, it is necessary that the acceleration of the centrifuge has a scale ratio equal to $r_a = 1/\lambda$.

A variable of possible interest is the hydraulic conductivity k, defined as

$$k = \frac{K \rho_f}{\mu}(g + a), \tag{8.28}$$

where K is the intrinsic permeability of the soil. The scale ratio of the hydraulic conductivity is equal to

$$r_k = \frac{r_K r_{\rho_f}}{r_\mu} r_a, \tag{8.29}$$

and using the same material and the same fluid of the prototype in the model results in $r_K = r_{\rho_f} = r_\mu = 1$ and therefore $r_k = r_a = 1/\lambda$. The velocity of the fluid can be expressed as

$$V_s = \frac{k\,i}{\eta},$$ (8.30)

where i is the hydraulic gradient and η is the porosity of the filtering medium. For a geometrically undistorted model (characterised by a single geometric scale), the gradient assumes the same value as in the prototype. Moreover, if the material is the same in the model and in the prototype, the porosity is also unchanged. Therefore,

$$r_{V_s} = \frac{r_k\,r_i}{r_\eta} \equiv r_k = \frac{1}{\lambda}.$$ (8.31)

The filtration time is equal to

$$t = \frac{l}{V_s},$$ (8.32)

and its scale ratio becomes

$$r_t = \frac{\lambda}{r_{V_s}} = \lambda^2.$$ (8.33)

The condition of similarity provides 8 equations on 11 unknown scales. On the basis of the dimensionless groups listed in Table 8.2, the set of equations is:

$$\begin{cases} r_C = r_{\rho_f}, \\ r_{V_s} = \lambda/r_t, \\ r_{D_m} = \lambda^2/r_t, \\ r_{l_\mu}\, r_a\, \lambda\, r_{\rho_f} = r_\sigma, \\ r_S = r_{\rho_f}, \\ r_{l_\mu}\, r_{V_s}\, r_{\rho_f} = r_\mu, \\ r_{V_s}\, r_{l_\mu} = r_{D_m}, \\ r_a\, r_t^2 = \lambda. \end{cases}$$ (8.34)

Finally, the additional 6 constraints

$$r_{\rho_f} = r_{D_m} = r_{l_\mu} = r_\sigma = r_S = r_\mu = 1,$$ (8.35)

lead to the following conditions of similarity:

$$\begin{cases} r_C = 1, \\ r_{V_s} = 1/\lambda, \\ r_t = \lambda^2, \\ r_a = 1/\lambda, \\ r_{V_s} = 1. \end{cases} \tag{8.36}$$

The second condition (derived from the similarity for the advection) and the fifth condition (derived from the similarity of Reynolds and Péclet) are incompatible. We can achieve a valid approximate similarity if Re < 1 and Pe < 1, neglecting the condition $r_{V_s} = 1$. This implies that, on a reduced geometric scale, the numbers of Reynolds and Péclet in the model assume a higher value than the prototype, with

$$\begin{aligned} \frac{\text{Re}_m}{\text{Re}_p} &\equiv r_{\text{Re}} = 1/\lambda, \\ \frac{\text{Pe}_m}{\text{Pe}_p} &\equiv r_{\text{Pe}} = 1/\lambda. \end{aligned} \tag{8.37}$$

Therefore, it is necessary to verify that in the prototype and in the model, these numbers are both small.

8.2.4 The Similarity in Dynamic Models in Centrifuges

We consider a simple oscillatory motion, with amplitude A and frequency n, which develops on a small geometric scale compared to the dominant geometric scales:

$$x = A \sin(2\pi n t). \tag{8.38}$$

Differentiating, we calculate the velocity

$$V = 2\pi n A \cos(2\pi n t), \tag{8.39}$$

and the acceleration

$$a = -4\pi^2 n^2 A \sin(2\pi n t). \tag{8.40}$$

The similarity conditions are:

$$\begin{cases} r_V = r_n \lambda, \\ r_a = r_n^2 \lambda. \end{cases} \tag{8.41}$$

Since $r_a = 1/\lambda$ in the centrifuge,

$$\begin{cases} r_n = \dfrac{1}{\lambda}, \\ r_V = 1. \end{cases} \qquad (8.42)$$

Therefore, in reproducing dynamic phenomena, the frequency in a model with a reduced geometric scale is amplified by a factor of $1/\lambda$; an oscillation at $10\,\text{Hz}$ in the prototype results in a model with $\lambda = 1/10$ at an oscillation at $100\,\text{Hz}$. For example, an earthquake of duration $18\,\text{s}$ with 36 cycles ($2\,\text{Hz}$) and amplitude $5\,\text{cm}$ in a $100\,g$ centrifuge with $\lambda = 1/100$ results in an oscillation with amplitude $5/100 = 0.5\,\text{mm}$, frequency $2 \times 100 = 200\,\text{Hz}$, and duration $18/100 = 0.18\,\text{s}$.

We observe that the time scale equal to λ does not coincide with the time scale equal to λ^2 calculated in filtration phenomena (see Eq. 8.36). Therefore, if the dynamic effects and filtration appear jointly (for example, the collapse of a bank with a stratum in the presence of an earthquake), to gain equal time scales for the two phenomena, it is necessary to recover a further degree of freedom.

In this regard, we observe that the time scale of the filtration can be controlled by changing the dynamic viscosity of the fluid in the model. The scale of hydraulic conductivity (see Eq. 8.28) for a fluid with the same density in the model and prototype and with identical filter material in the model and prototype ($r_K = 1$) is equal to

$$r_k = \frac{r_a}{r_\mu} \equiv \frac{1}{\lambda\,r_\mu}. \qquad (8.43)$$

The scale of filtration time for an undistorted model, with the same filter material in the model and prototype ($r_\eta = 1$), is equal to

$$r_t = \frac{\lambda}{r_{V_s}} = \lambda^2\,r_\mu. \qquad (8.44)$$

To obtain the same time scale λ of the dynamic phenomena, we need a fluid with $r_{\rho_f} = 1$ and $r_\mu = 1/\lambda$. For example, we can use silicone fluid with the same specific gravity as water but with a much higher dynamic viscosity than water (some orders of magnitude), which is nontoxic and nonhazardous. Recently, a nontoxic carboxymethylcellulose solution (CMC) has been used, currently used in pharmacology and in the food industry. It has a shear-thinning behaviour, but with minor effects if the range of shear rate is limited during filtration.

If $r_\mu = 1/\lambda$, the Reynolds number is unchanged since $r_{\text{Re}} = 1$. The Péclet number has a scale ratio that also depends on the molecular diffusivity of the contaminant in the new fluid:

$$r_{\text{Pe}} = \frac{r_{V_s}\,r_{l_\mu}}{r_{D_m}} = \frac{1}{\lambda\,r_{D_m}}. \qquad (8.45)$$

Molecular diffusivity depends on the speed of a molecule of contaminant moving in the fluid; it is directly proportional to the kinetic energy of the particle and inversely proportional to its size and to the viscosity of the fluid. Since $r_{D_m} = 1/r_\mu \equiv \lambda$, we have $r_{Pe} = 1/\lambda^2$. The amplification of the Péclet number in reduced scale models requires more attention to ensure that it is small enough to have negligible effects.

8.2.5 Similarity in Tectonic Processes

We consider the process of motion of rocky clusters or magma in flow conditions similar to those of a viscous fluid, possibly a Bingham fluid. The typical equation of the physical process is

$$f(\rho,\ V,\ l,\ \mu,\ g,\ \Delta p,\ \tau_c) = 0, \qquad (8.46)$$

where ρ is density, V and l are velocity and scale lengths, respectively, μ is dynamic viscosity, g is gravity acceleration, Δp is pressure difference, and τ_c is cohesion. The dimensional matrix of the 7 variables

$$
\begin{array}{c|ccccccc}
 & \rho & V & l & \mu & g & \Delta p & \tau_c \\
\hline
M & 1 & 0 & 0 & 1 & 0 & 1 & 1 \\
L & -3 & 1 & 1 & -1 & 1 & -1 & -1 \\
T & 0 & -1 & 0 & -1 & -2 & -2 & -2
\end{array}
\qquad (8.47)
$$

has rank 3; therefore, we can express the functional relationship by introducing $(7 - 3) = 4$ dimensionless groups. Applying the matrix method, we can calculate, for example, the following 4 groups:

$$\Pi_1 = \frac{\mu}{\rho\,V\,l}, \quad \Pi_2 = \frac{g\,l}{V^2}, \quad \Pi_3 = \frac{\Delta p}{\rho\,V^2}, \quad \Pi_4 = \frac{\tau_c}{\rho\,V^2}, \qquad (8.48)$$

that can be conveniently rearranged into groups of more immediate physical meaning:

$$
\begin{aligned}
& \mathrm{Re} = \frac{1}{\Pi_1} = \frac{\rho\,V\,l}{\mu}, \quad \mathrm{St} = \frac{\Pi_3}{\Pi_1} = \frac{l\,\Delta p}{\mu\,V}, \\
& \mathrm{Rm} = \frac{\Pi_2}{\Pi_1} = \frac{g\,l^2\,\rho}{\mu\,V}, \quad \mathrm{Rs} = \frac{\Pi_2}{\Pi_4} = \frac{g\,l\,\rho}{\tau_c},
\end{aligned}
\qquad (8.49)
$$

where Re is the Reynolds number, St is the Stokes number, Rm is the Ramberg number, and Rs is a number representing the relative effects of gradient pressure due to gravity and yield stress, similar to the Bingham number. To verify that the new groups are independent, we calculate the rank of their dimensional matrix,

$$
\begin{array}{c|ccccccc}
 & \rho & V & l & \mu & g & \Delta p & \tau_c \\
\hline
\text{Re} & 1 & 1 & 1 & -1 & 0 & 0 & 0 \\
\text{St} & 0 & -1 & 1 & -1 & 0 & 1 & 0 \\
\text{Rm} & 1 & -1 & 2 & -1 & 1 & 0 & 0 \\
\text{Rs} & 1 & 0 & 1 & 0 & 1 & 0 & -1
\end{array}
\tag{8.50}
$$

and check that it is equal to 4, equal to the number of rows (see Sect. 1.4.2.2).

The complete similarity requires satisfying the following 4 equations (the dimensionless groups must assume the same value in the model and in the prototype) in the 7 unknowns (the scales of the 7 quantities involved):

$$
\begin{cases}
r_\rho\, r_V\, \lambda = r_\mu, \\
\lambda\, r_{\Delta p} = r_\mu\, r_V, \\
r_a\, \lambda^2\, r_\rho = r_\mu\, r_V, \\
r_a\, \lambda\, r_\rho = r_{\tau_c}.
\end{cases}
\tag{8.51}
$$

There are still 3 degrees of freedom. Selecting the geometric scale λ, the scale of the density r_ρ and the scale of the dynamic viscosity r_μ, the system (8.51) admits the solution:

$$
\begin{cases}
r_{\Delta p} = \dfrac{r_\mu^2}{\lambda^2\, r_\rho}, \\[2ex]
r_{\tau_c} = \dfrac{r_\mu^2}{\lambda^2\, r_\rho}, \\[2ex]
r_a = \dfrac{r_\mu^2}{\lambda^3\, r_\rho^2}, \\[2ex]
r_V = \dfrac{r_\mu}{\lambda\, r_\rho}.
\end{cases}
\tag{8.52}
$$

However, full dynamic similarity is not necessary if, for example, the Reynolds number or the Ramberg number are very small (Weijermars and Schmeling 1986). In fact, we consider the equation of balance of the momentum for a viscous fluid that, for the ith component, assumes the following form in the scalar notation:

$$
\frac{\partial u_i}{\partial t} + u_j\, \frac{\partial u_i}{\partial x_j} - g_i + \frac{1}{\rho}\, \frac{\partial p}{\partial x_i} - \frac{1}{\rho}\, \frac{\partial \tau_{ij}}{\partial x_j} = 0,
\tag{8.53}
$$

where τ_{ij} is the nonisotropic component of the stress tensor. We can proceed to non-dimensionalise the equation, choosing a scale length l, a scale density ρ_0, and a dynamic viscosity scale μ_0. The time scale is equal to $\mu_0/(\rho_0\, g\, l)$, the velocity scale becomes $\rho_0\, g\, l^2/\mu_0$, and the pressure and tangential stress scales become $\rho_0\, g\, l$. In dimensionless form, Eq. (8.53) becomes

$$\frac{\rho_0^2 \, g \, l^3}{\mu_0^2} \, \tilde{\rho} \left(\frac{\partial \tilde{u}_i}{\partial \tilde{t}} + \tilde{u}_j \frac{\partial \tilde{u}_i}{\partial \tilde{x}_j} \right) - \tilde{\rho}_i + \frac{\partial \tilde{p}}{\partial \tilde{x}_i} - \frac{\partial \tilde{\tau}_{ij}}{\partial \tilde{x}_j} = 0, \tag{8.54}$$

where $\tilde{\rho}_i$ refers to the ith component of gravity. The dimensionless group that multiplies the inertia (the term in parentheses) is equal to Re·Rm. If Re $\ll$ 1 or Rm $\ll$ 1, then inertia is negligible, and the momentum balance becomes:

$$- \tilde{\rho}_i + \frac{\partial \tilde{p}}{\partial \tilde{x}_i} - \frac{\partial \tilde{\tau}_{ij}}{\partial \tilde{x}_j} = 0, \tag{8.55}$$

without dimensionless groups. Therefore, dynamic similarity simply requires geometric similarity (Eq. 8.53 is the momentum balance per unit volume and has no scale lengths), and some conditions of similarity are redundant. Equations (8.52) become

$$\begin{cases} \lambda \, r_{\Delta p} = r_\mu \, r_V, \\ r_a \, \lambda^2 \, r_\rho = r_\mu \, r_V, \\ r_a \, \lambda \, r_\rho = r_{\tau_c}, \end{cases} \tag{8.56}$$

and allow us to select 4 scales at will, for example, λ, r_ρ, r_μ and r_a. The 3 remaining scales are equal to

$$\begin{cases} r_{\tau_c} \equiv r_{\Delta p} = \lambda \, r_a \, r_\rho, \\ r_V = \dfrac{\lambda^2 \, r_a \, r_\rho}{r_\mu}. \end{cases} \tag{8.57}$$

The Reynolds number is not maintained but scales as follows:

$$\frac{\mathrm{Re}_m}{\mathrm{Re}_p} = \frac{\lambda^3 \, r_a \, r_\rho}{r_\mu^2}. \tag{8.58}$$

Example 8.2 To study the relationship between the deformation of the continental crust and the distribution of magma in the deep zones, a model in a centrifuge has been realised on a geometric scale $\lambda = 4.5 \cdot 10^{-7}$ (Corti et al. 2002). Since the Reynolds number is very small, the similarity is approximated, and the main dimensionless group is the Ramberg number, equal to the ratio between gravitational and viscous forces:

$$\mathrm{Rm} = \frac{\rho_d \, g \, h_d^2}{\mu \, V}, \tag{8.59}$$

where ρ_d, h_d and μ are the density, thickness and dynamic viscosity of the crust, respectively, g is the acceleration of gravity and V is the velocity of deformation. In the upper part of the crust, a relevant dimensionless group is given by the ratio between the forces of gravity and cohesion:

$$\mathrm{Rs} = \frac{\rho_b \, g \, h_b}{\tau_c}, \qquad\qquad \therefore \quad (8.60)$$

where ρ_b and h_b are the density and thickness of the upper crust, respectively, and τ_c is cohesion. On the basis of the analysis carried out in Sect. 8.2.1, we can fix 4 scales and calculate the 3 other scales. The upper crust, with fragile rupture, is reproduced with sand, the deeper crust is reproduced with a mixture of silicone and sand, and the magma is reproduced with glycerol.

8.3 Some Applications for the Solution of Classic Problems

Example 8.3 We wish to analyse the load-bearing capacity of an infinitely long foundation of width B on soil in drained conditions and in the presence of a uniform vertical load (see Fig. 8.8). We can assume that the maximum load per unit area q depends on the cohesion c, the angle of internal friction in drained conditions ϕ, the width B of the foundation, the depth D with respect to the average ground level (expressed as an equivalent load $q_0 = D\gamma$) and the specific gravity of the soil γ. The typical equation is

$$q = f(c, \ \phi, \ B, \ D, \ \gamma), \qquad\qquad (8.61)$$

with a dimensional matrix in an M, L, T system

$$
\begin{array}{c|cccccc}
 & q & c & \phi & B & D & \gamma \\
\hline
M & 1 & 1 & 0 & 0 & 0 & 1 \\
L & -1 & -1 & 0 & 1 & 1 & -2 \\
T & -2 & -2 & 0 & 0 & 0 & -2
\end{array}
\qquad (8.62)
$$

that has rank 2. For example, if we select B and γ as fundamental quantities, we can express the functional relation (8.61) as

$$\frac{q}{B\gamma} = \tilde{f}\left(\frac{c}{B\gamma}, \ \phi, \ \frac{D}{B}\right). \qquad\qquad (8.63)$$

This last relation is compared with the one by Terzaghi (1943),

$$\frac{q}{B\gamma} = \frac{c}{B\gamma} N_c + \frac{D}{B} N_q + 0.5 N_\gamma, \qquad\qquad (8.64)$$

where N_c, N_q and N_γ are dimensionless parameters depending on the friction angle ϕ.

If the foundation has a finite length l, this variable must be included and introduces an additional geometric scale. For a foundation of generic shape, there is also a shape factor. Equation (8.64) is redefined as

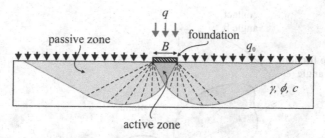

Fig. 8.8 Schematic of a long foundation. The depth D with respect to the average ground level is given by an equivalent load $q_0 = D\gamma$

$$\frac{q}{B\gamma} = \tilde{f}\left(\frac{c}{B\gamma}, \phi, \frac{D}{B}, \frac{l}{B}, \text{shape}\right), \tag{8.65}$$

comparable to Terzaghi's relationship (Terzaghi 1943),

$$\frac{q}{B\gamma} = \frac{c}{B\gamma} N_c s_c + \frac{D}{B} N_q s_q + 0.5 N_\gamma s_\gamma, \tag{8.66}$$

with s_c, s_q, $s_\gamma = f(l/B, \text{shape})$ and N_c, N_q, $N_\gamma = f(\phi)$.

Example 8.4 A class of rheological models for granular mixtures gives the following expressions of tangential and normal stresses:

$$\begin{cases} \tau = C_s\, \rho_s\, d^2 f_1(e,\ \beta)\, \dot{\gamma}^2, \\ \sigma = C_s\, \rho_s\, d^2 f_2(e,\ \beta)\, \dot{\gamma}\, |\dot{\gamma}|, \end{cases} \tag{8.67}$$

where C_s is the volumetric concentration of the sediment, ρ_s is their density, e is the elastic restitution coefficient for impact between the particles, $\dot{\gamma}$ is the shear rate, and β is the average contact angle between the grains (see the schematic in Fig. 8.9).

Under quasi-static conditions, with particles in long-lasting contact, a modified Mohr-Coulomb law can be adopted, namely,

$$\tau_f = \sigma_f \tan(\phi_0 + \beta), \tag{8.68}$$

where τ_f and σ_f are the stresses in almost static conditions, ϕ_0 is the true angle of friction, which depends on the surface characteristics of the grains, and β is the contact angle. We need a relationship to express β. We can assume here that

$$\beta = f(\rho_s,\ d,\ I_1,\ J_2,\ E), \tag{8.69}$$

where I_1 is the first invariant of the stress tensor, i.e., the sum of the diagonal elements (which is representative of the average normal stress) and where J_2 is the second invariant of the strain rate tensor (which is representative of the average angular

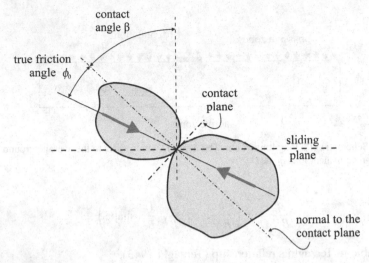

Fig. 8.9 Description of contact geometry (modified from Longo & Lamberti 2000)

strain rate). Six variables appear with a dimensional matrix of rank 3 (the contact angle β is dimensionless); if 3 independent variables are selected, it is possible to transform the typical equation (8.69) into a relation between 3 dimensionless groups, for example,

$$\beta = \tilde{f}\left(\frac{\rho_s\, d^2\, J_2^2}{E}, \frac{I_1}{E}\right). \qquad\qquad\therefore\quad (8.70)$$

The first group in the argument is the ratio between collisional and static stress (the grains have multiple contacts with deformability of the mixture due to deformation of the individual grains and small settlements). In the second group, a total normal stress estimator appears instead of the collisional component. We observe that gravity acceleration is not directly included but can cause stress in the granular material (for example, in the motion of sediment coming out of a silo or in a stony debris flow). The structure of $\tilde{f}$ can be inferred with experiments.

Example 8.5 We consider a structure-foundation-ground system stressed by a strong impulsive earthquake.

   In a linear approach, this scenario can be described as a system of concentrated elements with masses, springs and viscous dampers for the structure and the ground separately; see the schematic in Fig. 8.10. The variable of interest is the maximum displacement of the structure with respect to the foundation, $\max |x_s - x_f| = x_{max}$, for an impulse acceleration of amplitude a_p and duration T_p. The variables involved are the mass m_s, the stiffness k_s and the damping coefficient β_s of the structure, as well as the corresponding variables for the ground foundation, that is, m_f, k_f and β_f. The dimensional matrix is

$$
\begin{array}{c|ccccccccc}
 & x_{max} & a_p & T_p & m_s & m_f & k_s & k_f & \beta_s & \beta_f \\
\hline
M & 0 & 0 & 0 & 1 & 1 & 1 & 1 & 1 & 1 \\
L & 1 & 1 & 0 & 0 & 0 & 0 & 0 & 0 & 0 \\
T & 0 & -2 & 1 & 0 & 0 & -2 & -2 & -1 & -1
\end{array}
\qquad (8.71)
$$

and has rank 3. The variables a_p, T_p and m_s are independent and can be assumed to be fundamental variables. Buckingham's Theorem indicates that it is possible to express the physical process with only $(9 - 3) = 6$ dimensionless groups, for example:

$$
\Pi_1 = \frac{x_{max}}{a_p T_p^2}, \quad \Pi_2 = \frac{m_f}{m_s}, \quad \Pi_3 = \frac{k_s T_p^2}{m_s},
$$

$$
\Pi_4 = \frac{k_f T_p^2}{m_s}, \quad \Pi_5 = \frac{\beta_s T_p}{m_s}, \quad \Pi_6 = \frac{\beta_f T_p}{m_s}. \qquad (8.72)
$$

Not all the groups listed above have physical meaning, but they can be combined in such a way as to make their interpretation easier, for example, by defining the new groups

$$
\Pi_1' \equiv \Pi_1 = \frac{x_{max}}{a_p T_p^2}, \quad \Pi_2' \equiv \Pi_2 = \frac{m_f}{m_s}, \quad \Pi_3' = \sqrt{\Pi_3} = \omega_s T_p,
$$

$$
\Pi_4' = \frac{\Pi_4}{\Pi_3} = \frac{k_f}{k_s}, \quad \Pi_5' = \frac{\Pi_5}{2\sqrt{\Pi_3}} = \xi_s \quad \Pi_6' = \frac{\Pi_6}{\Pi_5} = \frac{\beta_f}{\beta_s}, \qquad (8.73)
$$

where $\omega_s = \sqrt{k_s/m_s}$ is the resonance frequency of the structure on an infinitely rigid foundation and $\xi_s = \beta_s/(2 m_s \omega_s)$ is the damping coefficient. The first group is the ratio between the relative displacement and the geometric scale of the impulsive load, the second is the ratio between the mass of the structure and the foundation, and the third is the relationship between the scale of the pulsation of the structure and that

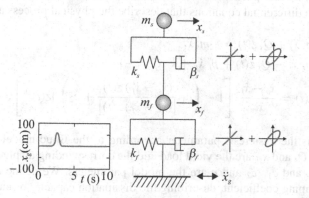

Fig. 8.10 Schematic of the structure-soil-foundation system with parameters concentrated in the linear case (modified from Zhang and Tang 2008)

of the load. In general, we can write

$$\frac{x_{max}}{a_p T_p^2} = \tilde{f}\left(\frac{m_f}{m_s}, \ \omega_s T_p, \ \frac{k_f}{k_s}, \ \xi_s, \ \frac{\beta_f}{\beta_s}\right). \tag{8.74}$$

Sometimes dimensionless groups can be identified on the basis of an algebraic or differential equation (or system of equations) to describe the physical process. In the present case, schematising the structure-ground-foundation system as a system with concentrated parameters with a single degree of freedom, we can write

$$\begin{cases} \ddot{x}_s + 2\,\xi_s\,\omega_s\,(\dot{x}_s - \dot{x}_f) + \omega_s^2\,(x_s - x_f) = -\ddot{x}_g, \\ \ddot{x}_f - 2\,\dfrac{m_s}{m_f}\,\xi_s\,\omega_s\,\dot{x}_s + \dfrac{m_s}{m_f}\left(1 + 2\,\dfrac{\beta_f}{\beta_s}\right)\xi_s\,\omega_s\,\dot{x}_f - \\ \qquad \dfrac{m_s}{m_f}\,\omega_s^2\,x_s + \dfrac{m_s}{m_f}\left(1 + \dfrac{k_f}{k_s}\right)\omega_s^2\,x_f = -\ddot{x}_g, \end{cases} \tag{8.75}$$

where x_s and x_f are the horizontal displacements of the structure and the foundation-ground system, $\ddot{x}_g$ is the acceleration imposed by the earthquake, and the dot and the double dot indicate the first and second temporal derivatives, respectively.

The system of Eq. (8.75) can be integrated by keeping 4 dimensionless groups constant, varying the fifth group and calculating the numerical value of the residual group. Figure 8.11 shows the results of the numerical integration performed for 3 different values of the imposed acceleration, with a constant pulse shape. As expected, the results in dimensionless form collapse onto a single curve. To realise a physical model, the scales should be selected using the 6 similarity conditions associated with the 6 dimensionless groups.

Example 8.6 A more realistic scheme than adopted in Example 8.5 includes plasticity and hysteresis phenomena. In a scheme with concentrated parameters, see Fig. 8.12, the differential equations that describe the physical process are:

$$\begin{cases} m_s\,\ddot{x}_s + Q_s\,z(t) = -m_s\,\ddot{x}_g, \\ m_f\,\ddot{x}_f - Q_s\,z(t) + \beta_f\,\dot{x}_f + k_f\,x_f = -m_f\,\ddot{x}_g, \\ \dot{z}(t) = \dfrac{\dot{x}_s - \dot{x}_f}{x_{sy}}\left[1 - \left(c_1\,\dfrac{(\dot{x}_s - \dot{x}_f)\,z(t)}{|(\dot{x}_s - \dot{x}_f)\,z(t)|} + c_2\right)|z(t)|^n\right], \end{cases} \tag{8.76}$$

where $z(t)$ is the hysteresis parameter according to the Bouc model (Bouc 1967; Wen 1976), Q_s and x_{sy} are the yield load and the corresponding displacement value, respectively, and c_1, c_2 and n are the model parameters. We define an equivalent viscous damping coefficient, describing the dissipation capacity of an elastoplastic structure with ductility μ, and the initial stiffness of the structure:

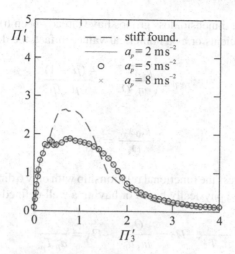

Fig. 8.11 System response for 3 different acceleration values that have been imposed. Integration performed for $\Pi_2' = 0.25$, $\log \Pi_4' = 0.25$, $\Pi_5' = 0.05$, $\log \Pi_6' = 0.25$ (modified from Zhang and Tang 2008)

$$
\begin{cases}
\beta_{s,equiv} = \dfrac{4\,m_s\,(\mu - 1)}{\pi\,\mu\,\sqrt{\mu}}\sqrt{\dfrac{Q_s}{m_s\,x_{sy}}}, \\[3mm]
k_{s0} = \dfrac{Q_s}{x_{sy}}.
\end{cases}
\tag{8.77}
$$

With these assumptions, the physical process is described by the typical equation

$$
x_{max} = f\left(a_p,\ T_p,\ m_s,\ m_f,\ k_{s0},\ k_f,\ Q_s,\ x_{sy},\ \beta_f,\ \beta_{s,equiv},\ c_1,\ c_2,\ n,\ \mu\right).
\tag{8.78}
$$

The dimensional matrix has rank 3, and applying Buckingham's Theorem, we can reduce the 11 variables and the 4 numerical coefficients to 8 dimensionless groups and 4 numerical coefficients. We observe that, based on the selected model and on

Fig. 8.12 Schematic structure-ground-foundation system with concentrated parameters in the nonlinear case (modified from Zhang and Tang 2008)

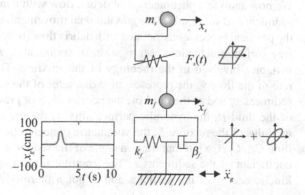

some assumptions, 2 dimensionless groups have already been fixed as a function of
the numerical coefficients or as a numerical value. In fact, the definitions of $\beta_{s,equiv}$
and k_{s0} imply that

$$\beta_{s,equiv}\sqrt{\frac{x_{sy}}{m_s\,Q_s}} = \frac{4\,(\mu-1)}{\pi\,\mu\,\sqrt{\mu}} \tag{8.79}$$

and

$$\frac{k_{s0}\,x_{sy}}{Q_s} = 1. \tag{8.80}$$

In short, we can express the functional relationship with only 6 dimensionless groups,
selected among those physically based or having a well-defined meaning, that is:

$$\Pi_1 = \frac{x_{max}}{a_p\,T_p^2}, \quad \Pi_2 = \frac{Q_s}{m_s\,a_p}, \quad \Pi_3 = \frac{x_{sy}}{a_p\,T_p^2},$$

$$\Pi_4 = \frac{m_f}{m_s}, \quad \Pi_5 = \frac{k_f}{k_{s0}}, \quad \Pi_6 = \frac{\beta_f}{\beta_{s,equiv}\,\mu}, \tag{8.81}$$

hence,

$$\frac{x_{max}}{a_p\,T_p^2} = \tilde{f}\left(\frac{Q_s}{m_s\,a_p}, \frac{x_{sy}}{a_p\,T_p^2}, \frac{m_f}{m_s}, \frac{k_f}{k_{s0}}, \frac{\beta_f}{\beta_{s,equiv}\,\mu}, c_1, c_2, n\right). \tag{8.82}$$

The numerical integration of the system of equations can be performed by modifying the value of the dimensionless groups. In practise, it is estimated that the
normalised yield strength and the normalised yield displacement are in the range
of $0.1 \le \Pi_2 \le 4.0$ and $0.01 \le \Pi_3 \le 1.00$. For the other dimensionless groups, it is
reasonable to assume $0.05 \le \Pi_4 \le 0.35$, $\Pi_5 \ge 0.1$, $\Pi_6 \le 1000$.

8.4 Dimensional Analysis of Debris Flows

We now analyse a phenomenon of debris flow with a mixture of high-concentration
sediment and water subject to gravitational movement. If we wish to analyse some of
the physical processes associated with debris flow (triggering process, arrest process,
free surface fluctuations, segregation of sediments), we first need to identify the
variables involved in the rheology of the mixture. These variables are the strain
rate of the flow $\dot{\gamma}$, the representative diameter of the sediment d, the density of the
sediment ρ_s and of the fluid ρ_f, the acceleration of gravity g, the dynamic viscosity
of the fluid μ, the hydraulic permeability k, the granular temperature θ, the bulk
modulus of the mixture E, the volumetric concentration of the sediments and of the
fluid C_s, C_f, the internal friction angle of the sediments ϕ, and the elastic restitution
coefficient of the sediments e. The granular temperature measures the fluctuating
kinetic energy of the particles and is not a thermodynamic variable. We wish to

study the dependence of normal stress on the other variables, with a typical equation

$$\sigma = f(\dot{\gamma},\ d,\ \rho_s,\ \rho_f,\ g,\ \mu,\ k,\ \theta,\ E,\ C_s,\ C_f,\ \phi,\ e). \tag{8.83}$$

The last four variables are dimensionless. In all, there are 14 variables with a dimensional matrix of rank 3, and applying Buckingham's Theorem, we calculate $(14 - 3) = 11$ dimensionless groups. The most straightforward fundamental quantities are d, the density ρ_s and the strain rate $\dot{\gamma}$, and they are independent. To identify the possible dimensionless groups, we can apply Rayleigh's method and express the other variables in power function with respect to the fundamental quantities, including the stress:

$$\sigma = d^{c_1}\ \rho_s^{c_2}\ \dot{\gamma}^{c_3}, \tag{8.84}$$

or

$$[\sigma] \equiv M\ L^{-1}\ T^{-2} = L^{c_1}\ (M\ L^{-3})^{c_2}\ (T^{-1})^{c_3}. \tag{8.85}$$

By equating the exponents of mass, length and time, we obtain a system of equations in the three unknowns c_1, c_2 and c_3. By performing calculations for all dependent variables, we can express the typical equation (8.83) as

$$\frac{\sigma}{\rho_s\ d^2\dot{\gamma}^2} = \tilde{f}\left(\frac{\rho_f}{\rho_s},\ \frac{\dot{\gamma}^2 d}{g},\ \frac{\dot{\gamma}\ d^2\ \rho_s}{\mu},\ \frac{k}{d^2},\ \frac{\theta}{\dot{\gamma}^2 d^2},\ \frac{E}{\dot{\gamma}^2\ d^2\ \rho_s},\ C_s,\ C_f,\ \phi,\ e\right), \tag{8.86}$$

where the 7 dimensionless groups and the 4 already-dimensionless variables appear.

The first dimensionless group in parentheses is the relative density of the sediments with respect to the liquid. The second dimensionless group is the Savage number, representing the role of gravity in granular dynamics. To take into account the presence of water in the pores, the Savage number can be expressed as

$$\text{Sa} = \frac{\rho_s\ \dot{\gamma}^2\ d^2}{(\rho_s - \rho_f)\ g\ h\ \tan\phi}. \tag{8.87}$$

In this last form, the Savage number is the ratio between the collisional stresses and the quasi-static stresses (frictional) due to the action of gravity in a flow field with a free surface in a uniform regime where h is the current depth, with a high concentration of sediments.

The third dimensionless group is the Bagnold number, usually expressed as

$$\text{Ba} = \frac{C_s\ \rho_s\ d^2\ \dot{\gamma}}{(1 - C_s)\ \mu}, \tag{8.88}$$

and which is the ratio between collisional and viscous stresses.

The fourth dimensionless group k/d^2 is representative of the role that the packing and grain size play in the liquid-solid interaction.

The fifth group is the granular temperature dimensionless with respect to the temperature source, that is, the strain rate.

The sixth dimensionless group is the ratio between the bulk modulus of the liquid-sediment mixture and the collisional stresses.

Other possible dimensionless groups derive from the relationship between the variables representative of different behaviours of the mixture. For example, the inertial behaviour of the mixture is described by the weighted average of the inertia of the granular component and the inertia of the liquid component (the inertia of the gas component, if the mixture is unsaturated, is negligible).

We define *mass number* as the ratio between the inertia of the grains and of the liquid:

$$N_{mass} = \frac{C_s}{1 - C_s} \frac{\rho_s}{\rho_f}. \tag{8.89}$$

This expression is valid for a saturated mixture, with a correction for an unsaturated mixture.

Other dimensionless groups of specific interest can be defined. For example, we consider the sediment-fluid interaction, with an inertial component and a quasi-static component due to viscosity. In general, the latter component is dominant, and we can write

$$\sigma_{s-f} \propto \frac{\dot{\gamma} \mu d^2}{k}. \tag{8.90}$$

The collisional component of the normal stress for sediments is

$$\sigma_c \propto C_s \rho_s \dot{\gamma}^2 d^2. \tag{8.91}$$

The ratio between the stress component of the interaction between the two phases and the collision component of the stress in the solid phase is defined as the Darcy number and is equal to

$$Da = \frac{\mu}{C_s \rho_s \dot{\gamma} k}. \tag{8.92}$$

From the ratio between Darcy number and the dimensionless group involving the bulk modulus of the mixture, we obtain a dimensionless group equal to the ratio between the time scale of pressure reduction by diffusion through the small channels of the porous medium and the time scale of pressure generation (equal to $1/\dot{\gamma}$):

$$\frac{\mu \dot{\gamma} d^2}{C_s k E}. \tag{8.93}$$

The ratio of the Bagnold number to the mass number is the Reynolds number of sediments:

$$Re_d = \frac{Ba}{N_{mass}} = \frac{\rho_f \dot{\gamma} d^2}{\mu}. \tag{8.94}$$

Fig. 8.13 Normal and
tangential stress for granular
mixtures (Bagnold 1954)
(modified from Fredsøe and
Deigaard 1992)

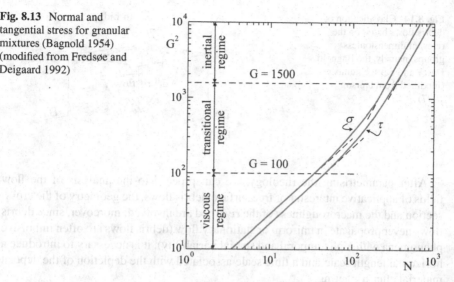

The ratio of the Bagnold number to the Savage number is the *frictional stress number*

$$N_{\text{frict}} = \frac{\text{Ba}}{\text{Sa}} = \frac{C_s}{1 - C_s} \frac{(\rho_s - \rho_f) g h \tan \phi}{\dot{\gamma} \mu}, \tag{8.95}$$

and represents the ratio between the stress component generated by the continuous contact of the grains and the stress component of viscous shear.

A great contribution to the knowledge of the rheology of granular mixtures is due to Bagnold (1954). His experiments led to the definition of an empirical equation correlating the tangential and normal stresses and the strain rate for granular mixtures in water:

$$\begin{cases} G = 0.114\,N & N > 450, \\ G = 1.483\,\sqrt{N} & N < 40, \end{cases} \tag{8.96}$$

where $G = \dfrac{d}{\nu}\sqrt{\dfrac{\tau}{\tilde{\lambda}}\dfrac{s}{\rho_w}}$, $N = \dfrac{\sqrt{\tilde{\lambda}}\,s\,d^2}{\nu}\,\dot{\gamma}$. The symbol $\tilde{\lambda}$ is the linear concentration of sediment, defined as

$$C_s = \frac{C_s^*}{(1 + 1/\tilde{\lambda})^3}, \tag{8.97}$$

where C_s^* is the maximum packing concentration. The experimental diagrams for normal and tangential stresses are shown in Fig. 8.13. Luckily, not all dimensionless groups are relevant at the same time, and the classification of a two-phase flow, with liquid and sediment, can be based on the range of variation of the groups and on the nature of the relevant groups. For example, Fig. 8.14 shows the classification of debris flows according to Iverson (1997).

Fig. 8.14 Classification of debris flows based on the relevant dimensionless groups, namely, the Bagnold, Darcy and Savage numbers (modified from Iverson 1997)

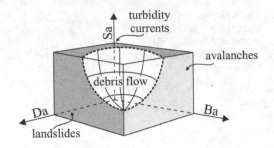

After parametrising the rheology, we can proceed to the analysis of the flow fields of applicative interest. For free surface debris flows, the geometry of the cross-section and the macroroughness of the riverbed are involved; moreover, since debris flows never propagate in uniform or stationary flow (debris flows are often impulsive phenomena with high temporal and spatial variability), it is necessary to introduce a horizontal length scale and a time scale associated with the depletion of the deposit material after triggering.

Example 8.7 To arrest a channelled debris flow, which propagates in channels incised by previous flows, metallic networks anchored in control sections are sometimes adopted, built in such a way as to dissipate the energy of the impact. Dissipation is essentially due to the plastic deformation of brakes, which are metal devices designed and installed for this purpose, although it is initially due to the change in mesh geometry and, to a lesser extent and in the subsequent phase, to the plastic deformation of the metal rod of the net.

The variables that characterise the current are the average density ρ_m, the average velocity V of the current of debris, the flow depth h and the width b of the riverbed. The variables that characterise the net are the Young's modulus of the material E and the yield strength σ_y of the net material and brakes, which is determined by the net mesh shape.

We distinguish between a first phase, in which the deformation controlled by the shape of the links dominates, and a second phase, in which plastic deformation, especially of the brakes, dominates. If we are interested in the maximum deformation δ of the net in the direction of the current, occurring in the second phase, the typical equation is

$$\delta = f\left(\rho_m, \ V, \ h, \ b, \ E, \ \sigma_y, \ \text{shape}\right). \tag{8.98}$$

The dimensional matrix of the 8 variables

	δ	ρ_m	V	h	b	E	σ_y	shape
M	0	1	0	0	0	1	1	0
L	1	−3	1	1	1	−1	−1	0
T	0	0	−1	0	0	−2	−2	0

$$\tag{8.99}$$

has rank 3, and applying Buckingham's Theorem, we can re-write the typical equation as a function of only $(8 - 3) = 5$ dimensionless groups, for example:

$$\Pi_1 = \frac{\delta}{b}, \ \Pi_2 = \frac{h}{b}, \ \Pi_3 = \frac{\sigma_y}{\rho_m V^2}, \ \Pi_4 = \frac{E}{\rho_m V^2}, \ \Pi_5 = \text{shape}. \qquad (8.100)$$

The conditions of similarity are

$$\begin{cases} r_\delta = r_b, \\ \lambda = r_b, \\ r_{\sigma_y} = r_{\rho_m} r_V^2, \\ r_E = r_{\rho_m} r_V^2, \end{cases} \qquad (8.101)$$

with the same mesh shape of the net in the model and in the prototype. The velocity scale from the Froude similarity of the current is $r_V = \sqrt{\lambda}$. Thus, if the material of the debris flow is the same in the model and in the prototype, the network must have a Young's modulus and yield strength that scale with λ.

In the second phase of plastic deformation, we can assume that the arrest process is so rapid as to render all other dissipative phenomena negligible. In this condition, it is necessary to guarantee that the kinetic energy per unit volume is the same in the model and in the prototype, with the consequent condition that the scale of the velocity is equal to the geometric scale (see Sect. 7.4.2), that is:

$$\frac{V_m}{V_p} = \lambda. \qquad (8.102)$$

This condition contrasts with the condition imposed by the Froude similarity of the current, and biases the deformation of the net and of the brakes with respect to λ. The distortion is in the sense of an excessive dissipation in the model compared to the prototype, with deformations of the net and of the brakes in the model that underestimate the real deformations. This aspect should be considered when rescaling measurements in the model to data in the prototype.

The arrest process is also favoured by the leakage of interstitial fluid from the current, which leads to an increase in frictional stresses compared to collision stresses, and the Savage number is reduced. This results in intense dissipation, which is added to the dissipation due to plastic deformation of the net and brakes. The similarity of Savage requires that

$$r_{\dot{\gamma}}^2 \, r_d^2 = r_s \, \lambda \, r_\phi, \qquad (8.103)$$

where $\dot{\gamma}$ is the average strain rate of the current, d is the diameter of the sediment, s is the specific gravity of the sediment, and ϕ is the internal friction angle of the granular mixture. If the same material is used in the model and in the prototype, with the diameter of the sediment on a geometric scale, since $\dot{\gamma} \approx V/h$, the condition of similarity of Savage becomes $r_V = \sqrt{\lambda}$ and coincides with the condition resulting from the similarity of Froude.

8.4.1   The Physical Process of Cliff Recession

Cliffs are steep rock faces in coastal areas, potentially subject to the action of waves (see Fig. 8.15). We wish to study, with the methods of dimensional analysis, their recess due to erosion of the foot.

The collapse of cliffs is a physical process in which the forcing is represented by the incident waves that, breaking on the face of the cliffs or rising the beach as bores after breaking in the surf zone, induce excavation that leads to the collapse of the overlying structure; see the schematic in Fig. 8.16.

The excavation provides loose material that is deposited forming a bar and then is redistributed with a beach profile that facilitates the rising of the bores. The system achieves equilibrium if solid transport along the coast does not displace the eroded material, but in most cases, long-shore currents mobilise the sediments, favouring further excavation. The initial cavern becomes progressively more widely engraved until it causes the collapse of the overlying structure and the consequent recession of the coastline. The physical process is three-dimensional, with foot erosion that does not occur along the entire front of the cliff but in local intermittent collapses.

We can assume that the speed of recession $R = l/t$ depends on the shear strength τ_s of the rock material, the intensity q_s of the solid transport along the coast (volume flow of sediment per unit of width), the density of the sediment ρ_s and of water ρ, the

Fig. 8.15 Cliffs in Ireland (courtesy of Helfrich 2004, https://upload.wikimedia.org/wikipedia/commons/d/d8/Ireland_cliffs_of_moher2.jpg)

height h of the cliff, the representative diameter of the sediment d, the acceleration of gravity g and the erosive action. This erosive action is proportional to the wave height, which depends on the parameter of surf-similarity ξ, defined as

$$\xi = \frac{\tan \beta}{\sqrt{H/l}}, \tag{8.104}$$

where β is the slope of the beach, H is the height of the wave, l is the length of the wave and H/l is the steepness of the wave.

The typical equation is

$$R = f\left(\tau_s,\ q_s,\ \rho_s,\ \rho_f,\ h,\ d,\ g,\ \xi\right). \tag{8.105}$$

The dimensional matrix of the 9 variables is

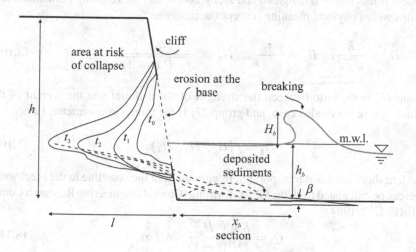

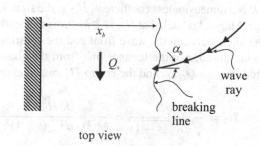

Fig. 8.16 Time evolution of excavation at the foot of a cliff

$$
\begin{array}{c|ccccccccc}
 & R & \tau_s & q_s & \rho_s & \rho & h & d & g & \xi \\
\hline
M & 0 & 1 & 0 & 1 & 1 & 0 & 0 & 0 & 0 \\
L & 1 & -1 & 2 & -3 & -3 & 1 & 1 & 1 & 0 \\
T & -1 & -2 & -1 & 0 & 0 & 0 & 0 & -2 & 0
\end{array}
\tag{8.106}
$$

and has rank 3. Selecting ρ_s, g and d as fundamental variables, the 6 following dimensionless groups are calculated:

$$
\Pi_1 = \frac{R}{\sqrt{g\,d}}, \quad \Pi_2 = \frac{\tau_s}{\rho_s\,g\,d}, \quad \Pi_3 = \frac{q_s}{\sqrt{g\,d^3}},
$$

$$
\Pi_4 = \frac{\rho}{\rho_s}, \quad \Pi_5 = \frac{h}{d}, \quad \Pi_6 = \xi. \tag{8.107}
$$

The experiments indicate that the number of groups needed to describe the physical process is less than 6 (Damgaard and Dong 2004), and the relevant 4 dimensionless groups with a physical meaning (except the first) are:

$$
\Pi_1' = \frac{R}{\sqrt{g\,d}}, \quad \Pi_2' = \frac{\tau_s}{\rho_s\,g\,h}, \quad \Pi_3' = \frac{q_s}{\sqrt{g\,d^3\,(s-1)}}, \quad \Pi_4' = \xi. \tag{8.108}
$$

Group Π_2' is the ratio between the strength of the material and the weight of the volume above the eroded cave, and group Π_3' is the transport parameter. Hence,

$$
\Pi_1' = \tilde{f}\left(\Pi_2',\ \Pi_3',\ \Pi_4'\right). \tag{8.109}
$$

The longshore sediment transport Q_s, integrated from the coastline to the breakwater line, can be calculated from the CERC formula (Coastal Engineering Research Center (CERC, US), 1984):

$$
Q_s = \frac{K\,g^{1/2}\,H_{sb}^{5/2}}{\xi^{1/2}(s-1)}\,\sin 2\alpha_b, \tag{8.110}
$$

where K is a dimensionless coefficient, H_{sb} is the significant breaking wave height, ξ is the breaking index such that $H_{sb} = \xi\,h_b$, h_b is the depth where breaking occurs, and α_b is the angle between the wave front and the coastline at breaking. If we indicate with x_b the distance of the breaker line from the coast, the sediment transport per unit width is $q_s = Q_s/x_b$, and the group Π_3' can be expressed as

$$
\Pi_3' = \frac{q_s}{\sqrt{g\,d^3\,(s-1)}} = \frac{K\,H_{sb}^{5/2}}{\xi^{1/2}x_b\,d^{3/2}\,(s-1)^{3/2}}\,\sin 2\alpha_b. \tag{8.111}
$$

The breaking depth is $h_b = x_b\,\tan\beta$ and thus results in

$$
\Pi_3' = \frac{K\,\xi^{1/2}}{(s-1)^{3/2}}\left(\frac{H_{sb}}{d}\right)^{3/2}\tan\beta\,\sin 2\alpha_b. \tag{8.112}
$$

The condition of similarity requires that dimensionless groups have the same value in the model and in the prototype. Assuming that $r_g = r_\beta = r_{\alpha_b} = 1$, the following system of equations for the unknown scales must be satisfied:

$$\begin{cases} r_R = \lambda^{1/2}, \\ r_{\tau_s} = r_{\rho_s} \lambda, \\ r_d^{3/2} r_{(s-1)}^{3/2} = r_K r_b \lambda^{3/2}, \\ r_\xi = 1. \end{cases} \quad (8.113)$$

If the material of the model has the same specific weight of the material used in the prototype, it is necessary that its resistance varies as the geometric scale, $r_{\tau_s} = \lambda$. Assuming that $r_b = r_K = 1$ and $r_{(s-1)} = 1$, the diameter scales with λ. This last condition is difficult to achieve if the geometric scale is too small. For this reason, the material in the model is usually the same as that of the prototype, accepting the corresponding scale effect, and the slope of the beach β, the breaking index ξ and the parameter K do not vary, with a unity scale ratio.

Summarising Concepts

- In geotechnical applications, it is advantageous to modulate mass forces by making physical models in a centrifuge. As always, scale effects are involved. In addition, the distortion of the acceleration field with respect to the gravity field must also be considered.
- If the physical process involves the structural characteristics of the soil and the flow in porous media, it is necessary to fix the scales so that all scaling times are equal.
- The simulation of tectonic processes takes advantage of the combined use of granular material and fluids to reproduce complex structures.

References

Bagnold, R. A. (1954). Experiments on a gravity-free dispersion of large solid spheres in a Newtonian fluid under shear. *Proceedings of the Royal Society London A, 225*(1160), 49–63.

Bairrao, R., & Vaz, C. (2000). Shaking table testing of civil engineering structures–the LNEC 3D simulator experience. In *Proceedings of the 12th World Conference on Earthquake Engineering. Auckland, New Zealand* (p. 2129).

Bouc, R. (1967). Forced vibrations of mechanical systems with hysteresis. In *Proceedings of the Fourth Conference on Nonlinear Oscillations, Prague, 1967.*

Bucky, P. (1931). The use of models for the study of mining problems. *Technical Publication 425*, Class A, Mining methods no. 44. American Institute of Mining and Metallurgical Engineers.

Coastal Engineering Research Center (CERC, US). (1984). *Shore Protection Manual* (Vol. I–II). United States Army Corps of Engineers.

Corti, G., Bonini, M., Mazzarini, F., Boccaletti, M., Innocenti, F., Manetti, P., Mulugeta, G., & Sokoutis, D. (2002). Magma-induced strain localization in centrifuge models of transfer zones. *Tectonophysics, 348*(4), 205–218.

Damgaard, J. S., & Dong, P. (2004). Soft cliff recession under oblique waves: Physical model tests. *Journal of Waterway, Port, Coastal, and Ocean Engineering, 130*(5), 234–242.

Fredsøe, J., & Deigaard, R. (1992). *Mechanics of coastal sediment transport* (Vol. 3). World Scientific Publishing Co. Pte. Ltd.

Iverson, R. M. (1997). The physics of debris flows. *Reviews of Geophysics, 35*(3), 245–296.

Longo, S., & Lamberti, A. (2000). Granular streams rheology and mechanics. *Physics and Chemistry of the Earth, Part B: Hydrology, Oceans and Atmosphere, 25*(4), 375–380.

Ramberg, H., & Stephansson, O. (1965). Note on centrifuged models of excavations in rocks. *Tectonophysics, 2*(4), 281–298.

Taylor, R. N. (1995). Centrifuges in modelling: Principles and scale effects. In R. N. Taylor (Ed.), *Geotechnical Centrifuge Technology* (19–33). Blackie Academic and Professional.

Terzaghi, K. (1943). *Theoretical Soil Mechanics*. Wiley.

Weijermars, R., & Schmeling, H. (1986). Scaling of Newtonian and non-Newtonian fluid dynamics without inertia for quantitative modelling of rock flow due to gravity (including the concept of rheological similarity). *Physics of the Earth and Planetary Interiors, 43*(4), 316–330.

Wen, Y.-K. (1976). Method for random vibration of hysteretic systems. *Journal of the Engineering Mechanics Division, 102*(2), 249–263.

Zhang, J., & Tang, Y. (2008). Dimensional analysis of soil-foundation-structure system subjected to near fault ground motions. In *Geotechnical Earthquake Engineering and Soil Dynamics Congress IV* (1–10).

Chapter 9
Applications in Wind Tunnel Technology

In recent decades, the applications of low-speed aerodynamics have increased, with the consequent need for extensive experimental tests for the development of vehicles, drones, and wind generators. Fifty years ago, many researchers predicted that computational fluid dynamics would replace experiments due to the availability of large computational capabilities at low cost. They were wrong, and physical experiments are still the only tool for comprehensive and effective planning of many devices. Turbulence still causes problems, and basic methods and theory have remained unchanged for several decades, without the quick rise in computational resources necessary for reducing time and costs. These experiments are routinely and extensively conducted in wind tunnels.

A broad classification of wind tunnels is in high-speed (M > 0.4) and low-speed wind tunnels (M < 0.4), where M is the Mach number, with a relatively recent new branch devoted to environmental wind tunnels, with a large cross-section, high Reynolds number and low Mach number, and devoted to modelling environmental flows.

There are also supersonic wind tunnels, up to Mach 5, hypersonic wind tunnels, up to Mach 15, which require heating of the flow. These are also special devices, such as the Ludwieg tube, a pipe with a converging/diverging nozzle at the end; the Ludwieg tube is operated through many cycles of expansion-wave reflection within the driver, thus providing a run time of a few seconds up to Mach 4 without heating.

Several details on the wind tunnel technology can be found in Barlow et al. (1999). Here we mainly focus on low-speed wind tunnels.

© The Author(s), under exclusive license to Springer Nature Switzerland AG 2021
S. G. Longo, *Principles and Applications of Dimensional Analysis and Similarity*,
Mathematical Engineering, https://doi.org/10.1007/978-3-030-79217-6_9

9.1 Classification of Wind Tunnels

Wind tunnels are divided into open-circuit and closed-circuit tunnels. In an open-circuit tunnel, the air follows a straight path from the inlet and passes through a contraction until it reaches the test section, followed by a diffuser and fan section before being released into the ambient environment. The flow can be free, as in the Eiffel tunnel, or confined between solid walls, as in most wind tunnels. Figure 9.1 shows an open-circuit tunnel at IISTA, Granada.

In a closed-circuit tunnel (of the Prandtl type), a circuit allows the current to be recirculated, since a service circuit is provided that injects the current back upstream of the test section. In most cases in the test section, the flow is confined, although for multiple applications, some wind tunnels provide the possibility of dual operation, confined or free flow. Figure 9.2 shows a closed-circuit wind tunnel at CET, Prague, with two different test sections. The choice of tunnel type is largely dependent on the

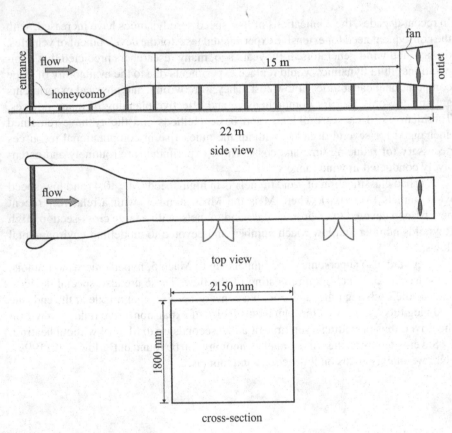

Fig. 9.1 Boundary layer open-circuit wind tunnel in the laboratory of Environmental Hydraulics at the Andalusian Institute for Earth System Research IISTA, Granada (modified from Jiménez-Portaz et al. 2020)

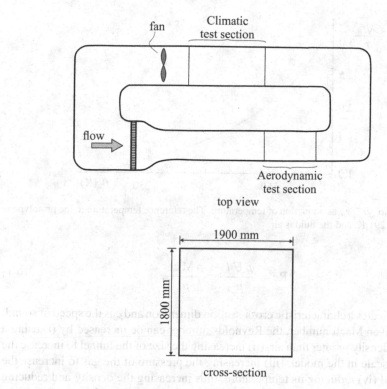

Fig. 9.2 Closed-circuit climatic boundary layer wind tunnel at the Center of Excellence Telč (CET) of the Institute of Theoretical and Applied Mechanics (ITAM), Prague, with a straight section 11 m long (modified from Jiménez-Portaz et al. 2020)

objectives and the availability of space and funds, with a number of advantages and disadvantages in the two schemes. The size of the measurement section is particularly important. In theory, it should be large enough to accommodate the model even at full scale, but for some models, such as aircraft, the size would be impractical. In any case, it is essential that the Reynolds number of the flow lies in the characteristic range of the real flow field to be simulated.

9.2 Aeronautical and Automobile Wind Tunnels

Most wind tunnels are designed and built for aeronautical purposes. Since it is difficult to handle tunnels with very large test sections, it may be convenient to limit the geometric scale by increasing the Reynolds number. The estimated maximum Reynolds number of a wind tunnel can be expressed as

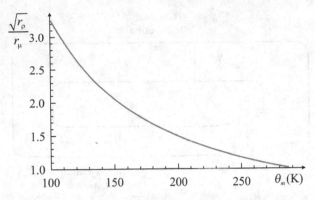

Fig. 9.3 Ratio $\sqrt{r_\rho}/r_\mu$ as a function of temperature. The reference temperature in the prototype is equal to $\theta = 291$ K, and the fluid is air

$$\mathrm{Re} \equiv \frac{\rho\, U\, l}{\mu} \equiv \frac{\rho\, \mathrm{M}\, c\, l}{\mu}, \qquad (9.1)$$

where l indicates a characteristic cross-section dimension and c is the speed of sound. With the given Mach number, the Reynolds number can be increased by (i) using a gas with a density greater than air, (ii) increasing the size of the tunnel to increase the geometric scale in the model, (iii) increasing the pressure of the gas to increase the density, and (iv) reducing its temperature, thus increasing the density and reducing the dynamic viscosity. Freon increases the Reynolds number by a factor of 3.6; in the past, Freon was not considered a threat to the environment, but today, it is and has been decommissioned almost everywhere. Increasing the size of the tunnels is quite expensive during building and for maintenance and operations. A pressurised tunnel at 1 MPa allows an increase in the Reynolds number of 10 compared to a tunnel with ambient pressure. The reduction in temperature increases the density and reduces the speed of sound; however, it is proportionally less than the kinematic viscosity μ/ρ, see Fig. 9.3.

Of all the solutions, this last one is the most advantageous, possibly coupled with pressure raising.

The operating costs of these tunnels, named *cryogenic wind tunnels*, are very high because steady-state conditions are reached after quite some time. For energy-saving reasons, these are closed-circuit tunnels that almost always operate by direct injection into the stream of liquefied gases (e.g., nitrogen), reaching temperatures below 150 K. The minimum operating temperature must be such that condensation is avoided in areas where the pressure has a minimum value.

The installed power, which is proportional to $P \propto Q\, U = Q\, \mathrm{M}\, c$, is reduced, as is the dynamic pressure:

$$p = \frac{\rho\, U^2}{2} \equiv \frac{\rho\, \mathrm{M}^2\, c^2}{2}. \qquad (9.2)$$

The reduction in dynamic pressure results in less stress in the model.

For the future, tunnels with a Reynolds number up to 10^8 and a Mach number just under 1 will be indispensable. The reduction in scale is due not only to economic reasons (for aircraft models) but also to the need to reduce energy costs, which are particularly high in the transonic regime.

For some very specific applications, such as modelling vertical or short take-off aircraft, (V/STOL, vertical/short take-off and landing) wind tunnels are characterised by some important variations, since a very large test section with limited wind speeds, up to 200 km h^{-1}, is required. A variant is represented by tunnels that allow free flight, in which the action of the air current is combined with the action of gravity. These tunnels, which present some level of complexity, provide data to be treated with extreme care since the Reynolds number is low and the results require attention in the extrapolation step.

Other variants are vertical tunnels and tunnels customised for specific purposes, such as the study of stability, with the presence of a series of blades able to generate swirling. Low-turbulence tunnels have a diffuser with a large contraction ratio and several screens to dampen turbulence.

Automobile wind tunnels are often full scale, although it is sometimes convenient to use a reduced scale, 0.5, but a sufficiently large Reynolds number. These tunnels can have temperature control to simulate environmental conditions, but the most relevant problem is the effect of the ground, which should move with the air stream velocity. In most cases, specific tools allow the generation of a boundary layer on fixed ground.

In water tunnels, the fluid is water, which allows a larger Reynolds number than air since the kinematic viscosity of water is an order of magnitude smaller than the air kinematic viscosity.

9.2.1 Environmental Wind Tunnels

Environmental wind tunnels are generally sized to guarantee a maximum speed in the test section of a few tens of meters per second. The suppression of edge effects and the need to test physical models at the largest possible scale require a suitably large measurement section. The requirements to be met are

- a high 'quality' of the generated flux, which depends on all the components of the tunnel and requires adequate technical arrangements, with the construction of screens to orient the current lines, diffusers, honeycombs to straighten the fluid trajectories and to suppress macrovortices;
- a reduced energy consumption. The installed power and the energy required for operation are directly proportional to the size of the test section and the length of the tunnel;
- the economy of the project. The size of all tunnel construction elements and the installation and operating costs are directly dependent on the size of the test section.

The construction elements required to ensure flow quality also aim to minimise pressure losses to reduce installation and operating costs. In wind tunnels, turbulence is not completely suppressed, but unless there are specific requirements, it is reduced to a minimum, with a turbulence index generally less than 5%. For some activities, the turbulence level is controlled with the adoption of roughness at the walls, spikes and similar devices.

The following requirements must be met to realise physical models in environmental tunnels:

– choice of an appropriate geometric scale;
– equal Reynolds number in the model (small scale) and in the prototype (full scale);
– equal Rossby number in the model and in the prototype;
– kinematic similarity of airflow, boundary layer velocity and turbulence;
– zero pressure gradient, as in the prototype.

The effects of the Reynolds number are generally modest, since the presence of sharp edges in most structures favours fully developed turbulence and consequent (asymptotic) viscosity independence. Some tests should be performed at increasing speeds to check the actual independence from Re, with Re calculated with a scale length equal to the width of the building. The Rossby number quantifies the effect of the Earth's rotation on the planetary winds. The Earth's rotation results in a change in wind direction of approximately $5''$ every 2 km, which is very modest and would be difficult to reproduce in the model if it were actually necessary. In the great majority of cases, the scale dimensions of the prototype are small enough to neglect Rossby similarity. The velocity distributions in the natural boundary layer should be reproduced as accurately as possible in the model. For example, at a geometric scale of 1/150, a building 90 m high in real life is 60 cm high in the model. The boundary layer should have a scale thickness of at least 90 cm and should ideally extend up to the ceiling of the test section. The velocity distribution of the boundary layer and the turbulence can be adequately reproduced by means of spires in the converging entrance section, with the addition of an artificial roughness (often realised with small cubic elements on the floor of the tunnel) covering a length of approximately 10–15 heights of the measurement section. Figure 9.4 shows the test section of the environmental wind tunnel at Laboratorio de Dinámica de Flujos Ambientales, Instituto Interuniversitario de Investigación del Sistema Tierra en Andalucía, Universidad de Granada.

The buildings or structures to be tested are usually placed on a turntable to allow testing with wind coming from different directions. As a rule, the boundary layer is not calibrated according to the direction of the wind (assuming that the characteristics of the environment in real life are isotropic) unless it is strictly necessary, as in the case of a structure facing a lake or the sea. In this case, if the wind blows over the lake or sea surface, it is necessary to reproduce a marine boundary layer with different characteristics than the boundary layer on land. The strong pressure gradient commonly found in a wind tunnel, exacerbated by the high thickness of the reproduced boundary layer, can be mitigated by a moving ceiling in the test section, suitably adjusted to ensure local expansion of the current. While in some experiments

Fig. 9.4 Section of tests of the environmental wind tunnel at IISTA, Granada. White blocks and lock nuts are necessary to generate the desired structure of the boundary layer

for pollutant diffusion models it is necessary to provide cooled or heated air and a free surface of adequate extent in the test section, in other experiments requiring the measurement of pressure, forces, vibrations, with sufficiently high wind speeds it is not necessary to reproduce temperature gradients which, even in real life, do not occur due to turbulence mixing.

9.2.1.1 Static Loads and Associated Tests on Buildings

Wind engineers can be asked to contribute to the improvement of existing structures or to the optimisation of structures at the design stage. Improvement includes eliminating swaying phenomena for very tall buildings, checking the strength of cladding, roofing, reducing noise, and reducing smoke emissions into the ventilation system.

Support in the design phase is often more difficult because of the designer's attitude to defend his project regardless of the plant requirements. Physical models are a useful tool for getting people to accept solutions that differ from the original solution, making the advantages of alternative solutions obvious. A typical wind tunnel test programme includes:

- Preliminary tests with tracers (smoke) to identify problematic sections where pressure taps should be installed.

- Tests with wind and structures in static conditions, which can be a useful guide for dynamic tests.
- Static wind loads, which can lead to dynamic experiments later.
- Studies on ventilation intake in the presence of smoke emitted from neighbouring factories or from the same building (higher wind speed at the top of the building results in higher dynamic pressure with unpredictable effects on smoke recirculation).
- Identification of areas with high local wind speeds that could cause problems for people.
- Studies with tracers to highlight the opportunity for changes in the structure.

9.2.1.2 Dynamic Loads on Buildings

In the presence of wind, buildings with an approximate ratio of more than 6 between the height and horizontal dimensions are subject to accelerations that are disturbing to users. There are two possible approaches:

- The structural engineer of the building knows the natural frequencies of the building, but not the frequencies of the forcing. Tests with a very rigid model on a scale with transducers with a high response frequency make it possible to estimate the unknown frequencies and to reinforce the structure to dampen the corresponding actions.
- An aeroelastic model can be made by reproducing the deformability characteristics of the prototype and measuring the accelerations and displacements resulting from the action of the wind at varying speeds and directions. The results are affected by some significant uncertainties, such as those associated with the dynamic characteristics of the prototype to be reproduced in the model (complete similarity is, in fact, unfeasible).

9.2.1.3 Study of Aerodynamic Instabilities

Wind can promote and induce structural oscillations and vibrations of various kinds:

Simple oscillations
All structures are characterised by a natural frequency spectrum that leads to an amplification of oscillations unless they are overdamped. The natural frequencies of many environmental or street furniture elements (trees, road signs, etc.) are very close to the frequencies of wind bursts, and a sequence of bursts can cause the structure to collapse. This is a response to a force with a frequency close to the resonance frequency of the structural element.

Transverse vibrations
Slender, simply wedged structures, such as chimneys and towers, tend to oscillate in the direction normal to the wind as a result of the Kutta-Jukowsky force associated

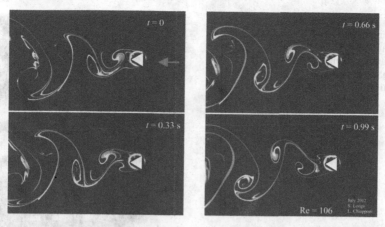

Fig. 9.5 Vortex shedding from a triangular cross-section rod

with the presence of vorticity in the flow field. The frequency of the forcing, equal to the frequency of detachment of the vortices, depends on the transverse dimension of the structural element and the average wind speed, moderately on the shape. The most representative dimensionless group is the Strouhal number St $= fd/V$, where f is the vortex shedding frequency, d is the scaled transverse dimension of the structural element and V is the average wind speed. In a wide range of conditions, a critical value St $= 0.2$ arises. This results in a forcing frequency of $f = 0.2V/d$, which may coincide with one of the resonance frequencies of the structure. Figure 9.5 shows vortex shedding in a water flume, and the flow is leftward directed.

Galloping and breathing
Another type of oscillation occurs when the lift curve of a body has a negative slope. Under certain conditions, rapid oscillations occur, which must be eliminated by changing the structural design or with other devices.

Sometimes large-diameter piles and other structures undergo deformation at the natural frequency, with such large deformations that the flow field is strongly affected. This phenomenon, called breathing, results in a nonlinear coupling between the structure and the wind field, which can limit the usability of the structure and can lead to collapse.

9.2.1.4 Study of Windbreaks for Agricultural/Civil Use and Application of Wind Tunnels in Agriculture

Windbreaks can be used to reduce loads on structures, to mitigate heat consumption in winter by reducing convective exchange, and to increase the yield of certain crops that do not like the action of the wind. Figure 9.6 shows the layout of a series of

Fig. 9.6 Section of tests of
the environmental wind
tunnel at IISTA, Granada.
Tree planting geometry is
investigated to optimise
pollination

experiments aiming to optimise the pollination of olive trees (Jiménez-Portaz et al.
2021).

Pollination is almost always due to the action of the wind and therefore depends on
the wind field modulated by the distribution of the plants. The planting of fruit trees
should be optimised according to the intensity and direction of the winds, favouring
flows in the case of areas with limited wind and reducing them in the case of intense
prevailing winds.

Soil erosion by wind causes a reduction in the vegetation layer. Road engineers,
on the other hand, would like to avoid roads in arid areas, with large amounts of
sediment available, being invaded by sand.

9.2.1.5   Dispersion of Pollutants

For pollutants released into the atmosphere, it is important not only to know where
they will be transported but also what distance from the source dispersion reduces
their concentration to values that render them harmless. The range of pollutants in the
air is particularly wide: industrial fumes, unrecovered gases emitted by chemical processes, and leaks from waste holding tanks at nuclear power stations. In other cases,
the dispersion of some substances or particles needs to be controlled: for example, it
is of practical interest to quantify the action of wind to disperse silver iodide particles

to facilitate rainfall; to quantify the distance from the LPG spill area needed to reduce its concentration to low enough to make the mixture noncombustible; to quantify the distance from the source needed to reduce the concentration of sulphur-based gases leaving geothermal power stations; and to estimate the dispersion efficiency of pesticides in crops. Ludwig and Skinner (1976), Skinner and Ludwig (1978) made a major contribution to wind tunnel dispersion testing methods, showing that (1) the dominant feature of plume mixing is turbulent diffusion and (2) viscous effects are negligible. There is a critical Reynolds number below which a viscous underlayer of excessive and intolerable thickness develops in the tunnel. If the tests are carried out with wind speeds such that a Reynolds number much higher than the critical value is induced, it is possible to implement very small geometric scale models by increasing the extension of the real domain of investigation, the more so as the average speed of the current increases. Furthermore, provided that the flow is in a sufficiently developed turbulent regime, it is possible to obtain the same dispersion in the model as is obtained on the prototype even by using cold gases (characterised by a higher density than the hot polluting gases), provided that the increase in the role of the body forces in the model, compared to the prototype, is compensated for by an increase in the tunnel speed. In addition to the advantages of physical modelling in wind tunnels, it should be remembered that all countries have guidelines and regulations on the dispersion of pollutants, which are sometimes mandatory.

9.3 Scale Effects in Wind Tunnels

The correct selection of the model scale is a relevant issue for obtaining reliable indications from wind tunnel experiments. For instance, for environmental wind tunnels, large errors in the modelling of the atmospheric boundary layer may result if the model scale is not properly selected. The relevant dimensionless groups are the Strouhal number $St = \delta/(Vt)$, the Euler number $Eu = p/(\rho V^2)$, the Reynolds number $Re = V\delta/\nu$, the Rossby number $Ro = V/(f\delta)$ and the Froude number $Fr = V^2/(g\delta)$, where δ is the thickness of the boundary layer, V is a reference velocity, t is a reference time, p is a reference pressure, g is the gravitational acceleration, ρ and ν are the density and the kinematic viscosity of air, respectively, and $f = 2\Omega \sin \theta$ is the Coriolis parameter with Ω being the rotational rate of the Earth and θ being the local latitude. Since a complete similarity cannot be achieved, due to some constraints on the nature of the fluid in the model and on g, scale effects are expected. The most relevant scale to be selected is λ_δ, the geometric scale of the boundary layer thickness. In most cases, the wind velocity profiles near the ground follow a logarithmic formula:

$$\frac{u}{u_*} = \frac{1}{k} \ln \frac{z - d_0}{z_0}, \tag{9.3}$$

where u is the mean velocity, u_* is the friction velocity, d_0 is the zero-plane displacement and z_0 is the roughness height. Defining the Jensen number as $Je = \delta/z_0$

and imposing an equal Jensen number in the model and in the prototype results in $\lambda_{z_0} = \lambda_\delta$. There are also several data sets retrieved in the field and in the laboratory indicating that $z_0 = 0.08K_s$, where K_s is the real roughness height (Wang et al. 1996); hence, $\lambda_{K_s} = \lambda_\delta$, indicating that the scale ratio of the roughness height should be the same as the scale ratio of the boundary layer. It also results that the velocity profile is unaffected by the distribution of the roughness elements for $z > 2K_s$; otherwise, the profile depends on the geometry of the elements and on their distribution. Other observations indicate that the best similarity of the velocity profile and turbulence intensity in the wind tunnel can be obtained with scales $1/200 - 400$ and $1/200 - 600$, respectively. As long as the roughness height is appropriately modelled, accurate results of the pressure coefficient on the building surface can be obtained even if the scale ratio of the body size is not correctly selected. The scale of the diffusion coefficient is the same as the scale of the boundary layer thickness, and the time scale for diffusion is $1/\lambda_\delta$.

9.4 Models for Multiphase Flows: An Application to Wind Waves

Similarity criteria can be difficult to meet, even when the process is well defined and the forces at play are unambiguous. In some processes, such as multiphase processes, which involve possible chemical reactions with a transition between solid, liquid and gas (or vapour) phases, the similarity criteria become increasingly complex. It is obvious that complete similarity, which is almost impossible to achieve in simple cases, is not achieved in practise in complex processes, and it is even more necessary to detail which forces are relevant and to what extent.

In the following, we analyse the generation of gravity waves at the interface of water and air due to the direct action of the wind. This kind of generation is dominant in the field (although for long tidal waves and tsunamis, the action is the gravitational attraction exerted by celestial bodies and the earthquake, respectively), while in the laboratory, the use of mechanical generators (paddles or plungers controlled by stepper motors or hydraulic actuators) that reproduce swell waves adequately well, with fairly good control even of the reflection, is preferred. This mitigates, to a large extent, interference from the laboratory. The purpose of direct wind generation in the laboratory can be the study of wave growth, a detailed analysis of processes at the interface, largely controlled by water and air turbulence, and many other investigations that would make little sense if the waves were generated mechanically.

To generate wind waves, wind tunnels are coupled over flumes, following both the construction rules and the management principles specific to the individual wind tunnel and flume. Almost always, the waves are unidirectional in a flume; very rarely, fans are used to generate three-dimensional wind waves in a tank. Figure 9.7 shows the schematic of a closed circuit wind-wave tunnel at Laboratorio de Dinámica de

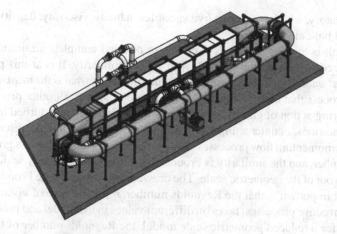

Fig. 9.7 Wind-wave tunnel at IISTA, Granada

Flujos Ambientales, Instituto Interuniversitario de Investigación del Sistema Tierra en Andalucía, Universidad de Granada.

Here, we are dealing with a two-phase system in which there are numerous scales that sometimes couple, more frequently overlap and coexist. In this respect, we aim to categorise what happens at the interface to establish the necessary criteria of similarity.

We start by considering only the presence of wind acting on a water surface initially at rest in the absence of currents (which, in a laboratory channel, are always present). On the water side, the physical process can be described using nine variables: the velocity, length, time, density, fluid viscosity, surface tension, pressure, compressibility and acceleration of gravity. We neglect, for the sake of simplicity, temperature and salinity, which play a fundamental role in many processes both at the ocean scale and in the shallow waters of coastal areas.

The problem is a dynamic problem involving 3 fundamental quantities, and by applying Buckingham's Theorem, we can group the nine variables into six dimensionless groups, already known in fluid mechanics: the Reynolds, Froude, Weber, Strohual, Euler and Mach numbers. The governed variables are the wave height, the period of the wave, or the shape of the wave spectrum if the analysis is more thorough, and we are not satisfied with a single significant wave, or rather, with a wave representative of the state of agitation of the water surface. These governed variables are made dimensionless with respect to the three quantities chosen as fundamental. Similarity requires that all dimensionless groups take on the same value in the model (in our laboratory) and in the prototype (in the field).

In theory, we have three degrees of freedom: we could set three scales arbitrarily, being confident that we could calculate the remaining six scales and the scales of all the governed variables. In practise, we do not enjoy such freedom since some of the variables are intensive (i.e., we cannot readily change them by an order of magnitude); moreover, for economic and safety reasons, it is convenient to use water in the laboratory, which constrains five variables, namely, viscosity, density, surface tension and bulk modulus.

The result is that the system is overconstrained and complete similarity cannot be guaranteed: we are forced to work with partial similarity. It is at this point that Bridgman's "*analysis within an analysis*" comes into play: what is the most important physical process that deserves special attention compared to all other processes? It is the restoring action of gravity, which tends to re-establish undisturbed horizontal surface conditions, counteracting the action of the wind and in the presence of all energy and momentum flow processes. The corresponding dimensionless group is the Froude number, and the similarity is Froude similarity, with a velocity scale equal to the square root of the geometric scale. The drawback of choosing the Froude number as the most important is that the Reynolds number, representative of viscous versus inertial convective processes, takes on different values in the model and prototype. In particular, for a reduced geometric scale model, the Reynolds number of the model is lower than that of the prototype.

Problems also arise with the Weber number, which indicates the importance of surface tension, and the Mach number, which has to do with the system's ability to propagate pressure waves. The Weber number is in any case negligible as long as the curvature of the waves, both in the model and in the prototype, is small (the effect of the surface tension is more pronounced the greater the curvature of the interface and is strictly null if the interface is flat); the Mach number is also negligible if the waves are not breaking (either in the laboratory or in the prototype) and the air is not trapped at all or is trapped in very low concentrations (transonic state, which involves the Mach number, are expected for fast movement of aerated breakers). With all these limitations, Froude similarity is valid and makes Strohual and Euler similarity valid, but it distorts the effects associated with the Reynolds number and Weber number.

For the water flow field in the presence of waves, we can define a Reynolds number based on the wave amplitude and orbital velocity:

$$\text{Re}_w = \frac{aV}{v_w} \equiv \frac{a^2\omega}{v_w}, \tag{9.4}$$

where $V = \omega a$ is the scale of the orbital velocity, a is the wave amplitude and v_w is the kinematic viscosity of the water. Substituting the dispersion relation yields

$$\text{Re}_w = \sqrt{2\pi g}\,\frac{a^2}{v_w\sqrt{L}}, \tag{9.5}$$

where L is the local wavelength. This Reynolds number decays vertically since a reduces almost exponentially in deep water with water depth $h > L/2$ and is constant in shallow water, where $h < 0.025L$. This implies a possible turbulent flow near the free surface, where breaking generates turbulence, and a transitional or even viscous flow beneath. This structure of the flow field in the vertical direction is highly distorted in the laboratory due to the reduction in the water Reynolds number, as indicated by Froude similarity.

Let us now analyse the air side, restricting the analysis only to the boundary layer and neglecting the large-scale pressure fluctuations and bursts that are always present in the turbulent wind flow field, neglecting also the compressibility of the air since the flow is isochoric.

Relevant time scales include the period of pressure fluctuations, which is considered important in the growth of wind waves (Teixeira and Belcher 2006). In the initial generation phase, for limited fetches, a time scale associated with the wind shear rate is important. Experimental measurements have found numerous correspondences between the wind boundary layer on water and that on a rough rigid surface. Therefore, it makes sense to select the classical boundary layer quantities, i.e., friction velocity $u_{*,a}$, apparent roughness z_0, density ρ_a and viscosity ν_a, and a time scale t that varies throughout the process. The most important dimensionless groups are the Reynolds number $\text{Re}_a = u_{*,a} z_0 / \nu_a$, the Euler number $\text{Eu}_a = \Delta p / (\rho_a u_{*,a}^2)$, and the Strohual number $\text{St}_a = z_0 / (u_{*,a} t)$, where the subscript 'a' stands for air. If we also use for the air side the geometric scale derived from the Froude similarity adopted for water, with $u_{*,a} \propto \lambda^{1/2}$, the Reynolds number of air scales like the Reynolds number of water, that is, $r_{\text{Re}_a} \propto \lambda^{3/2}$, with the same consequences as in water. Going into even more detail, let us consider the roughness scale for wind airflow. This is also defined as $z_0 = \alpha u_{*,a}^2 / g$, where $\alpha = 0.01 - 0.02$ is the Charnock parameter. The flow field is considered aerodynamically rough if $\text{Re}_a > \approx 2.5$ and smooth if $\text{Re}_a < \approx 0.13$. It is clear that a reduction in the Reynolds number of the airflow can result in the transition of the boundary layer from turbulent to transitional or viscous flow.

An analysis specifically aimed at reproducing the pressure fluctuations of the wind boundary layer on the water surface cannot disregard the fact that the Froude similarity time also scales with $\lambda^{1/2}$. This means that the pressure fluctuations of the air boundary layer in the laboratory may have an excessive frequency compared to the expected frequency, while the dynamic response of the interface remains the same in the laboratory and in the field. Therefore, resonance processes, especially for very small geometric scales of the model, can be radically cancelled out. See also Longo et al. (2013) for details on the free surface fluctuations and resonance induced by coherent structures on the water side.

Other time scales are induced by vorticity, with two facets: a first scale is closely linked to the existence of a dominant component of vorticity; a second scale is the period of intermittency, since vorticity, such as dissipation and the generation of turbulent kinetic energy itself, are processes characterised by a high level of intermittency, especially in the presence of breakers or micro breakers (Longo 2009). Given the nature of these scales, which are closely associated with the flow field through the Reynolds number and the Euler number, it is impossible to reproduce them adequately unless the Reynolds number is high enough to guarantee full turbulence even in the model. Many details of the process are hidden, also because the process of coupling the waves and wind flow field is still a hot topic of research, with the many variations due to wind blowing in the same direction or in the opposite direction to the wave propagation direction, or possibly not collinear with the waves.

Other details are known, such as the evidence that the first millimetres or tenths of a millimetre at the interface play a very important role in the initial phase of wave growth. However, we do not have the tools to include these important effects in our models, but all these limits should be considered when analysing and extrapolating the experimental results obtained in our wind-wave flumes.

Greater complexity immediately appears if, in addition to the action of wind generating waves or interacting with waves already present, the effect of currents is included. The velocity scale of the currents is added to the list of variables, leaving aside the velocity profile along the vertical and three-dimensional effects.

The typical equation is

$$f\left(H_{rms}, T_p, u_{*,a}, F, t_d, g, u_c, h\right) = 0, \tag{9.6}$$

where H_{rms} is the root-mean-square wave height, T_p is the peak period, $u_{*,a}$ is the friction velocity of the air flow, F is the fetch length, t_d is the duration of the wind, g is gravity acceleration, u_c is the velocity scale of the current, and h is the local depth. Neglecting other variables, we implicitly neglect the related effects. We are assuming partial similarity of Froude. The problem is purely kinematic in a space where we can select only two independent variables as fundamental variables. If we select g and $u_{*,a}$, Buckingham's Theorem indicates that no more than six dimensionless groups are needed to describe the problem, and the process is reduced to the following typical equation:

$$f\left(\frac{g H_{rms}}{u_{*,a}^2}, \frac{g T_p}{u_{*,a}}, \frac{u_c}{u_{*,a}}, \frac{g F}{u_{*,a}^2}, \frac{g t_d}{u_{*,a}}, \frac{g h}{u_{*,a}^2}\right) = 0. \tag{9.7}$$

In practise, the experimental results indicate that it is preferable to include the wave group velocity to scale the current velocity, and replacement by using the dispersion relation expressed as $f(c_g/u_{*,a}, g h/u_{*,a}^2, g T_p/u_{*,a}) = 0$ is possible. In deep water, the group $g h/u_{*,a}^2$ can be eliminated. Furthermore, if the wind duration is such that the fetch is saturated, the group $g t_d/u_{*,a}$ is also irrelevant. Therefore, the process can be expressed as

$$f\left(\frac{g H_{rms}}{u_{*,a}^2}, \frac{g T_p}{u_{*,a}}, \frac{u_c}{c_g}, \frac{g F}{u_{*,a}^2}\right) = 0. \tag{9.8}$$

The same procedure is followed for the other subcases.

A more important problem is the reduced effectiveness of trapping air bubbles in water and the limited presence of droplets in the air. This phenomenon is due to the excessive surface tension in the model, which significantly limits most of the physical phenomena at the interface, which are important in studies aimed at analysing the exchange of chemicals and gases in the marine environment. The problem is not amenable to resolution. Some naive attempts to reduce the surface tension of water by adding surfactants were doomed to failure since the scaling of the Weber number is $r_{We} = \lambda^2$: a geometric scale reduction $\lambda = 1/20$ requires a $1/400$ scale reduction of the surface tension, while the reduction obtainable by adding various substances (e.g., olive oil or ethyl alcohol) does not reach a single order of magnitude.

Summarising Concepts

- Wind tunnels were among the first devices designed to solve engineering problems. Similarity criteria require in most cases a geometric scale close to unity, which is impractical for physical models of many aircraft and rockets.
- The classification of wind tunnels refers to the Mach number in operation. The possibility of realising certain physical models in a wind tunnel is limited by the maximum Reynolds number of the flow field reproducible in the tunnel.
- The applications of wind tunnels are diversified in all sectors: civil engineering, environmental engineering, and aeronautics. As always, scale effects must be taken into account.
- Recently, wind tunnels coupled with water flumes for wind wave generation have been realised. The simulation of the processes is complicated by the presence of two phases: air and water. Similarity criteria are approximated and are developed on the basis of the specific physical process to be investigated.

References

Barlow, J. B., Rae, W. H., & Pope, A. (1999). *Low-speed wind tunnel testing*. Wiley.

Jiménez-Portaz, M., Chiapponi, L., Clavero, M., & Losada, M. A. (2020). Air flow quality analysis of an open-circuit boundary layer wind tunnel and comparison with a closed-circuit wind tunnel. *Physics of Fluids, 32*(12), 125120.

Jiménez-Portaz, M., Clavero, M., & Losada, M. A. (2021). A New Methodology for Assessing the Interaction between the Mediterranean Olive Agro-Forest and the Atmospheric Surface Boundary Layer. *Atmosphere, 12,* 658.

Longo, S. (2009). Vorticity and intermittency within the pre-breaking region of spilling breakers. *Coastal Engineering, 56*(3), 285–296.

Longo, S., Chiapponi, L., & Liang, D. (2013). Analytical study of the water surface fluctuations induced by grid-stirred turbulence. *Applied Mathematical Modelling, 37*(12–13), 7206–7222.

Ludwig, G. R., & Skinner, G. T. (1976). *Wind tunnel modeling study of the dispersion of sulfur dioxide in southern Allegheny County, Pennsylvania. Final report*. Calspan Corporation.

Skinner, G. T., & Ludwig, G. R. (1978). Physical modelling of dispersion in the atmospheric boundary layer. *Technical report 201*. Calspan Corporation.

Teixeira, M. A. C., & Belcher, S. E. (2006). On the initiation of surface waves by turbulent shear flow. *Dynamics of Atmospheres and Oceans, 41*(1), 1–27.

Wang, Z. Y., Plate, E. J., Rau, M., & Keiser, R. (1996). Scale effects in wind tunnel modelling. *Journal of Wind Engineering and Industrial Aerodynamics, 61*(2–3), 113–130.

Chapter 10
Physical Models in River Hydraulics

Rivers are characterised by numerous physical and chemical processes, all relevant and of great impact on the habitat, and with a level of complexity that is only partially addressed by numerical models. Processes such as (i) intense sediment transport, with cohesive and granular sediments, interfering with the anthropogenic works along the rivers and controlling the morphodynamics of the beaches; (ii) gas exchange with the atmosphere and affecting water quality; and (iii) floodings are some of the motivations for investigating rivers through physical models.

10.1 Similarity for a Non-prismatic Stationary (and Non-uniform) Stream

We consider a stream in a river of generic cross-section shape in a stationary and nonuniform fluid flow field. The physical process of flow can be described by the following typical equation:

$$f(\mu,\ \rho,\ R,\ k_s,\ U,\ i_b,\ g) = 0, \tag{10.1}$$

where μ is the dynamic viscosity, ρ is the density, R is the hydraulic radius in a representative section, k_s is the geometric scale of roughness, U is the cross-section average velocity of current in a representative section, i_b is the slope of the bottom and g is the acceleration of gravity. The process involves 3 fundamental quantities, and by virtue of Buckingham's Theorem, it is possible to describe it as a function of $(7 - 3) = 4$ dimensionless groups, for example:

$$\Pi_1 = \frac{\rho\, U\, R}{\mu}, \quad \Pi_2 = \frac{k_s}{R}, \quad \Pi_3 = i_b, \quad \Pi_4 = \frac{U^2}{g\, R}, \tag{10.2}$$

where Π_1 is the Reynolds number and Π_4 is the square of the Froude number.

The condition of similarity requires that the 4 dimensionless groups assume the same value in the model and in the prototype, equivalent to state that the scale ratios of the dimensionless groups are unitary:

$$r_{\Pi_1} = r_{\Pi_2} = r_{\Pi_3} = r_{\Pi_4} = 1. \tag{10.3}$$

This requires that the following system of 4 equations in 7 unknowns is satisfied:

$$\begin{cases} r_\rho\, r_U\, \lambda = r_\mu, \\ r_{k_s} = \lambda, \\ r_{i_b} = 1, \\ r_U^2 = r_g\, \lambda. \end{cases} \tag{10.4}$$

In addition, the 2 constraints are due to the presence of the same fluid in the model and prototype,

$$r_\mu = r_\rho = 1, \tag{10.5}$$

and an invariant acceleration of gravity,

$$r_g = 1. \tag{10.6}$$

The first and last equations in (10.4) are satisfied only for the trivial solution $\lambda = 1$.

We assume that the flow field is fully turbulent in the model and prototype. This happens if the Reynolds friction number satisfies the condition (see Table 5.2)

$$\mathrm{Re}_* = \frac{u_* k_s}{\nu} > 100, \tag{10.7}$$

where u_* is friction velocity. In this regime, the Reynolds number is physically irrelevant, the flow field no longer depends on the kinematic viscosity of the fluid, and Eq. (10.4) simplify as

$$\begin{cases} r_{k_s} = \lambda, \\ r_{i_b} = 1, \\ r_U^2 = r_g\, \lambda, \end{cases} \tag{10.8}$$

with the constraints represented by the two conditions:

$$r_g = r_\rho = 1, \tag{10.9}$$

where the second is inessential.

The system of 4 equations in 5 unknowns still has a degree of freedom and admits a parametric solution depending, for example, on the geometric scale; we select the geometric scale obtaining the following scales for the other variables:

$$\begin{cases} r_{k_s} = \lambda, \\ r_{i_b} = 1, \\ r_U = \sqrt{\lambda}. \end{cases} \tag{10.10}$$

This is a Froude similarity since the most representative number of the physical process is the Froude number, which assumes the same numerical value in the model and in the prototype. The roughness is related to the geometric scale in the model, and the slope in the model must be equal to the slope in the prototype, as can be deduced from the first and second in Eq. (10.10).

In order to verify the flow regime, we recall that Chézy's dimensionless coefficient is equal to

$$C\left(\text{Re}, \frac{k_s}{R}\right) \equiv \frac{U}{u_*} = \sqrt{\frac{8}{\lambda_{C\&W}\left(\text{Re}, \frac{k_s}{R}\right)}}, \tag{10.11}$$

where $\lambda_{C\&W}$ is the Darcy friction factor according to the Colebrook-White formula. In fully developed turbulence, C does not depend on the Reynolds number, but only on the relative roughness k_s/R. The relative roughness has a unit scale ratio, resulting in

$$r_{k_s/R} = \frac{r_{k_s}}{r_R} = \frac{r_{k_s}}{\lambda} = 1, \tag{10.12}$$

and $r_C = 1 \rightarrow r_U = r_{u_*}$. The condition of turbulent flow in the model

$$\text{Re}_{*,m} = \frac{u_{*,m}\, k_{s,m}}{\nu} > 100 \tag{10.13}$$

and in the prototype

$$\text{Re}_{*,p} = \frac{u_{*,p}\, k_{s,p}}{\nu} > 100, \tag{10.14}$$

for models on a reduced geometric scale ($\lambda < 1$) guarantees that

$$\text{Re}_{*,p} \equiv \frac{u_{*,p}\, k_{s,p}}{\nu} > \text{Re}_{*,m} \equiv \frac{u_{*,m}\, k_{s,m}}{\nu} > 100. \tag{10.15}$$

This means that, for the occurrence of the fully turbulent regime condition, the limiting factor is the model: if the flow regime in the model is fully turbulent and if $\lambda < 1$, the prototype is definitely in the same regime. In fact, the ratio between the Reynolds friction number in the model and the prototype is equal to

$$\frac{\text{Re}_{*,m}}{\text{Re}_{*,p}} = \frac{r_{u_*} r_{k_s}}{r_\nu} = \lambda^{3/2}, \tag{10.16}$$

since $r_{u_*} = r_U = \sqrt{\lambda}$, $r_{k_s} = \lambda$ and $r_\nu = 1$. As a consequence, the minimum allowed geometric scale to have turbulence in the model is:

$$\text{Re}_m = \text{Re}_p\, \lambda^{3/2} > 100 \;\rightarrow\; \lambda > \left(\frac{100\,\nu}{u_{*,p}\,k_{s,p}}\right)^{2/3}. \tag{10.17}$$

Example 10.1 We wish to reproduce a flow field in a riverbed with an average speed of the current $U_p = 2.5$ m s^{-1}, with Chézy coefficient $C_p = 15$ and geometric roughness $k_{s,p} = 5$ mm.

We first check that the flow field in the prototype is fully turbulent. The friction velocity is equal to

$$u_{*,p} = \frac{U_p}{C_p} = \frac{2.5}{15} = 0.17 \text{ m s}^{-1}. \tag{10.18}$$

Assuming a kinematic viscosity of water at temperature $\theta = 15$ °C equal to $\nu = 1.14 \cdot 10^{-6}$ m^{2}s^{-1} yields

$$\text{Re}_{*,p} = \frac{u_{*,p}\, k_{s,p}}{\nu} = \frac{0.17 \times 0.005}{1.14 \cdot 10^{-6}} = 731 > 100. \tag{10.19}$$

The flow is turbulent in the prototype. The minimum geometric scale in the physical model must be equal to

$$\lambda > \left(\frac{100\,\nu}{u_{*,p}\,k_{s,p}}\right)^{2/3} \equiv \left(\frac{100}{731}\right)^{2/3} \approx \frac{1}{3.8}. \tag{10.20}$$

Such a large geometric scale is impossible in most physical models: a relatively short river trunk 1000 m long would require a model almost 270 m long.

Suppose, instead, that the riverbed we wish to reproduce is very wide, with a hydraulic radius approximated by the depth of the current, $R_p \approx y_{0,p} = 3.1$ m, with width $b_p = 12$ m and a bottom slope $i_{b,p} = 0.002$. The average tangential stress at the wall is $\tau_p = \rho\, u_{*,p}^2 = \gamma\, R_p\, i_{b,p}$. The friction velocity is equal to

$$u_{*,p} = \sqrt{g\, R_p\, i_{b,p}} = \sqrt{9.806 \times 3.1 \times 0.002} = 0.25 \text{ m s}^{-1}. \tag{10.21}$$

Suppose also that the roughness has a geometric scale $k_{s,p} = 0.12$ m and that Chézy's coefficient has a value of $C_p = 13$. The average current velocity is equal to

$$U_p = C_p \sqrt{g\, R_p\, i_{b,p}} = 13 \times \sqrt{9.806 \times 3.1 \times 0.002} = 3.20 \text{ m s}^{-1}, \tag{10.22}$$

and the flow rate is

$$Q_p = U_p \, b_p \, y_{0p} = 3.20 \times 12 \times 3.1 = 119 \text{ m}^3 \text{ s}^{-1}. \tag{10.23}$$

The Reynolds friction number is given by

$$\text{Re}_{*,p} = \frac{u_{*,p} \, k_{s,p}}{\nu} = \frac{0.25 \times 0.12}{1.14 \cdot 10^{-6}} = 26\,300 \gg 100. \tag{10.24}$$

The minimum geometric scale in the model is

$$\lambda > \left(\frac{100 \, \nu}{u_{*,p} \, k_{s,p}} \right)^{2/3} \equiv \left(\frac{100}{26\,300} \right)^{2/3} \approx \frac{1}{41}. \tag{10.25}$$

Setting a geometric scale $\lambda = 1/40$, we calculate a section width in the model equal to

$$b_m = b_p \, \lambda = \frac{12}{40} = 30 \text{ cm}. \tag{10.26}$$

and an average speed of the current in the model equal to

$$U_m = U_p \, \sqrt{\lambda} = 3.20 \times \sqrt{\frac{1}{40}} = 0.51 \text{ m s}^{-1}, \tag{10.27}$$

with a flow rate

$$Q_m = Q_p \, \lambda^{5/2} = 119 \times \left(\frac{1}{40} \right)^{5/2} = 11.8 \text{ ls}^{-1}. \qquad \therefore \tag{10.28}$$

The roughness in the model should be $k_{s,m} = k_{s,p} \, \lambda = 0.12/40 = 3$ mm.

The results of the previous example indicate that models are only feasible, on a sufficiently small scale, if the roughness and slope in the prototype are high; otherwise, the minimum scale is excessive for most laboratories. This important limitation motivates a more thorough investigation on a possible distortion of the model, with a vertical scale greater than the horizontal scale. Distortion can also be a useful method to increase the Reynolds number in the model and ensure a fully developed turbulent regime, as is always the case in natural riverbeds. Finally, the distortion, with an increase in the vertical scale compared to the horizontal scale, makes the model less dissipative than it should be; this suggests incrementing the roughness to calibrate the dissipations correctly.

If we consider a natural stream with a sufficiently large width (compared to the depth of the current) and with uniform roughness along the wetted perimeter, it is possible to find a central zone where the flow field is mainly two-dimensional (see Fig. 10.1). This zone has a width of $B_c \approx B - 5y_0$ and therefore exists only if the width of the riverbed is at least $5y_0$ (Keulegan 1938).

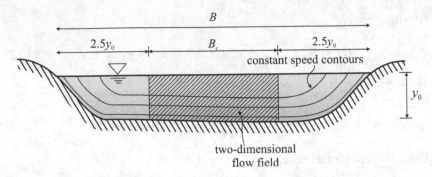

Fig. 10.1 Reference scheme for the identification of the zone with a two-dimensional flow field, which justifies a plano-altimetric distortion

If the roughness of the bottom is greater than the roughness of the walls (a condition almost always present in natural riverbeds), the width of the central zone is also greater than $(B - 5y_0)$. This is more evident if the cross-section is trapezoidal since in the trapezoidal section, the iso-speed curves are almost parallel to each other and to the walls at a shorter distance from the walls than for a cross-section with vertical walls.

The existence of a central core of the current, independent of the B width, allows the planimetric scales to be separated from the altimetric scales, adopting different scales in the horizontal plane and in the vertical plane:

$$\frac{B_m}{B_p} = \lambda_x, \quad \frac{y_m}{y_p} = \lambda_y, \quad \lambda_x \neq \lambda_y. \tag{10.29}$$

If we indicate with x the generic abscissa in the horizontal plane, the distortion ratio is defined as

$$n = \frac{\lambda_y}{\lambda_x}. \tag{10.30}$$

10.1.1 Distorted Models of Rivers and Canals in the Gradually Varied Flow Regime

We adopt direct analysis for detecting the similarity rules and consider the differential equation of the gradually varied flow:

$$\frac{1}{i_b}\frac{dy}{ds} = \frac{1 - \dfrac{\mathrm{Fr}^2}{i_b\,C^2}\left(1 - \alpha\,\dfrac{C^2}{P}\,\dfrac{\partial\Omega}{\partial s}\right)}{1 - \alpha\,\dfrac{B}{P}\,\mathrm{Fr}^2}, \tag{10.31}$$

where $\Omega = \Omega(s, y(s))$ is the area of the cross-section of the stream, s is the stream-wise curvilinear abscissa, α is the energy coefficient, P is the wetted perimeter, B is the top width, C is the dimensionless coefficient of Chézy, and Fr is the Froude number.

The equation, written for the variables referred to the prototype, is:

$$\frac{1}{i_{b,p}}\frac{\mathrm{d}y_p}{\mathrm{d}s_p} = \frac{1 - \dfrac{\mathrm{Fr}_p^2}{i_{b,p}\,C_p^2}\left(1 - \alpha_p\dfrac{C_p^2}{P_p}\dfrac{\partial\Omega_p}{\partial s_p}\right)}{1 - \alpha_p\dfrac{B_p}{P_p}\mathrm{Fr}_p^2}; \qquad (10.32)$$

the same equation, for the variables referred to the model, is:

$$\frac{1}{i_{b,m}}\frac{\mathrm{d}y_m}{\mathrm{d}s_m} = \frac{1 - \dfrac{\mathrm{Fr}_m^2}{i_{b,m}\,C_m^2}\left(1 - \alpha_m\dfrac{C_m^2}{P_m}\dfrac{\partial\Omega_m}{\partial s_m}\right)}{1 - \alpha_m\dfrac{B_m}{P_m}\mathrm{Fr}_m^2}. \qquad (10.33)$$

Equation (10.33) can be expressed introducing the scale ratios:

$$\frac{1}{i_{b,p}}\frac{\mathrm{d}y_p}{\mathrm{d}s_p}\left[\frac{\lambda_y}{r_{i_b}\lambda_x}\right] =$$

$$\frac{1 - \dfrac{\mathrm{Fr}_p^2}{i_{b,p}\,C_p^2}\left[\dfrac{r_{\mathrm{Fr}}^2}{r_{i_b}\,r_C^2}\right]\left(1 - \alpha_p\dfrac{C_p^2}{P_p}\dfrac{\partial\Omega_p}{\partial s_p}\left[r_\alpha\dfrac{r_C^2\,r_\Omega}{r_P\,\lambda_x}\right]\right)}{1 - \alpha_p\dfrac{B_p}{P_p}\mathrm{Fr}_p^2\left[r_\alpha\dfrac{r_B\,r_{\mathrm{Fr}}^2}{r_P}\right]}.$$

$$(10.34)$$

To guarantee dynamic similarity, Eqs. (10.32) and (10.34) must be equal; this requires that all expressions in square brackets assume unit values:

$$\begin{cases} \dfrac{\lambda_y}{r_{i_b}\lambda_x} = 1, \\[2mm] \dfrac{r_{\mathrm{Fr}}^2}{r_{i_b}\,r_C^2} = 1, \\[2mm] r_\alpha\dfrac{r_C^2\,r_\Omega}{r_P\,\lambda_x} = 1, \\[2mm] r_\alpha\dfrac{r_B\,r_{\mathrm{Fr}}^2}{r_P} = 1. \end{cases} \qquad (10.35)$$

The scale ratios of the top width and current cross-sectional area are $r_B = \lambda_x$ and $r_\Omega = \lambda_y \lambda_x$, respectively. In general, the wetted perimeter involves the two geometric scales, and its scale depends on the operating point. For example, considering a rectangular cross-section of width b and depth of the current y_0, we calculate:

$$P = b + 2y_0 \rightarrow \frac{P_m}{P_p} \equiv r_P = \frac{b_m + 2y_{0,m}}{b_p + 2y_{0,p}} =$$

$$\frac{b_m \left(1 + 2\dfrac{y_{0,m}}{b_m}\right)}{b_p \left(1 + 2\dfrac{y_{0,p}}{b_p}\right)} = \lambda_x \frac{1 + 2\dfrac{\lambda_y}{\lambda_x}\dfrac{y_{0,p}}{b_p}}{1 + 2\dfrac{y_{0,p}}{b_p}}. \qquad (10.36)$$

The r_P scale varies with varying flow conditions. However, for sufficiently large rectangular sections, the numerator and denominator of r_P in Eq. (10.36) tend to unity, and $r_P \rightarrow \lambda_x$.

We can also assume that $r_\alpha = 1$, in which case the system of Eq. (10.35) becomes

$$\begin{cases} \dfrac{\lambda_y}{r_{i_b} \lambda_x} = 1, \\[2mm] \dfrac{r_{\mathrm{Fr}}^2}{r_{i_b} r_C^2} = 1, \\[2mm] \dfrac{r_C^2 \lambda_y}{\lambda_x} = 1, \\[2mm] r_{\mathrm{Fr}} = 1, \end{cases} \qquad (10.37)$$

with the following solution:

$$r_{\mathrm{Fr}} = 1, \qquad r_{i_b} = \frac{\lambda_y}{\lambda_x}, \qquad r_C^2 = \frac{\lambda_x}{\lambda_y} \equiv \frac{1}{n}. \qquad (10.38)$$

By introducing distortion, we have recovered a further degree of freedom and can set, for example, the two geometric scales. We observe that the dependence of resistance on the Reynolds number and relative roughness is incorporated into the relationship expressing Chézy's coefficient.

10.1.2 The Scale Ratio of the Friction Coefficient and Roughness

The realisation of distorted models is limited to the case of sections large enough to allow the existence of a two-dimensional core of current, but it is also compulsory that the vertical and lateral accelerations of the fluid are negligible with respect to gravity. If the current profile is gradually varied, the velocity distribution remains logarithmic in all sections, and the average velocity of the current can be expressed as a function of the maximum velocity (assumed at the free surface) and the friction velocity, such that

$$\frac{U}{u_*} \equiv C = \frac{U_{max}}{u_*} - 2.5, \tag{10.39}$$

with

$$\frac{U_{max}}{u_*} = \frac{1}{\kappa} \ln \frac{y_0}{k_s} + B_s, \tag{10.40}$$

where κ is von Kármán's constant and B_s is a function of the Reynolds friction number. The experimental B_s function is shown in Fig. 10.2. Equations (10.39–10.40) can be merged obtaining

$$C = \frac{1}{\kappa} \ln \frac{y_0}{k_s} + (B_s - 2.5). \tag{10.41}$$

Using this last expression, the scale ratio of the friction coefficient is

$$r_C = \frac{C_m}{C_p} = \frac{\dfrac{1}{\kappa} \ln \dfrac{y_m}{k_{s,m}} + (B_{s,m} - 2.5)}{\dfrac{1}{\kappa} \ln \dfrac{y_p}{k_{s,p}} + (B_{s,p} - 2.5)}, \tag{10.42}$$

and after a few steps:

$$r_C = 1 + \frac{\ln \dfrac{\lambda_y}{r_{k_s}} + \kappa \, (B_{s,m} - B_{s,p})}{\ln \dfrac{y_p}{k_{s,p}} + \kappa \, (B_{s,p} - 2.5)}. \tag{10.43}$$

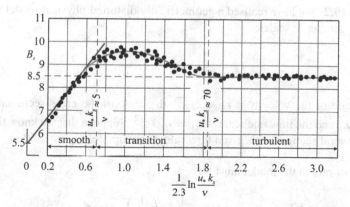

Fig. 10.2 Experimental diagram of the B_s function for varying friction Reynolds number (modified from Schlichting 1955)

If the Reynolds friction number in the model and in the prototype is greater than 100, it results in $B_{s,p} = B_{s,m} = 8.5$, and assuming $\kappa = 0.4$, we calculate:

$$r_C = 1 + \frac{\ln \dfrac{\lambda_y}{r_{k_s}}}{\ln \dfrac{y_p}{k_{s,p}} + 2.4}. \tag{10.44}$$

We observe that this result can also be used in the transition regime with the Reynolds friction number between 5 and 100 since in this regime, B_s assumes a maximum value of 9.5, only 10% greater than the asymptotic value, a fully acceptable deviation in the order of approximation required. For practical applications, the viscous regime with $Re_* < 5$ is irrelevant.

These analyses assume a logarithmic velocity profile in the section, which is correct if the relative roughness is less than 1/15 and incorrect for a relative roughness of $\approx 1/5.25$. This condition must be checked in the model and prototype. In particular, the choice of an absolute roughness ratio generally less than the vertical geometric scale results in a roughness in the model greater than the roughness in the prototype; it is always advisable to select the scales to limit the relative roughness in the model to 1/10.

The estimation of the roughness coefficient is indeed quite difficult in the prototype; the geometric characteristics of the roughness in the prototype are extremely varied, and almost always, the roughness of the natural riverbeds is estimated indirectly on the basis of the flow rate scale, possibly in several sections. Provided that the geometry of the riverbed is known in an adequate number of sections and that there are water level measurements for different flow rates, a numerical model is used to estimate the roughness coefficient with the best overlap between measurements and numerical results. Then, it is possible to proceed with a direct comparison between the water level measurements in the prototype and the expected water levels in the model.

Example 10.2 We have realised a geometrically distorted physical model with the following geometric scales:

$$\lambda_x = \frac{1}{120}, \quad \lambda_y = \frac{1}{40}. \tag{10.45}$$

The scale of the velocity is $r_U = \sqrt{\lambda_y}$, the scale of the cross-sectional area is $r_\Omega = \lambda_x \lambda_y$ and the flow rate scale is $r_Q = \lambda_x \lambda_y^{3/2}$. Suppose that we know the water level in several sections for a given flow rate; for example, $Q_p = 1850 \text{ m}^3\text{ s}^{-1}$; see Table (10.1).

The flow rate in the model must be

$$Q_m = Q_p \lambda_x \lambda_y^{3/2} \rightarrow Q_m = 1850 \times \frac{1}{120} \times \left(\frac{1}{40}\right)^{3/2} = 60.9 \text{ l s}^{-1}, \quad \therefore \tag{10.46}$$

Table 10.1 Water levels in the *prototype* corresponding to a flow rate of $Q_p = 1850 \text{ m}^3 \text{ s}^{-1}$

Section #	1	2	3	4
y_p (m)	5.50	5.20	5.62	6.05

Table 10.2 Water levels in the *model* corresponding to the water levels in the prototype listed in Table 10.1

Section #	1	2	3	4
y_m (cm)	13.8	13.0	14.1	15.1

and the expected water levels in the model are listed in Table 10.2, with values rounded to the nearest millimetre since the water level can seldom be measured with a lower uncertainty.

At this point, we proceeded by imposing the given flow rate on the physical model and modifying the roughness with an appropriate arrangement of wire mesh fixed to the walls, in number and area density to meet the required condition of correspondence between the actual and theoretical water levels. It may happen that the correspondence is guaranteed for a single value of flow rate and not for a higher value. This means that the roughness is not uniform along the contour and that it is necessary to refine the calibration, spatially changing the arrangement of the nets in the floodplains or embankments, that is, in the portion of the perimeter that is wetted only at higher water levels.

We now wish to assess the effect of the distortion ratio on the geometric roughness scale. Equation (10.43) can be inverted:

$$\frac{r_{k_s}}{\lambda_y} = \left(\frac{k_{s,p}}{y_p}\right)^{r_C - 1} \exp \kappa \left[(B_{s,m} - 2.5) - r_C (B_{s,p} - 2.5)\right]. \tag{10.47}$$

Since Eq. (10.38) results in $r_C = 1/\sqrt{n}$, we can write

$$\frac{r_{k_s}}{\lambda_y} = \left(\frac{k_{s,p}}{y_p}\right)^{\frac{1}{\sqrt{n}} - 1} \exp \kappa \left[(B_{s,m} - 2.5) - \frac{1}{\sqrt{n}}(B_{s,p} - 2.5)\right], \tag{10.48}$$

also available as a relationship between relative roughness in the model and prototype, depending on the distortion n:

$$\frac{k_{s,m}}{y_m} = \left(\frac{k_{s,p}}{y_p}\right)^{\frac{1}{\sqrt{n}}} \exp \kappa \left[(B_{s,m} - 2.5) - \frac{1}{\sqrt{n}}(B_{s,p} - 2.5)\right]. \tag{10.49}$$

Example 10.3 We wish to limit the relative roughness in the model to values between 1/8 and 1/10. Table 10.3 lists the relative roughness in the prototype for various distortion values.

Table 10.3 Relative roughness in the prototype corresponding to 1/8 and 1/10 relative roughness in the model, for different distortion coefficient n

$n =$	1	2	3	4	5
$\dfrac{k_{s,p}}{y_p} = \dfrac{1}{8}$	$\dfrac{1}{8}$	$\dfrac{1}{50}$	$\dfrac{1}{210}$	$\dfrac{1}{700}$	$\dfrac{1}{2000}$
$\dfrac{k_{s,p}}{y_p} = \dfrac{1}{10}$	$\dfrac{1}{10}$	$\dfrac{1}{70}$	$\dfrac{1}{310}$	$\dfrac{1}{1100}$	$\dfrac{1}{3300}$

Higher distortion values appear incompatible with the characteristics of natural streams. For example, if we consider a distortion ratio of $n = 5$ and we wish to limit the relative roughness ratio in the model to 1/8, from Table 10.3 we calculate a relative roughness of 1/2000. Assuming that the water depth is $y_p = 5.0$ m, the roughness must be $k_{s,p} = 5.0/2000 = 2.5$ mm. This results in a minimum friction velocity corresponding to $\text{Re}_* > 100$, equal to:

$$\frac{u_{*,p-min}\, k_{s,p}}{\nu} > 100 \rightarrow u_{*,p-min} > 100 \frac{\nu}{k_{s,p}} \equiv 100 \times \frac{1.14 \cdot 10^{-6}}{0.002\,5}$$

$$= 0.046 \text{ m s}^{-1}. \tag{10.50}$$

The friction velocity is equal to

$$u_{*,p} = \sqrt{g\, y_p\, i_{b,p}}, \tag{10.51}$$

where we have assumed a wide cross-section with the hydraulic radius equal to the water depth; hence:

$$i_{b,p} > \frac{u_{*,p-min}^2}{g\, y_p} \equiv \frac{0.046^2}{9.806 \times 5.0} = 4.3 \cdot 10^{-5}. \tag{10.52}$$

This is the lower limit of the slope in the prototype reproducible with the model.

However, the bottom inclination should not be so high as to mobilise the sediment. To perform this second verification, we assume that the size of the sediment is consistent with the geometric scale of the roughness, and we require that

$$\Theta < \Theta_{crit} \rightarrow \frac{\rho\, u_{*,p}^2}{\gamma_s\, k_{s,p}\, (s-1)} < 0.05, \tag{10.53}$$

where Θ is the Shields parameter, Θ_{crit} is the critical Shields parameter, assumed to be 0.05, and s is the relative specific gravity of the sediment. Depending on the characteristics of the riverbed, the following results are obtained:

$$i_{b,p} < 0.05 \frac{\gamma_s \, k_{s,p} \, (s-1)}{\gamma \, y_p} \equiv$$

$$0.05 \times \frac{27\,000 \times 0.0025 \times (2.7-1)}{9800 \times 5.0} = 1.17 \cdot 10^{-4}. \quad \therefore \quad (10.54)$$

This is the upper limit of bottom inclination, in the prototype, that can be reproduced with the present model.

In summary, the slope in the prototype must be within the range $4.3 \cdot 10^{-5} < i_{b,p} < 1.17 \cdot 10^{-4}$.

We observe that the upper limit of the slope is quite modest. In addition, in a riverbed with the characteristics assumed in the example, some bedforms are expected with an additional flow resistance, characterised by an equivalent roughness much greater than 2.5 mm.

10.1.3 Distorted Models of Rivers and Canals in the Generic Flow Regime

In the presence of concentrated energy losses, due to abrupt expansions or changes in the direction of the planimetric or altimetric axis, the energy balance equation of the current is modified as

$$-\frac{dH}{ds} = \frac{1}{C^2} \frac{U^2}{g\,R} + \sum \xi_i \frac{U^2}{2g\,L} \equiv \frac{1}{C^2} \frac{U^2}{g\,R} + \xi_L \frac{U^2}{g\,L} \equiv$$

$$\left(\frac{1}{C^2} + \xi_L \frac{R}{L} \right) \frac{U^2}{g\,R} \equiv E \frac{U^2}{g\,R}, \tag{10.55}$$

where L is the length over which the localised energy losses are distributed, ξ_L is the equivalent concentrated loss coefficient, and E is the equivalent total dissipation coefficient for distributed losses and for plano-altimetric variations of the stream axis. Equation (10.31) becomes:

$$\frac{1}{i_b} \frac{dy}{ds} = \frac{1 - \dfrac{\mathrm{Fr}^2}{i_b} \left[\left(\dfrac{1}{C^2} + \xi_L \dfrac{R}{L} \right) - \dfrac{\alpha}{P} \dfrac{\partial \Omega}{\partial s} \right]}{1 - \alpha \dfrac{B}{P} \mathrm{Fr}^2}. \tag{10.56}$$

For dimensional homogeneity, the scale ratio of the new term $\xi_L \, R/L$ must be equal to the scale ratio of the term $1/C^2$ and ultimately to the distortion ratio n:

$$r_{\xi_L} \frac{r_R}{r_L} = r_C^{-2} \equiv \frac{\lambda_y}{\lambda_x} \equiv n. \tag{10.57}$$

For very large sections resulting in $r_R = \lambda_y$, $r_L = \lambda_x$, and $r_{\xi_L} = 1$, the localised energy drop coefficients must have the same value in the model and prototype.

Localised energy drops are mainly due to local variations of the cross-section (mainly expansions) and depend on the Reynolds number, although this last dependence is lost in the fully turbulent regime. For the undistorted models, the cross-section variations are preserved, which is not true for distorted models; however, the effect of distortion is usually negligible, and we can assume that the loss coefficient is the same in the model and in the prototype for distorted models.

The energy balance for the prototype is

$$
\frac{1}{i_{b,p}} \frac{dy_p}{ds_p} = \frac{1 - \dfrac{Fr_p^2}{i_{b,p}} \left[\left(\dfrac{1}{C_p^2} + \xi_{L,p} \dfrac{R_p}{L_p} \right) - \dfrac{\alpha_p}{P_p} \dfrac{\partial \Omega_p}{\partial s_p} \right]}{1 - \alpha_p \dfrac{B_p}{P_p} Fr_p^2}
\tag{10.58}
$$

and, for the model:

$$
\frac{1}{i_{b,m}} \frac{dy_m}{ds_m} = \frac{1 - \dfrac{Fr_m^2}{i_{b,m}} \left[\left(\dfrac{1}{C_m^2} + \xi_{L,m} \dfrac{R_m}{L_m} \right) - \dfrac{\alpha_m}{P_m} \dfrac{\partial \Omega_m}{\partial s_m} \right]}{1 - \alpha_m \dfrac{B_m}{P_m} Fr_m^2}.
\tag{10.59}
$$

By introducing the scale ratios, Eq. (10.59) can be re-written as

$$
\frac{1}{i_{b,p}} \frac{dy_p}{ds_p} \left[\frac{\lambda_y}{r_{i_b} \lambda_x} \right] = \frac{1 - \dfrac{Fr_p^2}{i_{b,p} C_p^2} \left[\dfrac{r_{Fr}^2}{r_{i_b} r_C^2} \right]}{1 - \alpha_p \dfrac{B_p}{P_p} Fr_p^2 \left[r_\alpha \dfrac{r_B r_{Fr}^2}{r_P} \right]}
$$

$$
- \frac{\dfrac{Fr_p^2 \xi_{L,p} R_p}{i_{b,p} L_p} \left[\dfrac{r_{Fr}^2 r_{\xi_L} r_R}{r_{i_b} r_L} \right] - \dfrac{Fr_p^2}{i_{b,p}} \dfrac{\alpha_p}{P_p} \dfrac{\partial \Omega_p}{\partial s_p} \left[\dfrac{r_{Fr}^2}{r_{i_f}} \dfrac{r_\alpha r_\Omega}{r_P \lambda_x} \right]}{1 - \alpha_p \dfrac{B_p}{P_p} Fr_p^2 \left[r_\alpha \dfrac{r_B r_{Fr}^2}{r_P} \right]}.
\tag{10.60}
$$

The two equations (10.58) and (10.60) are coincident if the terms in square brackets are unity. The following system of equations should be satisfied:

$$\begin{cases} \dfrac{\lambda_y}{r_{i_b} \lambda_x} = 1, \\[2mm] \dfrac{r_{Fr}^2}{r_{i_b} r_C^2} = 1, \\[2mm] \dfrac{r_{Fr}^2 \, r_{\xi_L} \, r_R}{r_{i_b} \, r_L} = 1, \\[2mm] \dfrac{r_{Fr}^2 \, r_\alpha \, r_\Omega}{r_{i_b} \, r_P \, \lambda_x} = 1, \\[2mm] r_\alpha \, \dfrac{r_B \, r_{Fr}^2}{r_P} = 1. \end{cases} \tag{10.61}$$

Assuming that $r_{\xi_L} = 1$, the solution is identical to the one obtained in the case of gradually varying flow regimes. If the riverbed is very wide, with the wetted perimeter scale equal to λ_x, the solution of the system of Eq. (10.61) is simplified:

$$r_{i_b} = \frac{\lambda_y}{\lambda_x}, \quad r_C^2 = \frac{\lambda_x}{\lambda_y}, \quad r_{Fr}^2 = 1. \tag{10.62}$$

If local losses are much greater than distributed losses, with

$$\frac{Fr_p^2 \, \xi_{L,p} \, R_p}{i_{b,p} \, L_p} > \frac{Fr_p^2}{i_{b,p} \, C_p^2}, \tag{10.63}$$

the second equation in the system (10.61) is irrelevant, and with the same hypotheses adopted to derive the conditions of similarity for rivers and canals in a regime of gradually varied motion, we obtain the two conditions of similarity:

$$r_{Fr} = 1, \quad r_{i_b} = \frac{\lambda_y}{\lambda_x} \equiv n. \tag{10.64}$$

The limits to the maximum distortion ratio discussed in Sect. 10.1.2 are no longer relevant. This is the reason why many physical models of natural rivers and riverbeds, reproduced for short lengths in the presence of bridge piles, traverses, etc., with high local loss of energy, are often realised with high distortion ratios up to $n = 10$.

We observe that in the proposed scheme, localised and distributed energy losses have a common structure and can be interpreted as the result of a progressive transition. Localised losses are also due to abrupt plano-altimetric variations of the current axis and not only of the cross-section. When the plano-altimetric variations (for instance, meanders) reduce their size, they assume the dimensions of the bedforms and of the wall roughness. The expression of localised energy losses then becomes formally identical to that of distributed energy losses (Yalin 1971).

10.2   Models in the Unsteady Flow Regime

For currents in the unsteady flow regime, it is necessary to preserve the Strohual number, which must have the same value in the model and in the prototype:

$$St_m \equiv \frac{U_m\, t_m}{L_m} = St_p \equiv \frac{U_p\, t_p}{L_p}. \tag{10.65}$$

If the model is undistorted, it results in $r_t = r_U = \sqrt{\lambda}$, and Froude similarity is recovered. For a distorted model, the evaluation of the time scale requires further investigation.

Based on the schematic shown in Fig. 10.3, the displacement of a particle from A to B in a δt time interval can be recovered with a first displacement from A to C and a second displacement from C to B; hence:

$$\delta t = \frac{\delta l}{U} = \frac{\delta x}{U_x} = \frac{\delta y}{U_y}. \tag{10.66}$$

These equations, written for the model and prototype, become:

$$\begin{cases} \delta t_m = \dfrac{\delta l_m}{U_m} = \dfrac{\delta x_m}{U_{x,m}} = \dfrac{\delta y_m}{U_{y,m}}, \\[2mm] \delta t_p = \dfrac{\delta l_p}{U_p} = \dfrac{\delta x_p}{U_{x,p}} = \dfrac{\delta y_p}{U_{y,p}}. \end{cases} \tag{10.67}$$

Introducing scale ratios, we obtain:

$$r_t = \frac{\lambda_x}{r_{U_x}} = \frac{\lambda_y}{r_{U_y}}. \tag{10.68}$$

Assuming that the Strohual number, representative of the local inertia in the horizontal plane and in the vertical direction, has the same value in the model and

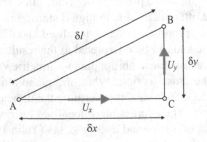

Fig. 10.3 Reference scheme for the calculation of time scaling in a distorted model with unsteady flow

prototype, yields:

$$\frac{r_U\, r_t}{\lambda} = \frac{r_{U_x}\, r_t}{\lambda_x} = \frac{r_{U_y}\, r_t}{\lambda_y} = 1. \tag{10.69}$$

The Froude number refers to the average velocity component in the streamwise direction and depends on the vertical length scale; hence, $r_{U_x} = \sqrt{\lambda_y}$. Therefore,

$$r_t = \frac{\lambda_x}{\sqrt{\lambda_y}} \equiv \frac{1}{n}\sqrt{\lambda_y}. \tag{10.70}$$

By substituting in Eq. (10.70), we obtain

$$r_{U_y} = n\sqrt{\lambda_y}. \tag{10.71}$$

This means that horizontal and vertical velocities have different scales, with a ratio equal to the distortion ratio:

$$\frac{r_{U_y}}{r_{U_x}} = n. \tag{10.72}$$

The same result was obtained in Sect. 4.1.6. Table 10.4 lists the scale ratios for physical models with a fixed bed, distorted and undistorted, in Froude similarity.

Example 10.4 We have realised a distorted physical model of a trunk of a river, and we wish to reproduce a flood. The horizontal scale is $\lambda_x = 1/120$, and the distortion ratio is $n = 4$. The riverbed has an average width of $B = 120$ m, and the depth of the current, for a flow rate of $Q = 580$ m^3 s^{-1}, is $y = 2.1$ m. During the flood, with a maximum flow rate of $Q_{max} = 820$ m^3 s^{-1}, the level rises and reaches a maximum of $y_{max} = 3.4$ m in a time interval $\Delta t = 5.5$ h. We wish to calculate the average rate of level rise in the model.

Based on the similarity rules, the vertical geometric scale is $\lambda_y = n\,\lambda_x = 4/120 \equiv 1/30$. The vertical velocity scale is $r_{U_y} = n\sqrt{\lambda_y} = 4\times\sqrt{1/30} = 1/1.37$.

The average rate of level rise in the prototype is $(3.4 - 2.1)/5.5 = 24$ cm h^{-1}, resulting in $24/1.37 = 17$ cm h^{-1} in the model. The time scale ratio is $r_t = (1/n)\sqrt{\lambda_y} = (1/4)\times\sqrt{1/30} = 1/21.9$, and the maximum level is reached after $5.5/21.9 = 15$ min.

10.3 Inclined Physical Models

The difficulty encountered in reproducing the dissipation level in small geometric scale models often suggests a model where only the equality $r_{i_b} = r_J$ is guaranteed; according to this equality, the slope of the bottom in the model is modified to ensure that the energy supplied by gravity balances dissipations. Since roughness in the models is always greater than necessary, an increase in slope is required to balance the extra dissipations.

If we use the Chézy formula with a dimensional coefficient according to Gauckler-Strickler, the similarity condition is

$$r_{i_b} = r_k^{-2} \lambda^{-1/3}, \tag{10.73}$$

where k is the roughness coefficient. If we use the Chézy formula with a dimensionless coefficient, it results in

$$r_{i_b} = r_C^{-2}. \tag{10.74}$$

In explicit form, the slope in the model must be equal:

$$i_{b,m} = i_{b,p} \frac{C_p^2}{C_m^2}. \tag{10.75}$$

Table 10.4 Summary of scale ratios for physical models with a fixed bed, distorted and undistorted, in Froude similarity

	Scale ratios	Model undistorted	Model distorted
Geometric			
Depth	λ	λ	λ_y
Length	λ	λ	λ_x
Width	λ	λ	λ_x
Cross-section area	r_Ω	λ^2	$\lambda_y \lambda_x$
Volume	r_V	λ^3	$\lambda_y \lambda_x^2$
Kinematic			
Time	r_t	$\lambda^{1/2}$	$\lambda_y^{-1/2} \lambda_x$
Horizontal velocity	r_{U_x}	$\lambda^{1/2}$	$\lambda_y^{1/2}$
Vertical velocity	r_{U_y}	$\lambda^{1/2}$	$\lambda_y^{3/2} \lambda_x^{-1}$
Friction velocity	r_{u_*}	$\lambda^{1/2}$	$\lambda_y \lambda_x^{-1/2}$
Flow rate	r_Q	$\lambda^{5/2}$	$\lambda_y^{3/2} \lambda_x$
Dynamic			
Mass	r_m	λ^3	$\lambda_y \lambda_x^2$
Pressure	r_p	λ	λ_y
Tangential stress	r_τ	λ	$\lambda_y^2 \lambda_x^{-1}$
Force	r_F	λ^3	$\lambda_y \lambda_x^2$
Dimensionless			
Inclination	r_{i_b}	1	$\lambda_y \lambda_x^{-1}$
Chézy C	r_C	1	$\lambda_y^{-1/2} \lambda_x^{1/2}$
Froude	r_{Fr}	1	1
Reynolds	r_{Re}	$\lambda^{3/2}$	$\lambda_y^{3/2}$

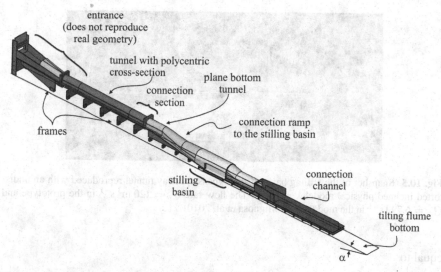

Fig. 10.4 Render of the bypass spillway tunnel reproduced with an undistorted inclined physical model (from Mignosa et al. 2010)

Since $C_m < C_p$ almost always holds, $i_{b,m} > i_{b,p}$ also holds.

The best way to select the slope value in the model is to realise the model on a tilting plane. This is possible for essentially rectilinear channels that can be reproduced in a variable slope channel. Figure 10.4 shows the physical model of a spillway on scale $\lambda = 1/50$, realised in polymethyl methacrylate (PMMA). The prototype is made of concrete.

From direct analysis, applying the following law of resistance

$$J = \frac{V^2}{k_s^2 \, R^{4/3}},$$
(10.76)

where J is the energy gradient, k_s is the Gauckler-Strickler coefficient, and R is the hydraulic radius, results in

$$r_J = \frac{r_V^2}{r_{k_s}^2 \, \lambda^{4/3}}.$$
(10.77)

To ensure the same level of dissipation, in the model and prototype, we must have $r_J = 1$. The ratio of the Gauckler-Strickler coefficient, considering that Froude similarity yields $r_V = \sqrt{\lambda}$, is equal to

$$r_{k_s} = \lambda^{-1/6}.$$
(10.78)

Concrete has a roughness with an estimated coefficient of $k_s = 70 \text{ m}^{1/3} \text{ s}^{-1}$. Therefore, the model should be made of rough material with a coefficient

Fig. 10.5 Snapshot of the stilling basin of the bypass spillway tunnel reproduced with an undistorted inclined physical model $\lambda = 1 : 50$. The flow rate $Q_p = 120 \, \text{m}^3 \, \text{s}^{-1}$ in the prototype and $Q_m = 6.79 \, \text{l} \, \text{s}^{-1}$ in the model (from Mignosa et al. 2010)

equal to

$$k_{s,m} = k_{s,p} \, \lambda^{-1/6} \rightarrow k_{s,m} = 70 \times \left(\frac{1}{50}\right)^{-1/6} = 135 \, \text{m}^{1/3} \, \text{s}^{-1}. \tag{10.79}$$

PMMA has a roughness of approximately $110 \, \text{m}^{1/3} \, \text{s}^{-1}$ and dissipates in excess, hence the need to tilt the model. The tilting ratio is:

$$r_{i_b} = r_J \rightarrow r_{i_b} = \frac{\lambda^{-1/3}}{r_{k_s}^2} = \frac{\left(\frac{1}{50}\right)^{-1/3}}{\left(\frac{110}{70}\right)^2} = 1.49. \tag{10.80}$$

The installation of the physical model in a variable slope channel has allowed the necessary adjustment to modify the slope to ensure the correct reproduction of the water levels in some significant cross-sections. The water levels in the prototype are calculated numerically.

We observe that the condition of Eq. (10.74) is valid only in uniform flow; in any other condition, the water profiles are distorted, and the model, in tranquil flow, over-estimates the accelerated current profile levels and underestimates the decelerated current profile levels.

Figure 10.5 shows a snapshot of the stilling basin of the physical model with rendering depicted in Fig. 10.4. The large eddies are evident, and it is also clear that the air bubbles are much larger than they should be as a consequence of the partial similarity, which neglects Weber number effects.

As a last remark, the slope of the bottom is precisely defined only for artificial channels and for long river trunks. Therefore, in the reproduction of a physical model of localised phenomena involving river trunks of modest length, the inclined model does not make much sense because the slope in the prototype is known with limited accuracy.

Summarising Concepts

- Physical models of rivers and planned works intersecting with rivers represent a good percentage of current modelling practise in hydraulics. They are used to simulate stationary or variable flood conditions.
- Wherever possible, the models should be geometrically undistorted. If the geometric scale is very small, a vertical distortion is preferred, with scales in the plane smaller than the scale along the vertical direction.
- The need for very small geometric scales to reproduce generally long trunks of the riverbed incurs the problem of correctly reproducing roughness in the model.
- Similarity is often achieved to have a correct reproduction of the dissipation of the stream, with the inclined models balancing the greater dissipation, in the model, due to a higher-than-theoretical roughness and the need to reproduce the turbulent regime.

References

Keulegan, G. H. (1938). Laws of turbulent flow in open channels. *Journal of Research of the National Bureau of Standard, 21*(RP1151), 708–741.

Mignosa, P., Longo, S., Chiapponi, L., D'Oria, M., & Mammí, O. (2010). *Prove su modello fisico della vasca di dissipazione al termine della galleria di bypass del Lago d'Idro (Physical model tests for the stilling basin at the end of the Lake Idro bypass tunnel) (in Italian)*. DICATeA, University of Parma.

Schlichting, H. (1955). *Boundary layer theory*. McGraw-Hill Inc.

Yalin, M. S. (1971). *Theory of hydraulic models*. Macmillan Publishers Ltd.

Chapter 11
Physical Models with Sediment Transport

To reproduce sediment transport phenomena occurring in rivers and in coastal zones or over land due to wind action, it is necessary to introduce modelling criteria for deriving the scales of the sediment transport variables, such as sediment flow rate, sediment diameter, and sediment specific weight. The physical models that reproduce sediment transport are named *movable-bed models* because the water or air stream is in contact with physical boundaries (the bed) susceptible to erosion or deposition of sediments. In most movable-bed models, the sediments are *non cohesive*, as in the case of sand and gravel. The case of *cohesive sediments*, such as clays and silt, is well different since their behaviour is difficult to reproduce in the model because it is highly dependent on the sequence of events. Clays, for example, are characterised by a wide variety of rheological parameters based on the degree of consolidation, which, in turn, may depend on the history of the loads. Much remains to be investigated to adequately frame the behaviour of cohesive sediments, which are prevalent in many natural environments.

11.1 Conditions of Similarity in Rivers in the Presence of a Movable Bed

In movable-bed physical models, we need to add all the variables involved in sediment transport, initially by assuming that the transport is essentially in the bed load regime. We assume that the relevant variables are the density of the sediments ρ_s, their representative diameter d, the friction velocity u_*, the density of water ρ, the kinematic viscosity of water v, and a scale length of the water flow field h (for example, the hydraulic radius). The physical process can be described with the typical equation

$$f(\rho,\ v,\ h,\ u_*,\ d,\ \rho_s,\ g) = 0, \tag{11.1}$$

© The Author(s), under exclusive license to Springer Nature Switzerland AG 2021
S. G. Longo, *Principles and Applications of Dimensional Analysis and Similarity*,
Mathematical Engineering, https://doi.org/10.1007/978-3-030-79217-6_11

with a dimensional matrix of rank 3. Applying Buckingham's Theorem and with a physically based selection of dimensionless groups, Eq. (11.1) reduces to:

$$\tilde{f}\left(\frac{u_* d}{\nu}, \frac{\rho u_*^2}{g d (\rho_s - \rho)}, \frac{\rho_s}{\rho}, \frac{h}{d}\right) = 0. \qquad (11.2)$$

The first two groups are the Reynolds number for the sediments,

$$\mathrm{Re}_* = \frac{u_* d}{\nu}, \qquad (11.3)$$

and the Froude number for the sediments, better known as number of Shields,

$$\Theta = \frac{\rho u_*^2}{g d (\rho_s - \rho)}. \qquad (11.4)$$

The remaining groups have immediate physical significance. We limit the analysis to a stream that has reached the transport capacity, with the maximum sediment load compatible with the kinematic characteristics of the water stream. The load capacity of the stream may not be saturated if sediments are scarce, as happens, for instance, in torrents. If we are interested in the spatial evolution of sediment transport, we have to include an initial concentration of sediments and the spatial coordinate of the cross-section, since the process is space varying.

The characteristics of the water stream are a function of the variables in Eq. (10.1) and can be described with 4 dimensionless groups: Reynolds and Froude numbers, bottom slope and relative roughness. As usual, in the turbulence regime, we neglect the Reynolds number and add the 4 dimensionless groups related to the sediment dynamics to obtain the following similarity equations:

$$\begin{cases} r_{u_*} = \dfrac{1}{r_d}, \\[2mm] r_{u_*}^2 = r_d \, r_{(\rho_s - \rho)}, \\[2mm] r_{\rho_s} = 1, \\[2mm] \lambda = r_d, \\[2mm] r_U = \sqrt{\lambda}, \\[2mm] r_{i_b} = 1, \\[2mm] r_{k_s} = \lambda, \end{cases} \qquad (11.5)$$

where we have assumed the same fluid in the model and in the prototype, with $r_\rho = 1$, $r_\nu = 1$, $r_g = 1$.

The second equation is reduced to $r_{u_*}^2 = r_d$, as opposed to the first equation, and the system has no solution. To force similarity, we could use a different fluid in the model and prototype, but handling a model where the circulating fluid is different from water is difficult, complex and expensive. In addition, the fourth equation may require the use of sediments so small that they fall into the field of silts and clays, which are cohesive sediments. We remind us that in cohesive sediments, the surface forces dominate the body forces and control the rheological behaviour: reducing the size of the sediments increases the relevance of the surface with respect to volume with a consequent enhancement of surface forces.

Therefore, partial similarity is compulsory, with only some of the dimensionless groups respecting the condition of similarity.

11.1.1 The Undistorted Models: Reynolds Number for the Sediments → ∞

When the Reynolds number for the sediments is very large, the typical equation (11.2) simplifies as

$$\tilde{f}\left(\frac{\rho\, u_*^2}{g\,(\rho_s - \rho)\, d},\ \frac{\rho_s}{\rho},\ \frac{h}{d}\right) = 0, \tag{11.6}$$

and the conditions of similarity (adding the conditions for the water stream) become

$$\begin{cases} r_{u_*}^2 = r_d\, r_{(\rho_s - \rho)}, \\ r_{\rho_s} = 1, \\ \lambda = r_d, \\ r_U = \sqrt{\lambda}, \\ r_{i_b} = 1, \\ r_{k_s} = \lambda; \end{cases} \tag{11.7}$$

hence,

$$\begin{cases} r_U = r_{u_*} = \sqrt{\lambda}, \\ r_d = r_{k_s} = \lambda, \\ r_\rho = r_{\rho_s} = r_{i_b} = 1. \end{cases} \tag{11.8}$$

This involves Froude similarity with the further assumption that roughness is controlled by the grain diameter only. The volumetric flow rate of water scales according to $\lambda^{5/2}$, and the volumetric flow rate of sediments per unit width scales with $\lambda^{3/2}$.

Example 11.1 We wish to plan a physical model with a movable bed to reproduce a very wide riverbed with water depth $h = 3.80$ m and bottom slope $i_b = 0.5\%$, with a median diameter of the sediments $d_{50} = 30$ mm. The relative specific gravity of the sediment is $s = 2.65$.

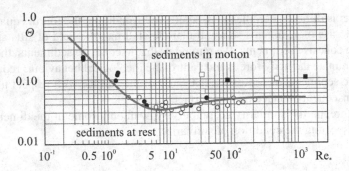

Fig. 11.1 Shields abacus

We first check the mobility of sediments in the prototype, which is guaranteed if the Shields parameter exceeds the critical value; see Fig. 11.1. The average tangential stress at the walls is equal to

$$\tau = \gamma_f \, R \, i_b = 9800 \times 3.80 \times 0.5/100 = 186 \text{ Pa.} \tag{11.9}$$

The friction velocity is

$$u_* = \sqrt{\frac{\tau}{\rho}} = \sqrt{\frac{186}{1000}} = 0.43 \text{ m s}^{-1}. \tag{11.10}$$

The Reynolds sediment number is $\text{Re}_* = u_* \, d/\nu = 0.43 \times 0.03/10^{-6} = 12\,900$, and the Shields parameter is $\Theta = u_*^2/(g \, d \, (s-1)) = 0.43^2/(9.806 \times 0.03 \times 1.65) = 0.38$. Since $\Theta > \Theta_c$, the sediments mobilise. The flow at the scale of the sediments is in a turbulent regime in the prototype, and the model must be designed with a length scale that guarantees the mobility of the sediments in the same regime.

The Reynolds number for the sediments has a scale ratio equal to

$$\frac{\text{Re}_{*,m}}{\text{Re}_{*,p}} = \frac{r_{u_*} \, r_d}{r_\nu} = \lambda^{3/2}. \tag{11.11}$$

By imposing that $\text{Re}_{*,m} > 70$,

$$\lambda > \left(\frac{70}{\text{Re}_{*,p}}\right)^{2/3} = \left(\frac{70}{12\,900}\right)^{2/3} = \frac{1}{32}. \tag{11.12}$$

Selecting a length scale $\lambda = 1/30$ yields:

$$\begin{cases} i_{b,m} = 0.5\%, \\ d_m = 30/30 = 1 \text{ mm}, \\ h_m = 380/30 = 12.7 \text{ cm}. \end{cases} \qquad \therefore \quad (11.13)$$

The volumetric flow rate of water scales as $r_Q = \lambda^{5/2} = (1/30)^{5/2} = 1/4930$, and the volumetric flow rate of sediment per unit width scales as $r_q = \lambda^{3/2} = (1/30)^{3/2} = 1/164$.

The analysis refers to a steady state regime for both the water stream and the sediment transport, assuming that the phenomena of practical interest can be considered as a sequence of stationary states neglecting inertia (quasi-stationary hypothesis). This approach hides the details of strongly nonstationary phenomena, such as sediment transport due to sea gravity waves in the surf and in the swash zone.

11.1.2 The Undistorted Models: Reynolds Number for the Sediments < 70

If $Re_* < 70$, the effect of viscosity is no longer negligible, and we cannot have a complete similarity. However, it is possible to simplify Eq. (11.2), in which the relative importance of the 4 dimensionless groups can be evaluated in relation to the dependent variable under analysis. For example, if we wish to analyse the motion of a single grain, eventually in the jumping regime, all the dimensionless groups are relevant; however, if we analyse an integral variable (the sediment flow rate, the average roughness of the bottom, the geometry of the bottom forms), the relative density does not intervene autonomously but is incorporated in the Shields parameter. In this last condition, Eq. (11.2) reduces to

$$\tilde{f}\left(\frac{u_* d}{\nu}, \frac{\rho u_*^2}{g\,(\rho_s - \rho)\,d}, \frac{h}{d}\right) = 0, \tag{11.14}$$

and the conditions of similarity (assuming, as usual, that $r_\rho = r_\mu = r_g = 1$) are:

$$\begin{cases} r_{u_*} = \dfrac{1}{r_d}, \\ r_{u_*}^2 = r_{(\rho_s-\rho)}\,r_d, \\ \lambda = r_d. \end{cases} \tag{11.15}$$

Unfortunately, some practical limitations prevent such a model from being realised. In fact, in uniform flow $r_{u_*} = \sqrt{r_{i_b}\,\lambda}$ and, therefore, $r_{i_b} = \lambda^{-3}$. In a model with $\lambda = 1/30$, the slope should be increased by a factor $r_{i_b} = 30^3 = 27\,000$. Even for large-scale models, we should increase the bottom slope dramatically, which renders useless any further investigation.

11.2   Hypothesis of Sediment Transport Independent of the Depth of the Water Stream

If we assume that the bed load sediment transport is essentially controlled by local parameters and does not depend on the depth of the stream h, it is possible to realise models in almost complete similarity as long as they are distorted. For sediment transport, we must have $r_\Theta = r_{Re_*} = 1$, and for the water stream, we must have $r_{Fr} = 1$ and $r_C^{-2} = r_{i_b} \equiv \lambda_y/\lambda_x$. If we assume the following resistance law in the presence of sediment (Julien 2002),

$$U = 5.75 \log_{10}\left(\frac{12.2R}{k_s}\right)\sqrt{g\,R\,i_b}, \tag{11.16}$$

approximating the logarithmic expression with a power function results in

$$\begin{cases} U = a\left(\dfrac{d}{h}\right)^m \sqrt{g\,R\,i_b} \equiv C\sqrt{g\,R\,i_b}, \\[2mm] m = 1/\ln\left(12.2\dfrac{h}{d}\right), \end{cases} \tag{11.17}$$

which, for $m = 1/6$, becomes the Manning-Strickler expression. The energy balance condition is

$$r_C = \frac{\lambda_y^m}{r_d^m}. \tag{11.18}$$

Substituting, we obtain:

$$\begin{cases} r_{u_*} = \dfrac{1}{r_d}, \\[2mm] r_{u_*}^2 = r_d\, r_{(\rho_s-\rho)}, \\[2mm] r_C = \dfrac{\lambda_y^m}{r_d^m}, \\[2mm] r_{U_x} = \lambda_y^{1/2}, \end{cases} \tag{11.19}$$

with the following solution:

$$r_{u_*} = \lambda_y^{\left(\frac{2m+2}{2m-1}\right)}, \quad r_d = \lambda_y^{\left(\frac{2m-1}{2m+2}\right)}, \quad r_{(\rho_s-\rho)} = \lambda_y^{\left(\frac{3-6m}{2m+2}\right)}, \quad r_{U_x} = \lambda_y^{1/2}, \tag{11.20}$$

and the horizontal length scale is

$$\lambda_x = \lambda_y^{\left(\frac{4m+1}{m+1}\right)}. \tag{11.21}$$

Table 11.2 lists the scales of other variables of interest: column (a) refers to a distorted model also in the plane, column (b) refers to a planimetrically undistorted model, and column (c) refers to a planimetrically undistorted model for $m = 1/6$.

From the data in the column (c) results $r_d > 1$ and $r_{(\rho_s - \rho)} < 1$; that is, the sediments in the model are of a larger size and lower specific gravity than the sediments in the prototype. This simplifies the realisation of models that reproduce very small incoherent granular materials. However, the use of sediments with a specific gravity too close to that of water should be avoided because it might lead to some practical difficulties in managing the model.

The scale of the volumetric flow rate of sediments in a channel with width b is calculated considering that

$$Q_s = b\,q \rightarrow r_{Q_s} = r_b\,r_q \equiv \lambda_z\,r_q. \qquad (11.22)$$

The dimensionless group in which the volumetric flow rate per unit of width appears is

$$\frac{q}{\sqrt{g\,d^3\,(\rho_s - \rho)/\rho}} \rightarrow r_q = r_d^{3/2}\,r_{(\rho_s - \rho)}^{1/2}\,r_\rho^{-1/2}; \qquad (11.23)$$

hence,

$$r_{Q_s} = \lambda_z\,r_d^{3/2}\,r_{(\rho_s - \rho)}^{1/2}\,r_\rho^{-1/2}. \qquad (11.24)$$

11.2.1 Hypothesis of Sediment Transport Independent of the Depth of the Water Stream and Reynolds Number for the Sediments $\rightarrow \infty$

If the Reynolds number for the sediments is so large to have negligible effects, the similarity equations are:

$$\begin{cases} r_{u_*}^2 = r_d\,r_{(\rho_s - \rho)}, \\[2mm] r_C = \dfrac{\lambda_y^m}{r_d^m}, \\[2mm] r_{U_x} = \lambda_y^{1/2}. \end{cases} \qquad (11.25)$$

The system admits the solution:

$$r_{u_*} = r_d^m\,\lambda_y^{\left(\frac{1-2m}{2}\right)}, \quad r_{(\rho_s - \rho)} = r_d^{2m-1}\,\lambda_y^{(1-2m)}, \quad r_{U_x} = \lambda_y^{1/2}, \qquad (11.26)$$

and the horizontal length scale is

$$\lambda_x = r_d^{-2m}\,\lambda_y^{(1+2m)}. \qquad (11.27)$$

Table 11.2 lists the scales for other variables: column *(d)* refers to a distorted model even in the plane, and column *(e)* refers to a planimetrically undistorted model. We can fix two scales, for example, the sediment scale and the vertical geometric scale. If we seek a planimetrically distorted model, the transverse geometric scale λ_z is arbitrary.

11.3 The Bottom in the Presence of Dunes, Ripples and Other Bedforms: The Calculation of the Equivalent Roughness

Under certain conditions, an initially flat bottom develops bedforms with characteristics also controlled by the stream depth. These bedforms increase the apparent roughness, and therefore, it is necessary to include their effect in the roughness calculation.

In the energy balance of the current, dissipation is due to (i) surface roughness (skin friction), (ii) bedforms (shape resistance), and (iii) plano-altimetric variations of the riverbed. Assuming that the total energy gradient is the sum of the three contributions, we can write:

$$J \equiv -\frac{dH}{dx} = J_1 + J_2 + J_3 \equiv (E_1 + E_2 + E_3)\frac{U^2}{g\,h}. \tag{11.28}$$

This is equivalent to expressing the tangential stress at the walls as the sum of the tangential stresses associated with the three distinct contributions:

$$\tau_0 = \tau_0' + \tau_0'' + \tau_0'''. \tag{11.29}$$

Tangential stresses can be expressed either as

$$\begin{cases} \tau_0' = \rho\,g\,J'\,h, \\ \tau_0'' = \rho\,g\,J''\,h, \\ \tau_0''' = \rho\,g\,J'''\,h, \\ J' + J'' + J''' = J, \end{cases} \tag{11.30}$$

or as

$$\begin{cases} \tau_0' = \rho\,g\,J\,h', \\ \tau_0'' = \rho\,g\,J\,h'', \\ \tau_0''' = \rho\,g\,J\,h''', \\ h' + h'' + h''' = h. \end{cases} \tag{11.31}$$

In terms of the friction velocity, we can assume

$$u_*^2 = u_*'^2 + u_*''^2 + u_*'''^2, \tag{11.32}$$

and, as a function of the Chézy dimensionless coefficient,

$$\frac{1}{C^2} = \frac{1}{C'^2} + \frac{1}{C''^2} + \frac{1}{C'''^2}. \tag{11.33}$$

For a flat bottom with limited sediment movement and in sheet-flow conditions, the contribution to the resistance of bedforms is zero, although in the latter regime, an additional contribution due to the intense momentum flux of the sediments should be added.

A problem arises in quantifying the different contributions to dissipation. Leaving aside the effects of the plano-altimetric variations of the riverbed, we restrict the analysis to the case in which only bedforms are present, with the wall tangential stress

$$\tau_0 = \tau_0' + \tau_0''. \tag{11.34}$$

According to Einstein & Barbarossa (1952), the dimensionless Chézy coefficient due to skin friction can be expressed as

$$\frac{U}{u_*'} \equiv C' = f\left(\frac{u_*' k_s'}{\nu}, \frac{h'}{k_s'}\right), \tag{11.35}$$

where k_s' is the geometric roughness associated with skin friction. For $u_*' k_s'/\nu \to \infty$, we can assume

$$\frac{U}{u_*'} \equiv C' = 7.66\left(\frac{h'}{k_s'}\right)^{1/6}, \tag{11.36}$$

which is rigorously valid in the presence of skin friction only. Pragmatically, we assume the same dependence in the presence of additional resistance due to bedforms.

Again, according to Einstein & Barbarossa (1952), the Chézy coefficient of the contribution of the bottom forms C'' is a function of the *number of mobility* of sediments (the reciprocal of the Shields parameter), defined as

$$\Psi_{35}' = \frac{\rho_s - \rho}{\rho} \frac{d_{35}}{h' J}, \tag{11.37}$$

where d_{35} is the sieve size at 35% passing in weight. The experimental function $C'' = f\left(\Psi_{35}'\right)$ is plotted in Fig. 11.2. The calculation procedure is iterative: given the values of h, d_{35} and J, we wish to calculate the average stream velocity.

We set a value $h' < h$ of the first attempt and then calculate the friction velocity u_*', the average velocity of the current U of the first attempt, the Chézy coefficient of the skin friction C', and the mobility parameter Ψ_{35}'; from the empirical diagram, we

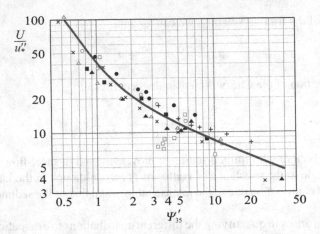

Fig. 11.2 Experimental data relating the dimensionless Chézy coefficient for bedform resistance and the mobility number due to skin friction (modified from Einstein & Barbarossa 1952)

estimate C'', the friction velocity u_*'', and the value of $h'' = u_*''^2/(g\,J)$. If the sum $h' + h''$ differs from h, another value is picked for h' until equality is achieved.

There are numerous variants of the method, and many authors detect a systematic deviation of the curve proposed by Einstein & Barbarossa (1952) compared to the experimental data, justified by the fact that C'' is a function of Ψ' and of other groups.

11.3.1 The Conditions of Similarity for Sediment and Water Streams in the Presence of Bedforms

The similarity conditions of the water stream (see Sect. 10.1.3) are

$$r_{i_b} = \frac{\lambda_y}{\lambda_x}, \quad r_E = \frac{\lambda_y}{\lambda_x}, \quad r_{\text{Fr}} = 1, \tag{11.38}$$

where E is the generalised energy loss in Eq. (10.55).

The second condition replaces and generalises the $r_C^2 = \lambda_x/\lambda_y$ condition of Eq. (10.62), and it can be rewritten as

$$r_E \equiv \frac{(E_1 + E_2 + E_3)_m}{(E_1 + E_2 + E_3)_p} = \frac{\lambda_y}{\lambda_x}, \tag{11.39}$$

which is certainly satisfied if the following conditions are met:

$$r_{E_1} = r_{E_2} = r_{E_3} = \frac{\lambda_y}{\lambda_x}. \tag{11.40}$$

As shown in Eq. (10.57), the condition $r_{E_3} = \lambda_y/\lambda_x$ is automatically satisfied. In addition, on a flat bottom, the contribution of the bedforms is null, and the similarity results in

$$r_{E_1} = \frac{\lambda_y}{\lambda_x} \equiv n, \tag{11.41}$$

or,

$$r_{C'} \equiv \frac{C'_m}{C'_p} = \sqrt{\frac{\lambda_x}{\lambda_y}} \equiv \frac{1}{\sqrt{n}}. \tag{11.42}$$

If bedforms are present, their contribution must be included. The coefficient of dissipation due to the bedforms can be expressed as follows:

$$E_2 = \frac{1}{2} \frac{\Delta^2}{\Lambda\, h}, \tag{11.43}$$

where Δ and Λ are the height and length of bed ripples, respectively. The condition of similarity $r_{E_2} = \lambda_y/\lambda_x$ requires that

$$r_{E_2} \equiv \frac{\Delta_m^2}{\Delta_p^2} \frac{\Lambda_p\, h_p}{\Lambda_m\, h_m} = \frac{\lambda_y}{\lambda_x} \equiv n. \tag{11.44}$$

Suppose the current is tranquil (or $Fr < 1$) and that the bedforms are ripples, assuming that these forms exist for $(u_* d/\nu) <\approx 20$. As long as $h/d <\approx 1000$, the geometrical characteristics of the ripples depend only on the diameter of the sediments and not on the water level. If, on the other hand, $h/d > 1000$, then only ripples develop for $(u_* d/\nu) <\approx 8$, and only dunes develop for $(u_* d/\nu) >\approx 24$; both ripples and dunes coexist in the range $8 <\approx (u_* d/\nu) <\approx 24$.

The geometrical characteristics of the bedforms are proportional to the diameter of the sediments, with coefficients of proportionality depending on the Reynolds number for the sediments and the Shield parameter. Since the model satisfies the similarity of these two parameters, it also results in

$$r_\Delta = r_\Lambda = r_d, \tag{11.45}$$

and Eq. (11.44) reduces to

$$r_{E_2} \equiv \frac{\Delta_m^2}{\Delta_p^2} \frac{\Lambda_p\, h_p}{\Lambda_m\, h_m} = \frac{r_d}{\lambda_y}. \tag{11.46}$$

To have $r_{E_2} = \lambda_y/\lambda_x$ requires $r_d = \lambda_y^2/\lambda_x$. Based on the values of r_d listed in Table 11.2, if the bedforms are ripples, the latter condition is never met, and the similarity of energy dissipation cannot be achieved.

On the other hand, if the bedforms are dunes (with $u_* d/\nu > \approx 20$), their height and length are proportional to the water depth and do not depend on the diameter of the sediments; hence,

$$r_{E_2} \equiv \frac{\Delta_m^2}{\Delta_p^2} \frac{\Lambda_p h_p}{\Lambda_m h_m} = 1, \tag{11.47}$$

which satisfies the condition of Eq. (11.44) only if the physical model is geometrically undistorted. This means that although dunes can grow in a distorted model, the energy balance of the bedforms is not properly scaled. In fact, in the presence of ripples,

$$r_{E_2} = \frac{r_d}{\lambda_y} = \lambda_y^{-1.286} > r_{i_b} \equiv \lambda_y^{-0.429} \quad \text{for} \quad m = 1/6, \tag{11.48}$$

and the model dissipates more than it should, with water depth in the model lower than expected.

In the presence of dunes,

$$r_{E_2} = 1 < r_{i_b} \equiv \lambda_y^{-0.429} \quad \text{for} \quad m = 1/6, \tag{11.49}$$

and the model dissipates less than it should, with water depth in the model higher than expected. However, the scale effect due to the lack of perfect similarity for energy dissipation is less important than expected. In fact, the maximum steepness of the bedforms is $\Delta/\Lambda < 1/10$, and the maximum relative height is $\Delta/h < 1/5$; therefore, the maximum dissipation coefficient is equal to:

$$E_{2,max} = \frac{1}{2} \frac{\Delta^2}{\Lambda h} = \frac{1}{100}. \tag{11.50}$$

This is a modest value, almost completely negligible compared to other localised dissipations, due to plano-altimetric variations and meanders.

11.4   Time Scales in Distorted Movable-Bed Models

When estimating time scales, it is necessary to refer to the significant variables involved. If we consider the jumping process of a single moving grain, the geometric scale is the grain diameter, and the velocity scale is the friction velocity. Therefore, for a complete similarity of the Reynolds sediment number and for $m = 1/6$,

$$r_{t_s} = \frac{r_d}{r_{u_*}} = \lambda_y^{-0.571}. \tag{11.51}$$

The time scale for the stream, see Table 11.2, is

$$r_t = \lambda_y^{0.928},$$ (11.52)

and the ratio between the two time scales is equal to

$$\frac{r_{t_s}}{r_t} = \frac{1}{\lambda_y^{3/2}}.$$ (11.53)

For example, if the model has a vertical length scale $\lambda_y = 1/20$, the time scale for the stream is $r_t = (1/20)^{0.928} \approx 1/16$, and the time scale for sediments is $r_{t_s} = (1/20)^{-0.571} = 5.5$. A physical process involving the motion of the sediments that in the prototype lasts 1 h, lasts 5.5 h in the model; if the physical process lasting 1 h in the prototype is related to the fluid motion, in the model, it lasts approximately 4 min. This means that the kinematics of the sediment are strongly delayed with respect to the kinematics of the fluid.

To estimate the time scales of evolution of the bedforms, we consider the equation by Exner (1925):

$$\frac{\partial \eta}{\partial t} = -\frac{1}{(1-\varepsilon)} \frac{\partial q}{\partial x},$$ (11.54)

where η is the bottom level, ε is the porosity of the sediments and q is the volumetric sediment flow rate per unit width. The time scale of the bottom evolution, indicated with $r_{t,bf}$, is equal to

$$r_{t,bf} = \frac{\lambda_y \lambda_x}{r_q} r_{(1-\varepsilon)}.$$ (11.55)

Assuming that the porosity is the same in the model and the prototype ($r_{(1-\varepsilon)} = 1$), a complete similarity of the sediments results in

$$r_{t,bf} = \lambda_y^{\left(\frac{2+5m}{1+m}\right)} \rightarrow r_{t,bf} = \lambda_y^{2.43} \quad \text{for } m = 1/6,$$ (11.56)

and the ratio between the evolutionary time scale of the bedforms and the time scale of the stream is equal to:

$$\frac{r_{t,bf}}{r_t} = \lambda_y^{3/2}.$$ (11.57)

In approximate similarity of the Reynolds sediment number,

$$r_{t,bf} = r_d^{-3m-1} \lambda_y^{\left(\frac{3+6m}{2}\right)}.$$ (11.58)

Table 11.2 lists the scale ratios for movable-bed physical models in Froude similarity.

Example 11.2 We wish to realise a physical model of a river trunk 450 m long, with an average riverbed width of 43 m and an average water depth of 1.00 m, having a useful space in the laboratory of 12 m maximum size. Suppose that the sediments in the prototype have a density of $\rho_{s,p} = 2700$ kg m^{-3} and that their representative diameter is $d_p = 10$ mm. The volumetric flow rate of the water is $Q_p = 90$ m^3 s^{-1}.

The planimetric length scale cannot be greater than $\lambda_x = 12/450 \equiv 1/37.5$. We set $\lambda_x = 1/40$, and with a distortion ratio $n = 2$, we calculate $\lambda_y = n\lambda_x = 2 \times 1/40 = 1/20$. The river trunk in the model will have a length of 11.25 m, a width of 1.08 m and a water depth of 5 cm.

We select a sediment ratio of 1/4; hence, $d_m = 10/4 = 2.5$ mm. The density of the sediments in the model must be equal to

$$r_{(\rho_s - \rho)} = \left(\frac{\lambda_y}{r_d}\right)^{2/3} = \left(\frac{1/20}{1/4}\right)^{2/3} \approx \frac{1}{3} \rightarrow$$

$$(\rho_s - \rho)_m = \frac{2700 - 1000}{3} = 565 \text{ kg m}^{-3} \rightarrow \tag{11.59}$$

$$\rho_{s,m} = 565 + 1000 = 1565 \text{ kg m}^{-3}.$$

Table 11.1 lists the data for some materials currently used for movable-bed physical models. When selecting the material, it is also necessary to check that, in the case of recirculation, the grains do not enter the impellers of the pumps. In fact, in many applications, it is necessary to guarantee a solid flow rate feed, normally by means of a hopper with an adjustable flow rate, which regulates the flow of sediment transferred from the closing section of the model, downstream, to the inlet section.

On the basis of the characteristics of the materials listed in Table 11.1, we can use Bakelite or coal. The volumetric flow rate of water in the model is equal to:

$$r_Q = \lambda_x \lambda_y^{3/2} = \frac{1}{40} \times \left(\frac{1}{10}\right)^{3/2} \approx \frac{1}{1265} \rightarrow$$

$$Q_m = Q_p r_Q = 90 \times \frac{1}{1265} = 71 \text{ l s}^{-1}. \tag{11.60}$$

The sediment mass flow rate, measured in the model, is equal to 0.27 kg s^{-1}, corresponding to a volumetric flow rate $Q_{s,m} = 0.27/1.565 = 0.172$ l s^{-1}.
The volumetric flow rate of the sediment in the prototype is equal to:

$$r_{Q_s} = \sqrt{n}\,\lambda_y^{5/2} = \sqrt{2} \times \left(\frac{1}{20}\right)^{5/2} \approx \frac{1}{1268} \rightarrow$$

$$Q_{s,p} = \frac{Q_{s,m}}{r_{Q_s}} = 0.172 \times 1268 \approx 0.219 \text{ m}^3 \text{ s}^{-1}. \tag{11.61}$$

Table 11.1 Some materials that can be used for the bed of movable-bed models

Material	Relative specific weight to water	Commercial diameter mm	Note
Polystyrene	1.035–1.05	0.5–3.0	Stable but tends to float and does not wet
Araldite (resin)	1.14	0.2–0.5	
Nylon	1.16	0.1–5.0	
PVC	1.14–1.25	1.5–4.0	Hydrophobic
Perspex	1.18–1.19	0.3–1.0	Dusty
Coal	1.2–1.6	0.3–4.0	Particle size and density are non-homogeneous
ABS	1.22	2.0–3.0	Sticky air bubbles
Nutshells	1.33	0.1–0.4	Deteriorates in a few weeks and dirty the water
Bakelite	1.38–1.49	0.3–4.0	Porous; swells and floats
Lytag®	1.52	1–8.0	Porous
Pumice	1.4–1.7		
Quartz sand	2.65	0.1–1.0	

The time ratio for water flow processes is equal to

$$r_t = \frac{\sqrt{\lambda_y}}{n} = \frac{\sqrt{1/20}}{2} = \frac{1}{8.95}, \qquad \therefore \quad (11.62)$$

while the evolution time ratio of the bedforms is equal to

$$r_{t,bf} = \frac{\lambda_y^{5/3}}{n\, r_d^{7/6}}\, r_{(1-\varepsilon)} = \frac{(1/20)^{5/3}}{2\times(1/4)^{7/6}} \approx \frac{1}{58}. \qquad \therefore \quad (11.63)$$

11.5 Localised Phenomena

The analysis of the behaviour of structures built in riverbeds, with the presence of localised phenomena of deposit or erosion, often requires a physical model. It is possible to reduce the number of variables involved, compared to the more general case, since the fluid velocity is generally high and the motion at the length scale of the sediments is almost always in a turbulent regime (at least in the initial phase of the phenomenon).

Table 11.2 Summary of scale ratios for movable-bed physical models in Froude similarity

Scale		$\lambda_z \neq \lambda_x$ (a)	Full $\lambda_z \equiv \lambda_x$ (b)	$\lambda_z \equiv \lambda_x$, $m=1/6$ (c)	Partial with $r_{Re_*} \neq 1$, $\lambda_z \neq \lambda_x$ (d)	$\lambda_z \equiv \lambda_x$ (e)
Geometric						
Depth	λ_y	λ_y	λ_y	λ_y	λ_y	λ_y
Length	λ_x		$\lambda_y^{\left(\frac{1+4m}{1+m}\right)}$	$\lambda_y^{1.43}$	$r_d^{-2m}\lambda_y^{1+2m}$	$r_d^{-2m}\lambda_y^{1+2m}$
Width	λ_z	λ_z	$\lambda_y^{\left(\frac{1+4m}{1+m}\right)}$	$\lambda_y^{1.43}$	λ_z	$r_d^{-2m}\lambda_y^{1+2m}$
Area of the cross-section	r_Ω	$\lambda_z\lambda_y$	$\lambda_y^{\left(\frac{2+5m}{1+m}\right)}$	$\lambda_y^{2.43}$	$\lambda_z\lambda_y$	$r_d^{-2m}\lambda_y^{2+2m}$
Volume	r_V	$\lambda_z\lambda_y\lambda_x^{\left(\frac{2+5m}{1+m}\right)}$	$\lambda_y^{\left(\frac{3+9m}{1+m}\right)}$	$\lambda_y^{3.86}$	$\lambda_z r_d^{-2m}\lambda_y^{2+2m}$	$r_d^{-4m}\lambda_y^{3+4m}$
Diameter of sediments	r_d	$\lambda_y^{\left(\frac{2m-1}{2+2m}\right)}$	$\lambda_y^{\left(\frac{2m-1}{2+2m}\right)}$	$\lambda_y^{-0.286}$	r_d	r_d
Kinematic						
Time (water stream)	r_t	$\lambda_y^{\left(\frac{1+7m}{2+2m}\right)}$	$\lambda_y^{\left(\frac{1+7m}{2+2m}\right)}$	$\lambda_y^{0.928}$	$r_d^{-2m}\lambda_y^{\left(\frac{1+4m}{2}\right)}$	$r_d^{-2m}\lambda_y^{\left(\frac{1+4m}{2}\right)}$
Time (sediments)	$r_{t,s}$	$\lambda_y^{\left(\frac{2m-1}{1+m}\right)}$	$\lambda_y^{\left(\frac{2m-1}{1+m}\right)}$	$\lambda_y^{-0.571}$	$r_d^{1-m}\lambda_y^{\left(\frac{2m-1}{2}\right)}$	$r_d^{1-m}\lambda_y^{\left(\frac{2m-1}{2}\right)}$
Time (bedforms)	$r_{t,bf}$	$\lambda_y^{\left(\frac{2+5m}{1+m}\right)}$	$\lambda_y^{\left(\frac{2+5m}{1+m}\right)}$	$\lambda_y^{2.429}$	$r_d^{-1-3m}\lambda_y^{\left(\frac{3+6m}{2}\right)}$	$r_d^{-1-3m}\lambda_y^{\left(\frac{3+6m}{2}\right)}$
Horizontal velocity	r_{U_x}	$\lambda_y^{1/2}$	$\lambda_y^{1/2}$	$\lambda_y^{1/2}$	$\lambda_y^{1/2}$	$\lambda_y^{1/2}$
Vertical velocity	r_{U_y}	$\lambda_y^{\left(\frac{1-5m}{2+2m}\right)}$	$\lambda_y^{\left(\frac{1-5m}{2+2m}\right)}$	$\lambda_y^{0.071}$	$r_d^{-2m}\lambda_y^{\left(\frac{1-m}{2}\right)}$	$r_d^{-2m}\lambda_y^{\left(\frac{1+4m}{2}\right)}$
Friction velocity	r_{u_*}	$\lambda_y^{\left(\frac{1-2m}{2+2m}\right)}$	$\lambda_y^{\left(\frac{1-2m}{2+2m}\right)}$	$\lambda_y^{0.286}$	$r_d^{m}\lambda_y^{\left(\frac{1-2m}{2}\right)}$	$r_d^{m}\lambda_y^{\left(\frac{1-2m}{2}\right)}$
Flow rate	r_Q	$\lambda_z\lambda_y^{3/2}$	$\lambda_y^{\left(\frac{5+11m}{2+2m}\right)}$	$\lambda_y^{2.929}$	$\lambda_z\lambda_y^{3/2}$	$r_d^{-2m}\lambda_y^{\left(\frac{5+4m}{2}\right)}$
Sediment flow rate per unit width	r_q	1	1	1	$r_d^{1+m}\lambda_y^{\left(\frac{1-2m}{2}\right)}$	$r_d^{1+m}\lambda_y^{\left(\frac{1-2m}{2}\right)}$

(continued)

Table 11.2 (continued)

Scale		$\lambda_z \neq \lambda_x$ (a)	Full $\lambda_z \equiv \lambda_x$ (b)	$\lambda_z \equiv \lambda_x\ m = 1/6$ (c)	Partial with $r_{Re_*} \neq 1$ $\lambda_z \neq \lambda_x$ (d)	$\lambda_z \equiv \lambda_x$ (e)
Dynamics						
Mass	r_m	$\lambda_z \lambda_y^{\left(\frac{2+5m}{1m}\right)}$	$\lambda_y^{\left(\frac{3+9m}{1+m}\right)}$	$\lambda_y^{3.857}$	$\lambda_z r_d^{-2m} \lambda_y^{2+2m}$	$r_d^{-2m} \lambda_y^{3+4m}$
Pressure	r_p	λ_y	λ_y	λ_y	λ_y	λ_y
Tangential stress	r_τ	$\lambda_y^{\left(\frac{1-2m}{1+m}\right)}$	$\lambda_y^{\left(\frac{1-2m}{1+m}\right)}$	$\lambda_y^{0.571}$	$r_d^{2m} \lambda_y^{1-2m}$	$r_d^{2m} \lambda_y^{1-2m}$
Density of sediments	$r(\rho_s-\rho)$	$\lambda_y^{\left(\frac{3-6m}{2+2m}\right)}$	$\lambda_y^{\left(\frac{3-6m}{2+2m}\right)}$	$\lambda_y^{0.857}$	$r_d^{2m-1} \lambda_y^{1-2m}$	$r_d^{2m-1} \lambda_y^{1-2m}$
Force	r_F	$\lambda_z \lambda_y^{\left(\frac{2+5m}{1+m}\right)}$	$\lambda_y^{\left(\frac{3+9m}{1+m}\right)}$	$\lambda_y^{3.857}$	$\lambda_z r_d^{-2m} \lambda_y^{2+2m}$	$r_d^{-2m} \lambda_y^{3+4m}$
Dimensionless						
Slope	r_{i_f}	$\lambda_y^{\left(\frac{-3m}{1+m}\right)}$	$\lambda_y^{\left(\frac{-3m}{1+m}\right)}$	$\lambda_y^{-0.429}$	$r_d^{2m} \lambda_y^{-2m}$	$r_d^{2m} \lambda_y^{-2m}$
C Chézy	r_C	$\lambda_y^{\left(\frac{3m}{2+2m}\right)}$	$\lambda_y^{\left(\frac{3m}{2+2m}\right)}$	$\lambda_y^{0.214}$	$r_d^{-m} \lambda_y^{m}$	$r_d^{-m} \lambda_y^{m}$
Froude	r_{Fr}	1	1	1	1	1
Reynolds	r_{Re}	$\lambda_y^{3/2}$	$\lambda_y^{3/2}$	$\lambda_y^{3/2}$	$\lambda_y^{3/2}$	$\lambda_y^{3/2}$
Reynolds of sediments	r_{Re_*}	$\lambda_y^{3/2}$	$\lambda_y^{3/2}$	$\lambda_y^{3/2}$	$\lambda_y^{3/2}$	$\lambda_y^{3/2}$
Shields	r_{Θ_*}	1	1	1	1	1

Fig. 11.3 Schematic
diagram for the analysis of
bottom erosion near a bridge
pile

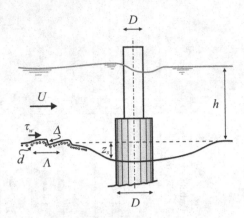

We consider the scouring of a bridge pile, according to the schematic shown in
Fig. 11.3. Numerous geometric scales are involved, such as the local depth and the
diameter of the pile, the geometry of the bedforms, the characteristic diameter of the
sediments and the grain size curve.

The physical process can be described with the typical equation

$$z_s = f\left(\rho, \ v, \ U, \ u_*, \ h, \ \rho_s, \ d_{50}, \ \sigma_g, \ D, \ K_1, \ K_2, \ K_3, \ g\right), \qquad (11.64)$$

where z_s is the depth of the excavation in equilibrium conditions, ρ is the density of
the water, v is the kinematic viscosity, U is the average velocity of the current, u_* is
the friction velocity, h is the water depth, ρ_s is the density of the sediment, d_{50} is the
median diameter of the grains, σ_g is the standard deviation of the grain size curve,
D is the size of the foundation pulvinus, K_1 and K_2 are form and alignment factors,
K_3 is a factor describing the geometry of the arrival channel and the characteristics
of the velocity field, and g is the acceleration of gravity.

Applying Buckingham's Theorem, the physical process can be expressed with 7
dimensionless groups and 4 parameters, for example, the following groups:

$$\frac{z_s}{D} = \tilde{f}\left(\frac{U^2}{g\,h}, \ \frac{\rho_s u_*^2}{g d_{50}(\rho_s - \rho)}, \ \frac{U d_{50}}{v}, \ \frac{\rho_s}{\rho}, \ \frac{h}{D}, \ \frac{d_{50}}{D}, \sigma_g, K_1, K_2, K_3\right). \qquad (11.65)$$

The Reynolds number for the sediments is always very high, and the associated
terms can be neglected. Considering the relative sediment density as constant, we
can restrict the analysis to less dimensionless groups, and ultimately, we obtain

$$\frac{z_s}{D} = \tilde{f}\left(\frac{U^2}{g\,h}, \ \frac{\rho_s\,u_*^2}{g\,d_{50}\,(\rho_s - \rho)}, \ \frac{h}{D}, \ \frac{d_{50}}{D}, \ \sigma_g, \ K_1, \ K_2, \ K_3\right). \qquad (11.66)$$

0

10 cm

Fig. 11.4 Physical model of scouring of bridge piles, realised with nutshells $d_{50} = 0.23$ mm, $\lambda = 1 : 150$, in Froude and Shields similarity, partial similarity with respect to other dimensionless groups (from Mignosa et al. 2017)

The similarity equations are:

$$\begin{cases} r_U = \sqrt{\lambda}, \\ r_{u_*}^2 = r_{d_{50}}, \\ r_{z_s} = r_{d_{50}} = r_D = \lambda, \\ r_{\sigma_g} = r_{K_1} = r_{K_2} = r_{K_3} = 1. \end{cases} \qquad (11.67)$$

The first condition is the classic Froude similarity for the water stream. The second condition derives from imposing the same dissipation level in the model and prototype. If the model is undistorted, this requires that the coefficient of Chézy assumes the same value in the model and prototype, including any bedform effects. In addition, sediments have a length scale equal to the geometric scale, which is easily achieved if they are large enough in the prototype. The limit of the geometric scale of the sediments derives from imposing that the Reynolds number for the sediments in the model is sufficiently high; see Eq. (11.12); hence,

$$r_{d_{50}} > \left(\frac{70}{\mathrm{Re}_{*,p}} \right)^{2/3}. \qquad (11.68)$$

Figure 11.4 shows the localised erosion around piles in a physical model in Froude-Shields similarity, partial with respect to other variables, and with $\lambda = 1 : 150$.

11.6   The Modelling of Sediment Transport in the Presence of Waves

In the analysis of coastal sediment transport, a class of approximate physical models respects only 3 dimensionless groups, with the following similarity equations:

$$
\begin{cases}
r_\rho \, r_{u_*}^2 = r_d \, r_{(\rho_s - \rho)}, \\
r_{\rho_s} = r_\rho, \\
\lambda = r_d.
\end{cases}
\tag{11.69}
$$

Using the same fluid in the model and prototype, it is necessary (i) to reduce the diameter of the sediments to a scale equal to the geometric scale of the model; (ii) to use sediments in the model with density equal to the density of the sediment in the prototype; and (iii) to scale the friction velocity according to $\lambda^{1/2}$. Since the similarity is approximate, the Reynolds number for the sediments is not preserved but is scaled according to the relationship

$$
r_{\frac{u_* d}{\nu}} \equiv r_{\mathrm{Re}_*} = \lambda^{3/2}.
\tag{11.70}
$$

In the presence of waves and currents, the analysis of sediment transport phenomena requires several additional variables associated with the unsteady flow field that governs the physical process, and the typical equation becomes

$$
q = f(\rho, \ \nu, \ l, \ \tau_b, \ d, \ \rho_s, g),
\tag{11.71}
$$

where q is the volumetric flow rate of sediment per unit width, l represents a geometric scale, and τ_b is the tangential stress at the bottom. The other variables have an immediate meaning. The dimensional matrix has rank 3, and it is possible to re-write Eq. (11.71) using only 5 dimensionless groups, for instance:

$$
\frac{q}{\sqrt{g \, d^3 \, (s-1)}} = \Phi \left(\frac{u_* \, d}{\nu}, \ \frac{u_*^2}{g \, d \, (s-1)}, \ \frac{\rho_s}{\rho}, \ \frac{l}{d} \right),
\tag{11.72}
$$

where $s = \rho_s / \rho$ and $u_* = \sqrt{\tau_b / \rho}$. In the presence of sediment transport in suspension, the sedimentation speed w of the grain becomes relevant, and we need to add the additional dimensionless group w/u_*. Equation (11.72) becomes

$$
\frac{q}{\sqrt{g \, d^3 \, (s-1)}} = \Phi \left(\frac{u_* \, d}{\nu}, \ \frac{u_*^2}{g \, d \, (s-1)}, \ \frac{\rho_s}{\rho}, \ \frac{l}{d}, \ \frac{w}{u_*} \right).
\tag{11.73}
$$

The scale ratio of the new group w/u_* is not preserved by Eq. (11.69). In fact, if the sediments in the model and prototype have a diameter between 0.13 mm and 1.0 mm,

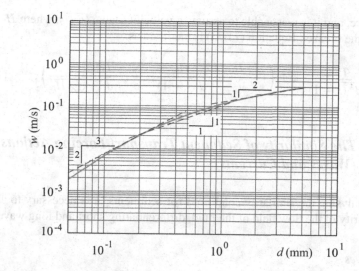

Fig. 11.5 Sedimentation speed for spherical grains

the sedimentation speed is approximately proportional to d (Fig. 11.5), and $r_w = \lambda$. Then, the scale ratio of the relative sedimentation speed is equal to

$$r_{\frac{w}{u_*}} = \lambda^{1/2}, \tag{11.74}$$

which for $\lambda < 1$ leads to a relative sedimentation speed in the model smaller than in the prototype.

The length scale l is controlled by the forcing of sediment motion and should coincide with the average wave height, in the presence of short waves, and with the local depth, in the presence of long waves.

According to Dalrymple (1989), it is appropriate to eliminate the l scale and to include the wave height and period in the list of variables, obtaining the following typical equation:

$$q = f(\rho, \, v, \, \tau_b, \, d, \, \rho_s, \, g, \, w, \, H, \, T). \tag{11.75}$$

However, only 5 of the 7 dimensionless groups are actually relevant:

$$\frac{q}{\sqrt{g\,d^3\,(s-1)}} = \Phi\left(\frac{u_* d}{v}, \; \frac{u_*^2}{g\,d\,(s-1)}, \; \frac{\rho_s}{\rho}, \; \frac{H}{w\,T}\right), \tag{11.76}$$

where the last dimensionless group is the *parameter of Dean*.

The two functional relations (11.72) and (11.76) are valid in the presence of bedload only and in the presence of an additional suspended load, respectively, and refer to a forcing action resulting from the interaction of the waves with the bottom. In the presence of breakers, the forcing action is also represented by the roller at the free surface, and it is reasonable to assume a velocity scale $\sqrt{g\,H_b}$, where H_b is the height of the breaker. Equation (11.73) is modified as:

$$\frac{q}{\sqrt{g\,d^3\,(s-1)}} = \Phi\left(\frac{\sqrt{g\,H_b}\,d}{\nu}, \frac{H_b}{d\,(s-1)}, \frac{\rho_s}{\rho}, \frac{H_b}{d}, \frac{w}{\sqrt{g\,H_b}}\right). \qquad (11.77)$$

11.6.1 The Similarity of Sediment Transport Forcing Actions (Waves and Currents)

Before analysing the similarity conditions for sediments, it is necessary to analyse the similarity of the flow field of the fluid, differentiating short and long waves, and currents.

Short waves

We first consider the presence of short waves, with a turbulent oscillating boundary layer at the bottom. We define the maximum tangential stress at the bottom:

$$\tau_{b,max} = \frac{\rho\,f_w\,U_{\delta,max}^2}{2}, \qquad (11.78)$$

and the friction factor:

$$f_w = 0.47\left(\frac{k_s}{a_\delta}\right)^{3/4}, \qquad (11.79)$$

where $U_{\delta,max}$ is the maximum fluid velocity at the upper edge of the boundary layer, k_s is the geometric roughness, and a_δ is the amplitude of the fluid particle oscillating motion at the upper edge of the boundary layer. Combining the two expressions results in

$$\tau_{b,max} = 0.24\,\rho\,U_{\delta,max}^2\left(\frac{k_s}{a_\delta}\right)^{3/4}. \qquad (11.80)$$

The condition of similarity requires that

$$r_{\tau_b} = r_\rho\,r_{U_\delta}^2\,\frac{r_{k_s}^{3/4}}{r_{a_\delta}^{3/4}}, \qquad (11.81)$$

or

$$\left(r_{\tau_b}\right)_{\text{short waves}} = \lambda^{1/4}\,r_{k_s}^{3/4} \rightarrow r_{u_*} = \lambda^{1/8}\,r_{k_s}^{3/8}. \qquad (11.82)$$

Long waves

In the presence of long waves, the structure of the boundary layer changes compared to that of short waves, and the expression of the maximum tangential stress at the bottom becomes (Yalin 1971)

$$\tau_{b,max} = \frac{\rho\, U_c^2}{\left[2.5 \ln \left(11 \frac{h_0}{k_s} \right) \right]^2},$$
(11.83)

where U_c is the vertically averaged speed of the current and h_0 is the thickness of the boundary layer (coincident with the depth of the current). Approximating the logarithmic law with a power function, we obtain:

$$2.5 \ln \left(11 \frac{h_0}{k_s} \right) \approx \text{const} \left(\frac{h_0}{k_s} \right)^{1/8},$$
(11.84)

which yields

$$\tau_{b,max} \propto \rho\, U_c^2 \left(\frac{h_0}{k_s} \right)^{-1/4}.$$
(11.85)

The condition of similarity requires

$$\left(r_{\tau_b} \right)_{long\ waves} = r_\rho\, r_{U_c}^2\, \frac{r_{k_s}^{1/4}}{\lambda^{1/4}} \rightarrow r_{u_*} = \lambda^{3/8}\, r_{k_s}^{1/8},$$
(11.86)

Under some assumptions (see Yalin 1971), the tangential stress scale ratio is equal to

$$r_\tau = \frac{\lambda_y^2}{\lambda_x}.$$
(11.87)

Short waves and currents (offshore models)

For this model, frequently adopted to reproduce sediment transport where oil platforms, submerged pipelines and short waves and currents are present, the horizontal velocity scale is calculated by equating (11.82) and (11.86):

$$r_{U_c} = \lambda^{1/4}\, r_{k_s}^{1/4} \rightarrow r_{U_c} = \lambda^{1/4}\, r_d^{1/4} \quad \text{(on a flat bottom)}.$$
(11.88)

Long waves and current (inshore models)

If long waves and currents in shallow water are simultaneously active in the mobilisation of sediments, a possible similarity criterion is computed by equating the scales of the bottom tangential stress for long waves (11.86) and for total stress (11.87), yielding the following relationship:

$$\frac{\lambda_y}{\lambda_x} = \frac{r_{k_s}^{1/4}}{\lambda_y^{1/4}}. \tag{11.89}$$

Alternatively, it is possible to create a model where the scale of the current is imposed. Thus, imposing that the scale of the tangential stress at the bottom due to the current,

$$\left(r_{\tau_b}\right)_{\text{current}} = r_{U_c}^2 \left(\frac{r_{k_s}}{\lambda_y}\right)^{1/4} \tag{11.90}$$

is equal to the scale of the tangential stress due to long waves,

$$\left(r_\tau\right)_{\text{long waves}} = \frac{\lambda_y^2}{\lambda_x}, \tag{11.91}$$

results in

$$r_{U_c} = \frac{\lambda_y^{9/8}}{\lambda_x^{1/2} r_{k_s}^{1/8}}. \tag{11.92}$$

In a geometrically undistorted model on a flat bottom and with $r_{k_s} = r_d$,

$$r_{U_c} = \frac{\lambda^{5/8}}{r_d^{1/8}}. \tag{11.93}$$

Table 11.3 provides a summary of scales dictated by the forcing actions in the four conditions.

11.6.2 The Hypothesis of a Dominant Bed Load

We now consider the additional constraints introduced by sediment dynamics.

Table 11.3 Summary of scales for the forcing actions of sediment transport

	r_{τ_b}	r_τ	r_{U_c}
Short waves	$\lambda_y^{1/2} r_{k_s}^{3/4}$	–	–
Long waves	$\lambda_y^{3/4} r_{k_s}^{1/4}$	$\lambda_y^2 \lambda_x^{-1}$	–
Inshore models (long waves and currents)	–	–	$\lambda_y^{9/8} \lambda_x^{-1/2} r_{k_s}^{-1/8}$
Offshore models (short waves and currents)	–	–	$\lambda^{1/4} r_{k_s}^{1/4}$

In addition to the conditions of similarity for the forcing actions listed in Table 11.3, it is necessary to impose the similarity conditions for sediment transport:

$$\begin{cases} r_{u_*} \, r_d = 1, \\ r_{u_*}^2 = r_d \, r_{(s-1)}, \\ r_{\rho_s} = r_\rho, \\ \lambda = r_d, \\ r_w = r_{u_*}. \end{cases} \tag{11.94}$$

The sedimentation speed can be expressed by the formula

$$w = \frac{\sqrt{(s-1)\,g\,d}}{A + 4B\,\nu/\left(d\,\sqrt{(s-1)\,g\,d}\right)}, \tag{11.95}$$

also depicted in Fig. 11.5; $A = 0.954$ and $B = 5.12$ are coefficients. For sediments with diameter $0.13 - 1.0$ mm, the speed variation is linear with the diameter and results in $r_w = r_d$, which is incompatible with the first equation (Reynolds sediment number similarity) and with the second equation (Shields parameter similarity) in (11.94). For this reason, only partial similarities are allowed by respecting some of the dimensionless groups that have generated the similarity conditions in (11.94), also adding similarity for one of the forcing actions listed in Table 11.3.

11.6.2.1 Partial Similarity: The "Best Model"

In the *best model*, the Reynolds number for the sediments and the relative sedimentation speed are neglected. For short waves, the similarity conditions are

$$\begin{cases} r_{u_*}^2 = \lambda^{1/4} \, r_{k_s}^{3/4}, \\ r_{u_*}^2 = r_d \, r_{(s-1)}, \\ r_{\rho_s} = r_\rho, \\ \lambda = r_d, \end{cases} \tag{11.96}$$

with the solution:

$$r_{k_s} = \lambda, \quad r_d = \lambda, \quad r_{u_*} = \lambda^{1/2}, \quad r_{(s-1)} = 1. \tag{11.97}$$

Once we have selected the geometric scale, all other scales can be calculated. For long waves, the system of equations to solve is:

$$\begin{cases} r_{u_*}^2 = \lambda^{3/4} \, r_{k_s}^{1/4}, \\ r_{u_*}^2 = r_d \, r_{(s-1)}, \\ r_{\rho_s} = r_\rho, \\ \lambda = r_d, \end{cases} \tag{11.98}$$

with the same solution (11.97).

For the offshore models, the similarity conditions (11.97) still apply, and it is necessary to reproduce the current with a velocity scale $r_{U_c} = \sqrt{\lambda}$. This also holds for inshore models that must be undistorted, and with a velocity scale of the current still equal to $r_{U_c} = \sqrt{\lambda}$. Table 11.4 lists the most relevant scales obtained by applying the criteria of the *best model*. In summary, discarding the similarity of the Reynolds number for the sediments and the similarity of the sedimentation speed, we obtain conditions valid for all possible combinations of forcing actions (short waves, long waves, currents, short waves and currents, long waves and currents), provided that the models are geometrically undistorted and in Froude similarity.

The scale effects resulting from the approximations are due to the viscosity, which occurs in phases where the boundary layer tends to relaminate. Particular care must be taken to avoid an excessive reduction of the geometric scale to avoid the need for sediments that are too small to fall into the category of cohesive materials. It is also necessary to check that the boundary layer in the model is turbulent with an adequate level of turbulence.

The sedimentation speed is not adequately reproduced. Observing the diagram in Fig. 11.5, results that for large sediments (for which it makes sense to make such a model), the ratio between the sedimentation speed and the friction velocity (which is the forcing agent of resuspension) is smaller than necessary, being equal to

$$\frac{r_w}{r_{u_*}} = \lambda^{1/2}, \tag{11.99}$$

whereas it should be one. However, this scale effect is of little importance because sedimentation processes are not as important in bed load conditions.

Finally, we highlight that similarity conditions for roughness have been obtained referring to a flat bottom, with skin friction only. In the presence of bedforms such as ripples, dunes and bars, the analysis of the similarity is much more complex, although the results obtained for flat bottoms are still sufficiently correct.

Table 11.4 Summary of the results of the *best model*. The n exponent for r_{w/u_*} is greater than 0.5

Forcing action	r_{Re_*}	r_{Fr_*}	$r_{\rho_s/\rho}$	$r_{l/d}$	r_{w/u_*}	$r_{(s-1)}$	r_d	r_{τ_b}	r_{U_c}
Short waves	$\lambda^{3/2}$	1	1	1	λ^n	1	λ	λ	–
Long waves	$\lambda^{3/2}$	1	1	1	λ^n	1	λ	λ	–
Offshore	$\lambda^{3/2}$	1	1	1	λ^n	1	λ	λ	$\lambda^{1/2}$
Inshore	$\lambda^{3/2}$	1	1	1	λ^n	1	λ	λ	$\lambda^{1/2}$

11.6.2.2 Partial Similarity: Similarity with "Light" Sediment

A second category of approximate models of sediment transport in the maritime environment is obtained by imposing Reynolds sediment number similarity and Shields similarity, neglecting the other dimensionless groups. The conditions for similarity for the sediments are

$$\begin{cases} r_{u_*} r_d = 1, \\ r_{u_*}^2 = r_d\, r_{(s-1)}, \end{cases} \tag{11.100}$$

which requires the selection of sediments according to

$$r_{(s-1)} = \frac{1}{r_d^3}. \tag{11.101}$$

If the forcing action is represented by short waves, the similarity conditions are:

$$\begin{cases} r_{u_*} = \lambda^{1/8}\, r_{k_s}^{3/8}, \\ r_{u_*} r_d = 1, \\ r_{u_*}^2 = r_d\, r_{(s-1)}. \end{cases} \tag{11.102}$$

Once the geometric scale is set, the other scales are:

$$r_{k_s} \equiv r_d = \lambda^{-1/11}, \quad r_{(s-1)} = \lambda^{3/11}. \tag{11.103}$$

For models with reduced geometric scale, the sediments in the model have a larger diameter and a lower specific weight than the prototype, which explains the name of this class of physical models.

For models where the forcing action is represented by long waves, the following conditions of similarity are obtained:

$$\begin{cases} r_{u_*} = \lambda^{3/8}\, r_{k_s}^{1/8}, \\ r_{u_*} r_d = 1, \\ r_{u_*}^2 = r_d\, r_{(s-1)}. \end{cases} \tag{11.104}$$

Additionally, in this case, when the geometric scale is set, the other scales are:

$$r_{k_s} \equiv r_d = \lambda^{-1/3}, \quad r_{(s-1)} = \lambda. \tag{11.105}$$

We observe that the sediments are very light, even for geometric scales not very small.

In the case of offshore models (short waves and currents), it is necessary to equalise the tangential stress of waves and current and consider it to be the stress acting on sediments. For the modelling of sediments, the scales are those already obtained for

Table 11.5 Synthesis of the results for a physical model with *light* sediments. The n exponent for r_{w/u_*} is greater than ≈ 0.66. The ratio $r_{\rho_s/\rho}$ is calculated based on $r_{(s-1)}$

Forcing action	r_{Re_*}	r_{Fr_*}	$r_{l/d}$	r_{w/u_*}	$r_{(s-1)}$	r_d	r_{τ_b}	r_{U_c}
Short waves	1	1	$\lambda^{12/11}$	λ^n	$\lambda^{3/11}$	$\lambda^{-1/11}$	$\lambda^{2/11}$	–
Long waves	1	1	$\lambda^{4/3}$	λ^n	λ	$\lambda^{-1/3}$	$\lambda^{2/3}$	–
Offshore	1	1	$\lambda^{12/11}$	λ^n	$\lambda^{3/11}$	$\lambda^{-1/11}$	$\lambda^{2/11}$	$\lambda^{5/22}$
Inshore	1	1	$\lambda^{4/3}$	λ^n	λ	$\lambda^{-1/3}$	$\lambda^{2/3}$	$\lambda_y^{25/22}\lambda_x^{-1/2}$

the action of short waves; see Eq. (11.97); the current cannot be reproduced in Froude similarity (that is, $r_{U_c} = \lambda^{1/2}$) but requires a scale equal to

$$r_{U_c} = \lambda^{1/4} r_{k_s}^{1/4} \rightarrow r_{U_c} = \lambda^{5/22}. \tag{11.106}$$

The same approach is valid for inshore models (long waves and currents): for the sediments, the scales derived from the action of long waves are valid; see Eq. (11.105); the current cannot be reproduced in Froude similarity but requires a scale

$$r_{U_c} = \frac{\lambda_y^{9/8}}{\lambda_x^{1/2} r_{k_s}^{1/8}} \rightarrow r_{U_c} = \frac{\lambda_y^{25/22}}{\lambda_x^{1/2}} \tag{11.107}$$

which, for a geometrically undistorted model, becomes

$$r_{U_c} = \lambda^{7/11}. \tag{11.108}$$

The most relevant scales are listed in Table 11.5.

The use of lighter sediments in the model, compared to sediments in the prototype, induces numerous scale effects. In particular, the inertia of the particles is underestimated, and the grains tend to be more easily suspended. The geometric scales of the bedforms are distorted due to the larger grain size, and the porosity in the model is excessive, making, for example, the beaches in the model more absorbent than they should be.

11.6.2.3 Partial Similarity: Froude Densimetric Models

In these models, only the number of Shields (also known as the sediment Froude number) is respected. For short waves, the similarity conditions are

$$\begin{cases} r_{u_*} = \lambda^{1/8} r_{k_s}^{3/8} \equiv \lambda^{1/8} r_d^{3/8}, \\ r_{u_*}^2 = r_d \, r_{(s-1)}, \end{cases} \quad \text{(flat bottom)} \qquad (11.109)$$

and we can select two scales and calculate the third.

For long waves,

$$\begin{cases} r_{u_*} = \lambda^{3/8} r_{k_s}^{1/8} \equiv \lambda^{3/8} r_d^{1/8}, \\ r_{u_*}^2 = r_d \, r_{(s-1)}, \end{cases} \quad \text{(flat bottom)} \qquad (11.110)$$

quite similar to the short wave case.

In the offshore models, the same scales for sediments apply as in the case of only short waves; for water, it is necessary to model the current with a scale ratio equal to that necessary to equalise the tangential stresses of the two joint forcing actions (short waves and current) on the sediments, that is,

$$r_{U_c} = \lambda^{1/4} r_d^{1/4}, \qquad (11.111)$$

and the model must be geometrically undistorted.

In the inshore models, the same scales for sediments apply as in the case of only long waves; for water, again it is necessary to model the current with a scale ratio equal to that necessary to equalise the tangential stresses of the two joint forcing actions (long waves and current) on the sediments, that is,

$$r_{U_c} = \frac{\lambda_y^{9/8}}{\lambda_x^{1/2} r_d^{1/8}}, \qquad (11.112)$$

and the model can also be geometrically distorted.

The results are summarised in Table 11.6.

Table 11.6 Summary of the results of a Froude densimetric model, $r_{Fr_*} = 1$. The n exponent for r_{w/u_*} is greater than ≈ 0.62. The ratio $r_{\rho_s/\rho}$ is calculated based on $r_{(s-1)}$

Forcing action	r_{Re_*}	$r_{l/d}$	r_{w/u_*}	$r_{(s-1)}$	r_{τ_b}	r_{U_c}
Short waves	$\lambda^{1/8} r_d^{11/8}$	λr_d^{-1}	$\lambda^{-1/8} r_d^n$	$\lambda^{1/4} r_d^{-1/4}$	$\lambda^{1/4} r_d^{3/4}$	–
Long waves	$\lambda^{3/8} r_d^{9/8}$	λr_d^{-1}	$\lambda^{-3/8} r_d^n$	$\lambda^{3/4} r_d^{-5/4}$	$\lambda^{3/4} r_d^{1/4}$	–
Offshore	$\lambda^{1/8} r_d^{11/8}$	λr_d^{-1}	$\lambda^{-1/8} r_d^n$	$\lambda^{1/4} r_d^{-1/4}$	$\lambda^{1/4} r_d^{3/4}$	$\lambda^{1/4} r_d^{1/4}$
Inshore	$\lambda^{3/8} r_d^{9/8}$	λr_d^{-1}	$\lambda^{-3/8} r_d^n$	$\lambda^{3/4} r_d^{-5/4}$	$\lambda^{3/4} r_d^{1/4}$	$\lambda_y^{9/8} \lambda_x^{-1/2} r_d^{-1/8}$

Table 11.7 Summary of the results of an unvaried sediment density model, $r_{\rho_s/\rho} = 1$. The n exponent for r_{w/u_*} is greater than ≈ 0.62

Forcing action	r_{Re_*}	r_{Fr_*}	$r_{l/d}$	r_{w/u_*}	r_{τ_b}	r_{U_c}
Short waves	$\lambda^{1/8}\,r_d^{11/8}$	$\lambda^{1/4}\,r_d^{-1/4}$	$\lambda\,r_d^{-1}$	$\lambda^{-1/8}\,r_d^{n}$	$\lambda^{1/4}\,r_d^{3/4}$	–
Long waves	$\lambda^{3/8}\,r_d^{9/8}$	$\lambda^{3/4}\,r_d^{-3/4}$	$\lambda\,r_d^{-1}$	$\lambda^{-3/8}\,r_d^{n}$	$\lambda^{3/4}\,r_d^{1/4}$	–
Offshore	$\lambda^{1/8}\,r_d^{11/8}$	$\lambda^{1/4}\,r_d^{-1/4}$	$\lambda\,r_d^{-1}$	$\lambda^{-1/8}\,r_d^{n}$	$\lambda^{1/4}\,r_d^{3/4}$	$\lambda^{1/4}\,r_d^{1/4}$
Inshore	$\lambda^{3/8}\,r_d^{9/8}$	$\lambda^{3/4}\,r_d^{-3/4}$	$\lambda\,r_d^{-1}$	$\lambda^{-3/8}\,r_d^{n}$	$\lambda^{3/4}\,r_d^{1/4}$	$\lambda_y^{9/8}\,\lambda_x^{-1/2}\,r_d^{-1/8}$

11.6.2.4 Partial Similarity: The Model with Unvaried Density

This model is based on the density value of the prototype and leaves 2 degrees of freedom, and the scales usually selected are the geometric scale and the sediment diameter scale. On the basis of these two scales, it is possible to calculate all the other scales, including the scales of the dimensionless groups not in similarity, summarised in Table 11.7.

The complexity of the phenomenon requires attention and caution in the realisation of physical models of sediment transport and in the interpretation of the results. We observe the presence of a vast number of dimensionless groups, with a relevance that cannot be easily assessed.

Among the many aspects still open to discussion and research, bedforms certainly change the scale of roughness and the intensity of solid transport. For the time scale in the processes of bed load sediment transport, the indications of Eq. (11.55) apply.

11.6.3 Hypothesis of Dominant Suspended Load

In many cases, near the coast and in the area of breakers, bed load is of minor importance with respect to suspended sediment transport, which is instead excited by the action of currents and breakers. Sediment transport can be modelled using Eq. (11.77):

$$\frac{q}{\sqrt{g\,d^3\,(s-1)}} = \Phi\left(\frac{\sqrt{g\,H_b}\,d}{\nu},\ \frac{H_b}{d\,(s-1)},\ \frac{\rho_s}{\rho},\ \frac{H_b}{d},\ \frac{w}{\sqrt{g\,H_b}} \right). \qquad (11.113)$$

The similarity criteria are expressed by the following equations:

$$\begin{cases} r_{H_b}^{1/2}\, r_d = 1, \\ r_{H_b} = r_{(s-1)}\, r_d, \\ r_{\rho_s} = r_\rho, \\ r_{H_b} = r_d, \\ r_w = r_{H_b}^{1/2}. \end{cases} \qquad (11.114)$$

Additionally, in this case, the constraint resulting from the use of water in the model reduces the number of degrees of freedom and prevents the realisation of a complete similarity. An alternative to the strict criteria of similarity deriving from dimensional analysis is provided by some *ad hoc* models developed to model sediment transport in suspension in some specific flow fields.

11.6.3.1 The Criteria of Similarity of Sediment Transport in Suspension, Without Respecting the Scale of Sedimentation Speed

Based on a series of experimental results in the laboratory, Noda (1972) proposed some similarity criteria to be used in very energetic areas, such as in the breaking zone, to adequately reproduce the equilibrium profiles of the beaches. The similarity criteria are:

$$r_d\, r_{(s-1)}^{1.85} = \lambda_y^{0.55}, \quad \lambda_x = \lambda_y^{1.32}\, r_{(s-1)}^{-0.386}, \qquad (11.115)$$

which, for $r_{(s-1)} = 1$, become:

$$r_d = \lambda_y^{0.55}, \quad \lambda_y/\lambda_x = 3.125. \qquad (11.116)$$

The distortion ratio for bathymetry is equal to

$$n = \lambda_y^{-0.32}\, r_{(s-1)}^{0.386}, \qquad (11.117)$$

whereas the flow field is undistorted. It is possible to avoid the distortion of the bathymetry by selecting light sediments to satisfy the following relation:

$$r_d = \lambda^{-0.984}, \quad r_{(s-1)} = \lambda^{0.83}. \qquad (11.118)$$

Summarising Concepts

- Sediment transport is among the most complex and, in many respects, unsolved problems due to the variety of forcing factors and the changing nature of sediments. Sediment transport models are often only "qualitative", providing indications of trends.

- Similarity conditions are differentiated according to the nature of the forcing (in rivers, in the sea), the characteristics of the site, the coexistence of different forcings (for instance, waves plus currents), and the type of sediment transport (bed load or suspended load).
- The number of dimensionless groups is very high, and partial similarities are currently adopted. The interaction between roughness due to skin friction and drag in the presence of dunes and ripples and the fluid stream requires special arrangements to ensure adequate coupling.

References

Dalrymple, R. A. (1989). Physical modelling of littoral processes. In R. Martins (Ed.), *Recent Advances in Hydraulic Physical Modelling* (567–588). Springer.

Einstein, H. A., & Barbarossa, H. L. (1952). River channel roughness. *Transactions of the American Society of Civil Engineers, 117*(1), 1121–1132.

Exner, F. M. (1925). Über Wechselwirkung zwischen Wasser und Geschiebe in Flüssen (On the interaction between water and sediment in streams). *Akademie der Wissenschaften in Wien, Mathematisch-Naturwissenschaftliche Klasse, 134*(2a), 165–204.

Julien, P. Y. (2002). *River mechanics*. Cambridge University Press.

Mignosa, P., Chiapponi, L., D'Oria, M. & Anelli, L. (2017). *Realizzazione del modello fisico delle pile del ponte sul fiume Po ex S.S 413 'Romana'* (*Physical model tests of the bridge piles in the riverbed of Po River ex S.S 413 'Romana'*)(*in Italian*). DICATeA, University of Parma.

Noda, E. K. (1972). Equilibrium beach profile scale-model. *Journal of Waterways, Harbors and Coastal Engineering Division, 98*(4), 511–528.

Yalin, M. S. (1971). *Theory of hydraulic models*. Macmillan Publishers Ltd.

Appendix A
Homogeneous Functions and Their Properties

A function is defined *homogeneous* of order k, if it results in $f(\alpha\,\mathbf{v}) = \alpha^k f(\mathbf{v})$. A linear combination of monomials of the type

$$f(x_1,\ x_2,\ \ldots,\ x_r) = \sum_{i=1}^{N} a_i x_1^{n_{1i}} x_2^{n_{2i}} \cdots x_{ri}^{n_{ri}}, \tag{A.1}$$

in which the coefficients are constant and the sum of the exponents of each term is constant and equal to k,

$$n_{1i} + n_{2i} + \cdots + n_{ri} = k \quad \forall i, \tag{A.2}$$

is a homogeneous function of order k.

A homogeneous function of r variables equal to zero is equivalent to a homogeneous function of $(r-1)$ variables equal to zero.

To prove this, simply divide all the terms by any variable raised to the order k, e.g. by x_q^k. It is possible to re-write the function in the new variables like:

$$x_1' = \frac{x_1}{x_q}, \quad x_2' = \frac{x_2}{x_q}, \quad \ldots, \quad x_{r-1}' = \frac{x_{r-1}}{x_q}. \tag{A.3}$$

For example, given the following homogeneous function of degree 11, in 3 variables, equal to zero

$$3\,x^2 y^5 z^4 + 12\,x y^2 z^8 = 0, \tag{A.4}$$

S. G. Longo, *Principles and Applications of Dimensional Analysis and Similarity*, Mathematical Engineering, https://doi.org/10.1007/978-3-030-79217-6

dividing by z^{11} we get

$$3 \left(\frac{x}{z}\right)^2 \left(\frac{y}{z}\right)^5 + 12 \left(\frac{x}{z}\right) \left(\frac{y}{z}\right)^2 = 0. \tag{A.5}$$

This is a new function homogeneous in the two variables $x' = x/z$ and $y' = y/z$, equalised to zero.

The condition that the function must be equal to zero is not necessary if the order of the function is zero.

For example, the following homogeneous function in 4 variables of order zero

$$3 \, x^2 y^2 z t^{-5} + 11 \, x y^{-3} z^2, \tag{A.6}$$

can be re-written as

$$3 \left(\frac{x}{t}\right)^2 \left(\frac{y}{t}\right)^2 \left(\frac{z}{t}\right) + 11 \left(\frac{x}{t}\right) \left(\frac{y}{t}\right)^{-3} \left(\frac{z}{t}\right)^2 \tag{A.7}$$

and becomes a new function in the 3 variables $x' = x/t$, $y' = y/t$ and $z' = z/t$.

Homogeneous functions satisfy Euler's Theorem: if f is a homogeneous function of order k, then it results:

$$\mathbf{x} \cdot \nabla f = kf \rightarrow x_1 \frac{\partial f}{\partial x_1} + \cdots + x_r \frac{\partial f}{\partial x_r} = kf. \tag{A.8}$$

For example, if the homogeneous function is

$$3 \, x^2 y^5 z^4 + 12 \, x y^2 z^8 = 0, \tag{A.9}$$

then it results:

$$x \frac{\partial f}{\partial x} + y \frac{\partial f}{\partial y} + z \frac{\partial f}{\partial z} = x \, (6 \, x y^5 z^4 + 12 \, y^2 z^8) +$$
$$y \, (15 \, x^2 y^4 z^4 + 24 \, x y z^8) + z \, (12 \, x^2 y^5 z^3 + 96 \, x y^2 z^7) =$$
$$11 \, (3 \, x^2 y^5 z^4 + 12 \, x y^2 z^8). \tag{A.10}$$

A consequence of Euler's Theorem is that the solution of a partial derivative equation of the kind

$$x_1 \frac{\partial f}{\partial x_1} + \cdots + x_r \frac{\partial f}{\partial x_r} = 0, \tag{A.11}$$

is the homogeneous function $f(x_1, \, x_2, \, \ldots, \, x_r) = 0$.

The definition of a homogeneous function given in (A.1) can be extended to include the case where the coefficients are replaced by arbitrary homogeneous functions of order zero. Thus, if the equation (A.6) is homogeneous, then the function

$$x^2 y^2 z t^{-5} \sin\left(\frac{x^3 t}{z^4}\right) + 5\, x y^{-3} z^2 \, \sinh\left(\frac{z}{t}\right) \tag{A.12}$$

is homogeneous and satisfies Euler's Theorem and all other properties.

Appendix B
Relevant Dimensionless Parameters (or Groups or Numbers)

Below is a list of dimensionless numbers and groups frequently used in the physical sciences. Many of the numbers are named after the scientists and researchers who first identified them. Sometimes, the name given to the numbers is the natural choice in the light of the quantities involved. A larger selection of numbers is available in Massey, B. S., 1971. *Units, dimensional analysis and physical similarity.* Van Nostrand Reinhold.

Absorption number, $Ab = k_L \sqrt{\dfrac{z}{D \overline{V}}}$

k_L = individual liquid coefficient of absorption, z = length of the surface covered by the liquid film (starting from the inlet section), D = diffusion coefficient of gas in liquid, $\overline{V}$ = mean speed of liquid film over wetted wall column.

Acceleration number, $Ac = \dfrac{\varepsilon^3}{\rho \, g^2 \, \mu^2}$

ε = bulk modulus of the fluid, ρ = density of the fluid, g = acceleration of gravity, μ = dynamic viscosity. It occurs in the flow of rapidly accelerated fluids.

Advance ratio (of propellers), $J = \dfrac{V}{\omega D}$

V = forward speed, ω = angular velocity, D = diameter of the propeller.

Aeroelasticity number, $Ae = \dfrac{\rho \, V^2}{E}$

ρ = density of the fluid, V = speed of the fluid, E = Young's modulus of the elastic material. It is equal to the ratio between the acting aerodynamic load and the elastic stresses in the structure.

© The Editor(s) (if applicable) and The Author(s), under exclusive license to Springer Nature Switzerland AG 2021
S. G. Longo, *Principles and Applications of Dimensional Analysis and Similarity,*
Mathematical Engineering, https://doi.org/10.1007/978-3-030-79217-6

Archimedes number, $\mathrm{Ar} = \dfrac{d^3 \, g \, (\rho_s - \rho_f) \, \rho_f}{\mu^2}$

d = diameter of sediments, g = acceleration of gravity, ρ_s, ρ_f = density of sediments, of fluid, μ = dynamic viscosity. It is the ratio (*Inertial force* × *Gravity force* / *Viscous force*^{2}).

Arrhenius number, $\dfrac{E}{RT}$

E = activation energy per unit mass, R = gas constant, from the relation $p = \rho RT$, T = absolute temperature. It is the ratio (*Activation energy* / *Potential energy of the gas*).

Atwood number, $\mathrm{A} = \dfrac{\rho_1 - \rho_2}{\rho_1 + \rho_2}$

ρ_1, ρ_2 = density of the denser (1) and less dense fluid (2). It occurs in the instability of stratified fluids.

Bagnold number, $\mathrm{Ba} = \dfrac{3 \, C_D \, \rho_f \, V^2}{4 \, d \, \rho_s \, g}$

C_D = drag coefficient, ρ_s, ρ_f = density of sediments, of fluid, V = speed, d = diameter of sediments, g = acceleration of gravity. It is the ratio (*Drag force* / *Gravity force*) and occurs in sediment transport due to a stream.

Bagnold number (second definition), $\mathrm{Ba} = \dfrac{C_s \, \rho_s \, d^2 \, \dot{\gamma}}{(1 - C_s) \, \mu}$

C_s = void ratio of sediments, ρ_s = density of sediments, $\dot{\gamma}$ = strain rate, μ = dynamic viscosity. It is the ratio (*Collisional stress* / *Viscous stress*) and it occurs in sediment transport at high concentration in the presence of a viscous interstitial fluid.

Bagnold number (third definition), $\dfrac{\rho_w \, k_w \, u_0^2}{p_{atm} \, D}$

ρ_w = density of water, k_w = thickness of the cushion of water considered as active during impulse, D = initial size of the air pocket between the vertical wall and the breaking wave, u_0 = speed, p_{atm} = absolute atmospheric pressure. It occurs in waves breaking on a vertical rigid wall.

Bansen number, $\dfrac{h_r \, A_w}{\dot{m} \, c}$

h_r = coefficient of heat transfer by radiation, A_w = exchange surface area, $\dot{m}$ = mass flow rate, c = specific heat. It is the ratio (*Heat transferred by radiation* / *Thermal capacity of fluid*).

Béranek number, $\mathrm{Be} = \dfrac{V_t^3\, \rho_f^2}{\mu\, g\, (\rho_s - \rho_f)}$

V_t = terminal falling velocity of solid particle, ρ_s, ρ_f = density of sediments, of fluid, μ = dynamic viscosity, g = acceleration of gravity. It is the ratio *Inertia force*^{2} / (*Viscous force × Gravity force*).

Bingham number (or Plasticity number), $\mathrm{Bm} = \dfrac{\tau_y\, L}{\mu_p\, V}$

τ_y = yield stress, L = length scale, μ_p = apparent dynamic viscosity, V = velocity. It is the ratio (*Yield stress / Viscous stress*) in a Bingham or Herschel-Bulkley fluid.

Biot number, $\mathrm{Bi} = \dfrac{h\, l}{k_s}$

h = heat transfer coefficient, l = length scale, k_s = thermal conductivity. It is the ratio (*Internal thermal resistance of solid body / Surface resistance*). It is similar, but not identical, to the Nusselt number.

Biot number for mass transfer, $\mathrm{Bi}_m = \dfrac{k_m\, L}{D_{\mathrm{int}}}$

k_m = mass exchange coefficient (mass flow rate per unit area and per unit concentration difference), L = layer thickness, D_{int} = molecular diffusivity at interface. It is the ratio (*Mass flow at the solid-fluid interface / Internal mass flow through the layer of thickness L*).

Blake number, $\mathrm{Bl} = \dfrac{V}{\nu\, (1 - e)\, S}$

V = speed, ν = kinematic viscosity, e = porosity, S = specific area, ratio between external surface area and volume. It is the ratio (*Inertia force / Viscous force in flow through granular materials*).

Bodenstein number, $\mathrm{Bd} = \dfrac{V\, L}{D_a}$

V = speed, L = axial length, D_a = effective axial diffusivity. Describes diffusion in a bed of granular material, it is a special case of Péclet number.

Bond (Eötvös) number, $\mathrm{Bo} = \dfrac{(\rho - \rho_f)\, d^2\, g}{\sigma}$

ρ, ρ_f = density of bubbles or droplets, of surrounding fluid, d = diameter of bubbles or droplets, g = acceleration of gravity, σ = surface tension. It is the ratio (*Gravity force / Surface tension force*).

Boussinesq number, $\text{Bq} = \dfrac{V}{\sqrt{2\,g\,h}}$

V = speed, h = hydraulic mean depth of open channel, g = acceleration of gravity. It is similar to the Froude number and is the square root of the ratio (*Inertia force / Gravity force*).

Brinkman number, $\dfrac{\mu\,V^2}{k\,\Delta\theta}$

μ = dynamic viscosity, V = speed, k = thermal conductivity, $\Delta\theta$ = temperature difference. It is the ratio (*Heat generated by viscous action / Heat transferred by conduction*).

Bulygin number, $\text{Bu} = \dfrac{\lambda\,c_b\,\Delta p}{c\,(\theta_a - \theta_0)}$

λ = latent heat of evaporation, c_b = mass of vapour per unit mass of dry gas per unit pressure change, Δp = pressure variation, c = specific heat of wet body, θ_a = boiling temperature of the liquid, θ_0 = initial temperature. It is the ratio (*Heat needed to vapourize that part of liquid which is removed by seepage / Heat needed to raise temperature of moist body to boiling point of liquid*).

Camp number, $\text{Ca} = \sqrt{\dfrac{P\,W}{\mu\,Q^2}}$

P = power dissipated by viscous action, W = volume, μ = dynamic viscosity, Q = flow rate. *Elapsed time × average rate of shear in fluid*. Criterion for degree of flocculation of suspended particles.

Capillarity number, $\dfrac{\mu^2\,\varepsilon}{\rho\,\sigma^2}$

μ = dynamic viscosity, ε = bulk modulus, ρ = density, σ = surface tension. Concerns action of surface tension in flowing fluid.

Capillarity-buoyancy number, $\dfrac{g\,\mu^4}{\rho\,\sigma^3}$

g = acceleration of gravity, μ = dynamic viscosity of the surrounding fluid, ρ = density of the surrounding fluid, σ = surface tension. Concerns effects of surface tension, viscosity and acceleration when liquid globules move through another fluid.

Capillary number, $\dfrac{\mu\,V}{\sigma}$

μ = dynamic viscosity, V = speed, σ = surface tension. It is the ratio (*Viscous force / Surface tension force*). Concerns atomization processes and two-phase flow through porous media.

Carnot number, $\dfrac{T_2 - T_1}{T_2}$

T_2 = absolute temperature of the hot source, T_1 = absolute temperature of cold source. It represents the theoretical efficiency of a machine with Carnot cycle operating between the two heat reservoirs.

Cauchy number, $\dfrac{\rho V^2}{\varepsilon}$, $\dfrac{\rho V^2}{E}$

ρ = density of the fluid, V = speed, ε = bulk modulus for a fluid, E = Young's modulus for an elastic body. It is the ratio (*Inertia force / Elastic force*).

Cavitation number, $\dfrac{p - p_v}{\frac{1}{2}\rho V^2}$

p = pressure, p_v = vapour pressure, ρ = density of the fluid, V = speed. Quantifies the distance from cavitation.

Clausius number, $Cl = \dfrac{V^3 l \rho}{k_f \, \Delta\theta}$

V = speed, l = length scale, ρ = density, k_f = thermal conductivity, $\Delta\theta$ = temperature difference. It is used in the study of heat transfer in the presence of forced convection.

Colburn J factor, $J = \dfrac{h}{\rho \, c_p \, V} \left(\dfrac{c_p \, \mu}{k_f} \right)^{2/3}$

h = heat transfer coefficient, ρ = density of the fluid, c_p = specific heat at constant pressure, V = speed, μ = dynamic viscosity, k_f = thermal conductivity of the fluid. It also results in $J = St \cdot Pr^{2/3} = Nu \cdot Re^{-1} \cdot Pr^{-1/3}$.

Condensation number, $Co = \dfrac{h}{k_f} \left(\dfrac{v^2}{g} \right)^{1/3}$

h = heat transfer coefficient, k_f = thermal conductivity of the fluid, v = kinematic viscosity, g = acceleration of gravity, k_f = thermal conductivity of the fluid.

Condensation number for vertical walls, $Co = \dfrac{l^3 \rho^2 g \lambda}{k_f \, \mu \, \Delta\theta}$

l = length scale, ρ = density, g = acceleration of gravity, λ = latent heat of condensation, μ = dynamic viscosity, $\Delta\theta$ = temperature variation.

Crispation number, $\dfrac{\mu \, \alpha}{\sigma * l}$

μ = dynamic viscosity, $\alpha = k/\rho \, c_p$ = thermal diffusivity, $\sigma *$ = undisturbed surface tension, l = length scale. Concerns convective currents with cell formation caused by surface tension gradient.

Crocco number, $Cr = \dfrac{V}{V_{max}} = \left[1 + \dfrac{2}{(\gamma - 1)M^2}\right]^{-1/2}$

V = speed, V_{max} = maximum possible velocity of a perfect gas expanding isentropically, $\gamma = c_p/c_v$ = ratio between specific heat at pressure and at constant volume, M = Mach number.

Darcy coefficient, $f = \dfrac{2gRh_f}{l\overline{V}^2}$

g = acceleration of gravity, R = hydraulic radius, h_f = head loss, l = pipe length, $\overline{V}$ = cross-section average speed of the fluid. In US more usual is $f = \dfrac{8gRh_f}{l\overline{V}^2}$.

Dean number, $Dn = Re\sqrt{\dfrac{r}{R}}$

Re = Reynolds number, $r = d/2$ = channel (pipe) semi-width (diameter), R = channel (pipe) curvature radius. Expresses effect of centrifugal force in curved channels (pipes).

Dean number (second definition), $\dfrac{H}{wT}$

H = wave height, w = sedimentation speed, T = wave period. Expresses the effects of sea waves on the suspension/sedimentation of sediments.

Deborah number, $De = \dfrac{t_r}{t_0}$

t_r = relaxation time of material , t_0 = time scale of the observer/process.

Generalized Deborah number, $(I_e - I_w)^{1/2}t_n$

I_e = invariant of rate of strain tensor, I_w = invariant of vorticity tensor, t_n = natural time of visco-elastic material.

Dimensionless specific speed, $K_n = \dfrac{NP^{1/2}}{\rho^{1/2}(gH)^{5/4}}$ (for turbines), $K_n = \dfrac{NQ^{1/2}}{(gH)^{3/4}}$ (for rotodynamic pumps)

N = speed of rotation (in revolutions per unit time), P = power, ρ = density of the fluid, g = acceleration of gravity, H = head difference across the machine, Q = volumetric flow rate. Concerns the characteristics of a rotating hydraulic machine.

Drag coefficient, $C_D = \dfrac{Drag\ force\ on\ body}{\frac{1}{2}\rho V^2 A}$

ρ = density of the fluid, V = velocity, A = maximum cross-sectional area of body perpendicular to the velocity.

Eckert number, $Ec = \dfrac{V_\infty^2}{c_p\,\Delta\theta}$

V_∞ = asymptotic speed, c_p = specific heat at constant pressure, $\Delta\theta$ = temperature variation between moving gas and adiabatic wall.

Ekman number, $Ek = \sqrt{\dfrac{\nu}{2\,\omega\,l^2}}$ (the '2' is sometimes omitted)

ν = kinematic viscosity, ω = angular velocity, l = length scale. It is the square root of the ratio (*Viscous force / Coriolis force*).

Elasticity number, $\dfrac{\mu\,t_r}{\rho\,r^2}$

μ = dynamic viscosity, t_r = relaxing time of the fluid, ρ = density of the fluid, r = pipe radius. It is the ratio (*Elastic force / Inertia force*) in visco-elastic flows.

Elasticity number (second definition), $\dfrac{\rho\,c_p}{\beta\,\varepsilon}$

ρ = density, c_p = specific heat at constant pressure, β = coefficient of thermal expansion of volume at constant pressure, ε = bulk modulus of the fluid. Concerns effect of elasticity of fluid on flow processes.

Ellis number, $El = \dfrac{\mu_0\,V}{\tau_{1/2}\,d}$

μ_0 = limiting value of viscosity as shear rate $\to 0$, V = speed, $\tau_{1/2}$ = shear stress when $\mu = \mu_0/2$, d = length scale.

Euler number, $Eu = \dfrac{\Delta p}{\rho\,V^2}$

Δp = pressure variation, ρ = density, V = speed of the stream.

Evaporation number, $\dfrac{V^2}{\lambda}$

V = speed, λ = specific enthalpy of evaporation (latent heat of evaporation).

Evaporation number (second definition), $\dfrac{c_p}{\beta\,\lambda}$

c_p = specific heat at constant pressure, β = coefficient of thermal expansion of volume at constant pressure, λ = specific enthalpy of evaporation (latent heat of evaporation).

Evaporation-elasticity number, $\dfrac{\varepsilon}{\lambda\,\rho}$

ε = bulk modulus of the fluid, λ = specific enthalpy of evaporation (latent heat of evaporation), ρ = density of the fluid.

Expansion number, $\dfrac{g\,d}{V^2}\left(\dfrac{\rho_l - \rho_g}{\rho_l}\right)$

d = diameter of gas bubbles in the liquid, g = acceleration of gravity, V = speed, $\rho_{l,\,g}$ = liquid/gas density. It is the ratio (*Buoyancy force / Inertia force*).

Fedorov number (see Archimedes number), $\mathrm{Fe} = \left[\dfrac{4\,g\,d_p^3\,(\rho_s - \rho_f)\,\rho_f}{3\,\mu^2}\right]^{1/3} \equiv$

$\left(\dfrac{4}{3}\times\mathrm{Ar}\right)^{1/3}$

Flow coefficient, $\dfrac{Q}{N\,D^3}$

Q = flow rate, N = rotational rate of the blade (in revolutions per unit time), D = diameter of the impeller. Concerns turbomachines.

Fluidization number, $\dfrac{V}{V_0}$

V = velocity scale, V_0 = fluid velocity at start of fluidization.

Fourier criterion for heat diffusion, $\mathrm{Fo}' = \dfrac{D_v\,t}{2\,\pi\,l^2}$

D_v = molecular diffusivity, t = time scale, l = length scale.

Fourier flow number, $\mathrm{Fo}_f = \dfrac{\nu\,t}{l^2}$

ν = kinematic viscosity, t = time scale, l = length scale.

Fourier number, $\mathrm{Fo} = \dfrac{\alpha\,t}{l^2}$

$\alpha \equiv k/(\rho\,c_p)$ = thermal diffusivity, t = time scale, l = length scale.

Fourier number for mass transfer, $\mathrm{Fo}_m = \dfrac{k_m\,t}{l}$

k_m = mass exchange coefficient, t = time scale, l = length scale.

Frequency parameter, $\dfrac{\omega l}{V} \equiv 2\pi \times$ Strouhal number

ω = pulsation, l = length scale, V = speed.

Frössling number for heat transfer, $\mathrm{Fr_h} = \dfrac{\mathrm{Nu}}{\mathrm{Re}^{1/2}}$ (for laminar flow over a flat plate), $\mathrm{Fr_h} = \dfrac{\mathrm{Nu} - 2}{\mathrm{Re}^{1/2} \cdot \mathrm{Pr}^{1/3}}$ (for turbulent flow round sphere)

Nu = Nusselt number, Pr = Prandtl number, Re = Reynolds number.

Frössling number for mass transfer, $\mathrm{Fr_m} = \dfrac{\mathrm{Sh} - 2}{\mathrm{Re}^{1/2} \cdot \mathrm{Sc}^{1/3}}$ (for mass transfer from sphere)

Sh = Sherwood number, Re = Reynolds number, Sc = Schmidt number.

Froude number, $\mathrm{Fr} = \dfrac{V}{\sqrt{gl}}$

V = speed, g = acceleration of gravity, l = length scale. It is the square root of the ratio (*Inertia force / Gravity force*).

Froude number (rotating), $\dfrac{D\omega^2}{g}$

D = diameter of the rotating flow, ω = angular velocity (in revolutions per unit time), g = acceleration of gravity. It is a modification of the Froude number for use with stirred liquids.

Galileo number, $\mathrm{Ga} = \dfrac{l^3 g}{\nu^2}$

l = length scale, g = acceleration of gravity, ν = kinematic viscosity. It is (*Inertia force × Gravity force / Viscous force^2*).

Gay-Lussac number, $\mathrm{Gc} = \dfrac{1}{\beta \, \Delta\theta}$

β = coefficient of thermal expansion of volume at constant pressure, $\Delta\theta$ = temperature variation.

Goertler parameter, $\mathrm{Gl} = \dfrac{V\theta}{\nu} \left(\dfrac{\theta}{r}\right)^{1/2}$

V = speed, θ = momentum thickness of boundary layer, ν = kinematic viscosity, r = radius of curvature of boundary.

Goucher number, $\mathrm{Go} = r \left(\dfrac{\rho\, g}{2\,\sigma} \right)^{1/2}$

r = radius of the wall or wire on which the fluid coating is deposited, ρ = density of the fluid, g = acceleration of gravity, σ = surface tension. It is the square root of the ratio (*Gravity force / Surface tension force*).

Graetz number, $\mathrm{Gz} = \dfrac{\dot{m}\, c_p}{k_f\, l}$

$\dot{m}$ = mass flow rate, c_p = specific heat at constant pressure, k_f = thermal conductivity of the fluid, l = length of heat transfer path. It is the ratio (*Thermal capacity of the fluid / Heat transferred by conduction*).

Grashof number, $\mathrm{G} = \dfrac{l^3\, g\, \beta\, \Delta\rho}{\rho\, \nu^2}$

l = length scale, g = acceleration of gravity, β = coefficient of thermal expansion of volume at constant pressure, ρ = density, $\Delta\rho$ = density variation, ν = kinematic viscosity. It is (*Inertia force × Buoyancy force / Viscous force²*).

Gukhman number, $\mathrm{Gu} = \dfrac{(T_0 - T_m)}{T_0}$

T_0 = absolute temperature of hot gas stream, T_m = absolute temperature of the wet surface. It is used in the analysis of convective heat transfer with constant pressure evaporation.

Gümbel number, $\mathrm{G\ddot{u}} = \dfrac{F\, b^2}{2\, \mu\, U\, r^2}$

F = force per unit of length of bearing, b = meatus height, μ = dynamic viscosity of the lubricant fluid, U = relative speed respect to the bearing, r = shaft radius.

Gümbel number (second definition), $\dfrac{\mu\, \omega\, D}{F}$

μ = dynamic viscosity of the lubricant fluid, ω = angular velocity of shaft, D = diameter of shaft, F = force per unit of length of bearing.

Hadamard number, $\mathrm{Ha} = \dfrac{3\, \mu_b + 3\, \mu_f}{3\, \mu_b + 2\, \mu_f}$

μ_b = dynamic viscosity of the fluid in bubble, μ_f = dynamic viscosity of the surrounding fluid.

Harrison number, $\text{Ha} = \dfrac{6\mu U L}{p_a h_0^2}$

μ = dynamic viscosity of the lubricant fluid, U = relative speed speed between bearing pad and wall, L = length of bearing pad in the direction of motion, p_a = ambient or supply pressure, h_0 = thickness of lubricant film at exit.

Hatta number, $\text{Ha} = \dfrac{\gamma}{\tanh \gamma}$, $\gamma = l\sqrt{\dfrac{k_n C_B^{(n-1)}}{D_A}}$

k_n = reaction constant for chemical reaction of order n, C_B = average molar concentration of the component B, D_A = diffusion coefficient of component A through the other components, l = length of diffusion path (or $(v^2/g)^{1/3}$ for process in packed towers).

Head coefficient, $\dfrac{g H}{N^2 D^2}$

H = head difference across turbomachine, g = acceleration of gravity, N = rotational rate (in revolutions per unit time), D = impeller diameter.

Heat transfer factor, $\dfrac{q}{V^3 l^2 \rho}$

q = thermal power, V = speed, l = length scale, ρ = density.

Hedström number, $\text{He} = \dfrac{\tau_y l^2 \rho}{\mu_p^2}$

τ_y = yield stress for shear rate $\to 0$, l = length scale, ρ = density, μ_p = apparent viscosity. Concerns Herschel-Bulkley fluid flows.

Helmoltz resonator group, $\text{Hh} = \dfrac{\sqrt{d^3/W}}{\text{M}}$

d = pipe diameter, W = volume, M= Mach number. It is (*Frequency of pulsating combustion* × *Residence time*).

Hersey number, $\text{Hs} = \dfrac{F}{\mu U}$

F = load per unit length, μ = dynamic viscosity, U = relative speed.

Hodgoson number, $\text{Ho} = \dfrac{W f \, \Delta p^*}{\overline{p}\,\overline{Q}}$

W = volume of system, f = frequency of pulsation of gas flow, Δp^* = drop of piezometric pressure $(p + \rho g z)$ due to friction and losses past obstructions, $\overline{p}$ = average pressure, $\overline{Q}$ = average flow rate.

Hydraulic resistance group, $\Gamma c = \dfrac{\Delta p}{\rho_l g L}$

Δp = pressure drop in the distillation line, ρ_l = density of the liquid, g = acceleration of gravity, L = depth of liquid layer on tray.

Ilyushin number, $\dfrac{4 \operatorname{Re} \tau_D}{3 \rho V^2}$

Re = Reynolds number, τ_D = maximum dynamic slip stress, ρ = density, V = speed. For flow of visco-plastic liquid in circular pipe.

Jacob modulus, $\operatorname{Ja} = \dfrac{c_l \, \rho_l \, \Delta\theta}{\lambda \, \rho_v}$

c_l = specific heat of the liquid, ρ_l = density of the liquid, $\Delta\theta$ = excess temperature of the hot surface above the boiling temperature of the liquid, λ = specific enthalpy of evaporation (latent heat of evaporation), ρ_v = density of the vapour. It is the ratio (*Maximum bubble radius / Thickness of superheated liquid on hot surface*).

Johnson (or damage) number , $\operatorname{Dn} = \dfrac{\rho_c \, V_0^2}{\sigma_0}$

ρ_c = density of the hitting body, V_0 = speed of the hitting body, σ_0 = yield strength of the material. Concerns impact phenomena.

von Kármán number, $\operatorname{Ka} = f^{1/2} \operatorname{Re}$

f = friction factor, Re = Reynolds number.

von Kármán number (second definition), $\dfrac{k}{v} \sqrt{\dfrac{\tau_w}{\rho}}$

k = mean height of surface roughness, v = kinematic viscosity, τ_w = shear stress at the wall, ρ = density of the fluid. It is the ratio (*Roughness length / Viscous length*).

Kirpichev heat transfer number, $\operatorname{Ki_h} = \dfrac{h(\theta_s - \theta_a) l}{k_s \, \Delta\theta}$

h = heat transfer coefficient, θ_s = temperature at surface of body, θ_a = ambient temperature, $\Delta\theta$ = drop temperature over length l within the body. It is the ratio (*Heat flux across surface of body / Heat flux within body*).

Kirpichev mass transfer number, $\operatorname{Ki_m} = \dfrac{\dot{m} \, l}{\lambda_m (\theta_0 - \theta_p)}$

$\dot{m}$ = mass flow rate, λ_m = mass conductivity coefficient, θ_0 = initial mass transfer potential, θ_p = mass transfer potential at equilibrium. It is the ratio (*Mass flux through the lateral surface / Mass flux within the system*).

Knudsen number, $Kn = \dfrac{l_p}{l}$

It is the ratio (*Mean free path of molecules in gas / Length scale of the domain of interest*).

Knudsen number for diffusion, $Kn_D = \dfrac{e\,D_{AB}}{q_D\,D_{KA}}$

e = porosity, D_{AB} = binary bulk diffusion coefficient for the system AB, q_D = diffusion tortuosity, D_{KA} = Knudsen diffusion coefficient. It is the ratio (*Bulk diffusion / Knudsen diffusion in granular bed*).

Kondrat'ev number, $\dfrac{h}{k\,S}\left(\dfrac{\theta_a - \theta_s}{\theta_a - \overline{\theta}}\right)$

h = thermal conductivity coefficient, S = specific surface, k = thermal conductivity, θ_a = ambient temperature, θ_s = temperature at surface of body, $\overline{\theta}$ = mean temperature of body.

Kossovich number, $Ko = \dfrac{\lambda\,\Delta u}{c_p\,\Delta\theta}$

λ = latent heat of evaporation, Δu = mass of moisture/mass of dry medium, c_p = specific heat at constant pressure, $\Delta\theta$ = temperature variation. It is the ratio (*Heat required for evaporation / Heat used in raising temperature of body*).

Kozeny function, $k = \dfrac{\Delta p^*}{\mu\,l}\,\dfrac{e^3}{(1 - e^2)}\,\dfrac{1}{\overline{V}\,S^2}$

Δp^* = piezometric pressure drop by flow through a layer of thickness l of porous material, μ = dynamic viscosity, e = porosity, $\overline{V}$ = mean component of fluid velocity, S = specific area, ratio of surface area to volume.

Lagrange number, $Lg = \dfrac{P}{\mu\,l^3\,N^2}$

P = power supplied to an agitator of characteristic size l, μ = dynamic viscosity, N = rotational rate of the agitator (in revolutions per unit time).

Lagrange number (second definition), $\dfrac{\Delta p^*\,r}{\mu\,\overline{V}}$

Δp^* = piezometric pressure drop ($p^* = p + \rho\,g\,z$), r = pipe radius, μ = dynamic viscosity of the fluid, $\overline{V}$ = average speed of the fluid.

Laplace number, $La = \dfrac{\Delta p\,L}{\sigma}$

Δp = pressure drop across the interface between two fluids, L = characteristic length of interface curvature, σ = interface tension.

Laval number, $\text{Lv} = \dfrac{V}{\left(\dfrac{2\gamma}{\gamma+1}RT\right)^{1/2}}$

V = speed of the gas stream, γ = ratio between specific heat at constant pressure and at constant volume, R = gas constant, T = absolute temperature.

Leverett function, $\text{j} = \left(\dfrac{k}{e}\right)^{1/2}\dfrac{p_c}{\sigma}$

k = permeability of porous material, e = porosity, p_c = capillary pressure (pressure difference at the interface between two immiscible fluids), σ = surface tension at the interface. It is the ratio (*Characteristic radius of interface curvature / Characteristic pore size*).

Lewis number, $\text{Le} = \dfrac{\rho\, c_p\, D_v}{k}$

ρ = density, c_p = specific heat at constant pressure, D_v = molecular diffusivity, k = thermal conductivity.

Turbulent Lewis number, $\text{Le}_T = \dfrac{\rho\, c_p\, \varepsilon_D}{k_T}$

ρ = density, c_p = specific heat at constant pressure, ε_D = turbulent diffusivity, k_T = eddy thermal conductivity.

Lock number, $\text{Lk} = \dfrac{d\,C_L}{d\,\alpha}\dfrac{\rho\, c\, r^4}{I}$

C_L = lift coefficient, α = angle of attack for blades of helicopter rotor, $\rho ='$ density of air, c = chord length of blade , r = radius of rotor, I = moment of inertia of blade about flapping hinge.

Lorentz number, V/c

V = velocity of the body, c = speed of light.

Luikov number, $\text{Lu} = \dfrac{k_m\, l}{\alpha}$

k_m = mass exchange coefficient, l = length scale, α = thermal diffusivity. It is the ratio (*Mass diffusivity / Thermal diffusivity*).

Lyashchenko number, $\dfrac{V^3\, \rho_f^2}{\mu\, g\, (\rho_s - \rho_f)}$

V = speed, ρ_f = fluid density, ρ_s = sediment density, μ = dynamic viscosity, g = acceleration of gravity. It is the ratio (*Inertia force^2 / Viscous force × Gravity force*).

McAdams group, $\mathrm{Mc} = \dfrac{h^4 \, l \, \mu_l \, \Delta\theta}{k_l^3 \, \rho_l^2 \, g \, \lambda}$

h = heat transfer coefficient, l = length scale, μ_l = dynamic viscosity of the liquid, $\Delta\theta$ = temperature difference, k_l = thermal conductivity of the liquid, ρ_l = density of the liquid, g = acceleration of gravity, λ = latent heat of condensation.

Mach number, $\mathrm{M} = \dfrac{V}{c}$

V = speed, c = sound speed.

Marangoni number, $\mathrm{Mr} = \dfrac{\Delta\sigma}{\Delta\theta} \, \dfrac{\Delta\theta}{\Delta L} \, \dfrac{L^2}{\mu \, \alpha}$

$\Delta\sigma$ = surface tension variation, $\Delta\theta$ = temperature variation, ΔL = change in layer thickness, μ = dynamic viscosity, α = thermal diffusivity.

Mass number, $\mathrm{N}_{mass} = \dfrac{C_s}{1 - C_s} \, \dfrac{\rho_s}{\rho_f}$

C_s = void concentration of sediments, ρ_s, ρ_f = density of sediments, of fluid. It is the ratio (*Inertia of grain / Inertia of fluid*).

Merkel number, $\mathrm{Me} = \dfrac{k_m A}{\dot{m}_g}$

k_m = mass exchange coefficient, A = surface area of water in contact with the gas, $\dot{m}_g$ = dry gas mass flow rate. It is the ratio (*Mass of water transferred in cooling per unit humidity difference / Mass of dry gas*).

Miniovich number, $\mathrm{Mn} = \dfrac{S \, r}{e}$

S = specific area, ratio of surface area to volume, r = radius of particles in packed bed, e = porosity.

Mobility parameter, $\psi = \dfrac{1}{Shields \ number}$.

Newton number, $\mathrm{Ne} = \dfrac{F}{\rho \, V^2 \, l^2}$

F = Drag force, ρ = density of the fluid, V = relative speed, l = length scale of the body.

Nusselt film thickness number, $L \left(\dfrac{g}{v_l^2} \right)^{1/3}$

L = film thickness, g = acceleration of gravity, v_l = kinematic viscosity of the liquid.

Nusselt number, $\mathrm{Nu} = \dfrac{h\,l}{k_f}$

h = heat transfer coefficient, l = length scale, k_f = thermal conductivity. It is the ratio (*Actual heat transfer in forced convection / Heat transfer which would occur by conduction across stationary fluid layer of thickness l*). Number similar, but not identical to Biot number.

Ocvirk number, $\mathrm{Oc} = \dfrac{F}{\mu\,U}\left(\dfrac{2\,b}{L}\right)^2$

F = load force, μ = dynamic viscosity of the liquid, U = speed of the moving wall, b = thickness of the meatus, L = axial length of bearing. It is the ratio (*Load force on bearing / Viscous force*).

Ohnesorge number, $Z = \dfrac{\mu}{\sqrt{\rho\,l\,\sigma}} = \dfrac{1}{(Suratman\ number)^{1/2}}$

μ = dynamic viscosity, ρ = density of fluid, l = length scale, σ = surface tension. It is the square root of the ratio (*Viscous force^2 / Inertia force × surface tension*).

Péclet number, $\mathrm{Pe} = \dfrac{l\,V\,\rho\,c_p}{k_f} = \dfrac{l\,V}{\alpha} = \mathrm{Re} \cdot \mathrm{Pr}$

l = length scale, V = speed of the fluid, ρ = density of the fluid, c_p = specific heat at constant pressure, k_f = thermal conductivity, α = thermal diffusivity. It is the ratio (*Bulk transport of heat in forced convection / Heat transfer by conduction*).

Péclet number for mass transfer, $\mathrm{Pe_m} = \dfrac{l\,V}{D_v}$

l = length scale, V = speed of the fluid, D_v = molecular diffusivity. It is the ratio (*Bulk mass transport / Mass transport by diffusion*).

Plasticity number = *Bingham number*.

Poiseuille number, $\mathrm{Ps} = \dfrac{V\,v}{(\rho_s - \rho_f)\,g\,d_p^2}$

V = speed, v = kinematic viscosity, ρ_s = density of sediments, ρ_f = density of fluid, d_p = diameter of sediment. It is the ratio (*Viscous force / Gravity force*).

Poisson's ratio, v

It is the ratio (*Lateral strain / Longitudinal strain*).

Posnov number, $\text{Pn} = \dfrac{\delta \, \Delta\theta}{\Delta n}$

δ = Soret thermogradient coefficient, $\Delta\theta$ = temperature difference, Δn = difference in specific moisture content (mass of moisture per unit mass of completely dry gas).

Power number, $\dfrac{P}{l^5 \, \rho \, N^3}$

P = power, l = length scale, ρ = density, N = rotational rate (in revolutions per unit time). Usually l = diameter of the blade/impeller. It is the ratio (*Drag force on rotating impeller / Inertia force*).

Prandtl dimensionless distance, $y^+ = \dfrac{y}{\nu}\left(\dfrac{\tau_b}{\rho}\right)^{1/2}$

y = distance from wall, ν = kinematic viscosity, τ_b = tangential stress at the wall, ρ = density.

Prandtl number, $\text{Pr} = \dfrac{c_p \, \mu}{k_f} = \dfrac{\nu}{\alpha}$

c_p = specific heat at constant pressure, μ = dynamic viscosity, k_f = thermal conductivity of the fluid, ν = kinematic viscosity, α = thermal diffusivity. It is the ratio (*Momentum diffusivity / Thermal diffusivity*). It depends only on the properties of the fluid.

Prandtl number for mass transfer = *Schmidt number*

Diffusion Prandtl number, $\dfrac{\nu}{D_\nu}$

ν = kinematic viscosity, D_ν = molecular diffusivity.

Turbulent Prandtl number, $\text{Pr}_T = \dfrac{\varepsilon_M}{\varepsilon_T}$

ε_M = turbulent diffusivity of momentum, ε_T = turbulent heat diffusivity.

Total Prandtl number, $\dfrac{\varepsilon_M + \nu}{\varepsilon_T + \alpha}$

ε_M = turbulent diffusivity of momentum, ε_T = turbulent heat diffusivity, ν = kinematic viscosity, α = thermal diffusivity. It is the ratio (*Total momentum diffusivity / Total thermal diffusivity for heat transfer in combined laminar and turbulent flows*).

Prandtl velocity ratio, $u^+ = \dfrac{u}{(\tau_b/\rho)^{1/2}}$

u = speed, τ_b = tangential stress at the wall, ρ = density. It is the ratio (*Inertia force / Shear force at boundary*).

Predvoditelev number, $\mathrm{Pd} = \dfrac{\Gamma\, l^2}{\alpha\, T_0}$

Γ = maximum rate of change of ambient temperature, l = length scale, α = thermal diffusivity, T_0 = initial temperature. It is the ratio (*Rate of change of ambient temperature / Rate of change of temperature of body*).

Pressure coefficient, $\dfrac{\Delta p}{\rho\, V^2}$

Δp = pressure variation, ρ = density, V = speed.

Pressure number, $\dfrac{p}{\sqrt{g\,\sigma\,(\rho_l - \rho_g)}}$

p = pressure, g = acceleration of gravity, σ = surface tension, ρ_l = density of the liquid, ρ_g = density of gas. It is the ratio (*Absolute pressure in system / Pressure jump at liquid surface*).

Ramberg number, $\mathrm{Rm} = \dfrac{g\, l^2\, \rho}{\mu\, V}$

g = acceleration of gravity, l = length scale, ρ = density, V = speed, μ = dynamic viscosity.

Psychrometric ratio (for wet- and dry-bulb thermometry), $\dfrac{h_c}{k_m\, s}$

h_c = heat transfer coefficient for convection, k_m = mass exchange coefficient, s = heat required for a unit rise in temperature to a unit mass of dry air plus contained water vapour.

Radiation number, $\dfrac{k\,\varepsilon}{\sigma\, s\, T^3}$

k = thermal conductivity, ε = bulk modulus of the fluid, σ = surface tension, s = Stefan-Boltzmann constant, T = absolute temperature.

Radiation parameter, $\Phi = \dfrac{\zeta\, s\, T_w^3\, R}{k_f}$

ζ = coefficient expressing the average emissivity of the channel walls, s = Stefan-Boltzmann constant, T_w = absolute temperature of wall, R = hydraulic radius of the channel, k_f = thermal conductivity of the fluid. It expresses the influence of radiation on convective heat transfer in the channel. It is a variant of Stefan number.

Rayleigh number, $Ra = \dfrac{l^3 \rho^2 g \beta c_p \Delta\theta}{\mu k_f} \equiv Gr \cdot Pr$

l = length scale, ρ = density, g = acceleration of gravity, β = coefficient of thermal expansion of volume at constant pressure, c_p = specific heat at constant pressure, $\Delta\theta$ = temperature variation, μ = dynamic viscosity, k_f = thermal conductivity.

Reaction enthalpy number, $\dfrac{(\Delta h)_A \Delta n_A}{c_p \Delta T}$

$(\Delta h)_A$ = enthalpy of reaction/mass of A produced, n_A = mass fraction of A, c_p = specific heat at constant pressure, ΔT = temperature variation. It is the ratio (*Change in reaction energy / Change in thermal energy*).

Recovery factor, $RF = \dfrac{2 c_p \Delta\theta}{V^2}$

c_p = specific heat at constant pressure, $\Delta\theta$ = temperature difference between the gas in motion and the adiabatic wall, V = speed of the gas. It is the ratio (*Actual temperature recovery / Theoretical temperature recovery for an ideal gas*).

Reech number, $\dfrac{V^2}{g\,l} \equiv$ Froude number.

Reynolds number, $Re = \dfrac{\rho V l}{\mu}$

ρ = density, V = speed, l = length scale, μ = dynamic viscosity. It is the ratio (*Inertia force / Viscous force*).

Generalized Reynolds number for non-Newtonian fluids, $\dfrac{8 \rho \overline{V}^2}{\tau_w}$

ρ = density, $\overline{V}$ = average speed, τ_w = tangential wall stress. Applies for non-Newtonian fluid flow in circular cross-section.

Reynolds number (rotating), $Re_R = \dfrac{\rho \omega D^2}{\mu}$

ρ = density, ω = angular velocity, D = length scale of the rotating body, μ = dynamic viscosity.

Richardson number, $Ri = -\dfrac{g}{\rho} \left(\dfrac{d\rho}{dz}\right) \Big/ \left(\dfrac{dV}{dz}\right)_w^2$

g = acceleration of gravity, ρ = density, z = height of layer (measured vertically upwards), $(dV/dz)_w$ = shear rate at the wall. It is the ratio (*Gravity force / Inertia force*).

Romankov number, $\mathrm{Ro} = \dfrac{T_0 - T_{pr}}{T_0}$

T_0 = absolute temperature of the hot gas stream used in a drying process, T_{pr} = absolute temperature of product being dried.

Rossby number, $\mathrm{Ro} = \dfrac{V}{\omega\, l}$

V = speed, ω = angular velocity , l = length scale.

Rossby number (second definition), $\mathrm{Ro} = \dfrac{V}{2\, \omega\, l\, \sin \alpha}$

V = speed, ω = angular velocity of the Earth, l = length scale, α = angle between the direction of laminar flow and the Earth's rotation axis. It is the ratio (*Inertia force / Coriolis force*).

Savage number, $\mathrm{Sa} = \dfrac{\rho_s\, \dot{\gamma}^2\, d^2}{(\rho_s - \rho_f)\, g\, h\, \tan \phi}$

ρ_s, ρ_f = density of sediments, of fluid $\dot{\gamma}$ = shear rate, d = diameter of sediments, g = acceleration of gravity, h = height of the stream, ϕ = internal friction angle of sediments. It is the ratio (*Collisional stress / Quasi-static (frictional) stress due to gravity*).

Serrau number = *Mach number.*

Schiller number, $\left(\dfrac{\mathrm{Re}}{C_D} \right)^{1/3}$

Re = Reynolds number, C_D = drag coefficient.

Schmidt number (molecular), $\mathrm{Sc} = \dfrac{\nu}{D_v}$

ν = kinematic viscosity, D_v = molecular diffusivity. It is the ratio (*Momentum diffusivity / Molecular diffusivity*).

Schmidt number (turbulent), $\mathrm{Sc}_T = \dfrac{\varepsilon_M}{\varepsilon_D}$

ε_M = turbulent diffusivity of momentum, ε_D = turbulent diffusivity of mass.

Total Schmidt number, $\dfrac{\varepsilon_M + \nu}{\varepsilon_D + D_v}$

ε_M = turbulent diffusivity of momentum, ε_D = turbulent diffusivity of mass, ν = kinematic viscosity, D_v = molecular diffusivity. It is the ratio (*Total momentum diffusivity / Total mass diffusivity in combined laminar and turbulent flows*).

Semenov number = *Lewis number*.

Sherwood number, $\mathrm{Sh} = \dfrac{k_m \, l}{D_v}$

k_m = mass exchange coefficient, l = length scale, D_v = molecular diffusivity. It is the ratio (*Mass diffusivity / Molecular diffusivity*).

Shields number, $\Theta = \dfrac{\rho_f \, u_*^2}{g \, d \, (\rho_s - \rho_f)}$

ρ_s, ρ_f = density of sediments, of fluid, u_* = friction velocity, g = acceleration of gravity, d = diameter of sediments. It is the ratio (*Tangential destabilizing stress / Tangential stabilizing stress*).

Size number (of turbomachine), also termed Specific diameter, $\dfrac{D \, (g \, H)^{1/4}}{Q^{1/2}}$

D = diameter, g = acceleration of gravity, H = difference of head across turbomachine, Q = flow rate.

Smoluchowski number$= \dfrac{1}{Knudsen \; number}$

Sommerfeld number, $\mathrm{Sm} = \dfrac{F \, b^2}{\mu \, U \, r^2}$

F = force on bearing, b = thickness of the meatus, μ = dynamic lubricant viscosity, U = relative speed, r = shaft radius.

Spalding number, $\mathrm{Sp} = -\dfrac{\partial \theta}{\partial u^+}$

$\theta = (T - T_\infty)/(T_w - T_\infty)$, $u^+ = V/(\tau_w/\rho)^{1/2}$, T = absolute temperature, T_w = absolute wall temperature, T_∞ = absolute temperature far from the body, V = speed, τ_w = tangential stress at the wall, ρ = density. It is the temperature gradient at the wall expressed in dimensionless form.

Spalding number (second definition), $\dfrac{h \, v}{k \, (\tau_w/\rho)^{1/2}}$

h = heat transfer coefficient, v = kinematic viscosity, k = thermal conductivity, τ_w = tangential stress at the wall, ρ = density.

Spalding number (third definition), $\dfrac{c_p \, \Delta T}{\lambda - (q_r/\dot{m})}$

c_p = specific heat at constant pressure, ΔT = temperature variation, λ = specific enthalpy of evaporation (latent heat of evaporation), q_r = radiant heat flow, $\dot{m}$ =

mass flow rate. It is the ratio (*Change in thermal energy / Latent heat for evap-ourated material*).

Stanton number, $St = \dfrac{h}{\rho \, c_p \, V} \equiv \dfrac{Nu}{Re \cdot Pr}$

h = heat transfer coefficient, ρ = density, c_p = specific heat at constant pressure, V = speed. It is the ratio (*Quantity of heat actually transferred / Thermal fluid capacity*).

Stanton number for mass transfer, $St_m = \dfrac{k_m}{V} \equiv \dfrac{Sh}{Re \cdot Sc}$

k_m = mass exchange coefficient, V = speed.

Stark number = *Stefan number*.

Stefan number, $Sf = \dfrac{s \, T^3 \, l}{k}$

s = Stefan-Boltzmann constant, T = absolute temperature, l = length scale, k = thermal conductivity. It is the ratio (*Rate of energy radiation / Rate of heat conduction*).

Stokes number, $Sk = \dfrac{v \, t}{l^2} \equiv \dfrac{1}{Sr \cdot Re}$

v = kinematic viscosity, t = vibration time of the particle in the fluid, l = characteristic particle size.

Stokes number (second definition), $\dfrac{\omega \, l^2}{v}$

ω = pulsation of vibration of particle, l = characteristic particle size, v = kinematic viscosity.

Stokes number (third definition), $\dfrac{l \, \Delta p}{\mu \, V}$

l = characteristic particle size, Δp = pressure variation, μ = dynamic viscosity, V = speed. It is the ratio (*Pressure force / Viscous force*).

Strohual number, $Sr = \dfrac{f \, l}{V}$

f = vibration frequency, l = length scale, V = speed.

Suratman number, $Su = \dfrac{\rho \, l \, \sigma}{\mu^2} \equiv \dfrac{1}{Ohnesorge \; number}$

ρ = density of fluid, μ = dynamic viscosity, l = length scale, σ = surface tension. It is the ratio (*Inertia force × Surface tension force / Viscous force^2*).

Surface elasticity number, $\dfrac{\Gamma'}{D_s} L \dfrac{\partial \sigma}{\partial \Gamma'}$

Γ' = concentration at the surface of a surfactant in undisturbed state, D_s = surface diffusivity, L = liquid layer thickness, σ = surface tension.

Surface viscosity number, $\dfrac{\mu_s}{\mu L}$

μ_s = 'surface viscosity', μ = dynamic viscosity, L = liquid layer thickness.

Taylor number, $\mathrm{Ta} = \dfrac{\omega \, \bar{r}^{1/2} \, b^{3/2}}{\nu}$

ω = angular velocity of the internal cylinder, $\bar{r}$ = mean radius of annulus surrounding rotating cylinder, b = gap, ν = kinematic viscosity. It is used in Taylor's vortex instability criterion.

Taylor number (second definition), $\dfrac{2 \, \omega \, L^2 \cos \theta}{\nu^2} = \dfrac{1}{(Ekman \; number)^4}$

ω = angular velocity, L = length scale, θ = angle between rotation axis and vertical, ν = kinematic viscosity. It expresses the effect of rotation on free convection and is the square of the ratio (*Coriolis force / Viscous force*).

Taylor number (third definition),= *Sherwood number.*

Frictional stress number, $\mathrm{N}_{frict} = \dfrac{C_s}{1 - C_s} \dfrac{(\rho_s - \rho_f) \, g \, h \, \tan \phi}{\dot{\gamma} \, \mu} \equiv \dfrac{\mathrm{Ba}}{\mathrm{Sa}}$

C_s = void concentration of sediments, ρ_s, ρ_f = density of sediments, of fluid, g = acceleration of gravity, h = height of the stream, ϕ = internal friction angle for sediments, $\dot{\gamma}$ = strain rate, μ = dynamic viscosity.

Thoma number, $\sigma = \dfrac{p - p_v}{\Delta p}$

p = absolute pressure, p_v = vapour pressure, Δp = total pressure variation.

Thoma number (second definition), $\sigma = \dfrac{p - p_v}{\rho \, V^2}$

p = absolute pressure, p_v = vapour pressure, ρ = density, V = speed. Indicates the onset of cavitation in a system in which the pressure scales according to the velocity of the fluid.

Thomson number = *Marangoni number.*

Thomson number (second definition), $Th = \dfrac{V\,t}{l}$

V = speed, t = time scale, l = characteristic size. It the time scale si $t = f^{-1}$, then $Th = Sr$.

Thring number, $Tg = \dfrac{\rho\,c_p\,V}{\varepsilon\,s\,T^3}$

ρ = density, c_p = specific heat at constant pressure, V = speed, ε = surface emissivity, s = Stefan-Boltzmann constant, T = absolute temperature. It is the ratio (*Bulk transport of heat / Heat transferred by radiation*).

Thrust coefficient (of propellers), $T_c = \dfrac{T}{\rho\,V^2\,D^2}$

T = thrust, ρ = density, V = forward speed, D = diameter of propeller.

Torque coefficient (of propellers), $M_c = \dfrac{M}{\rho\,V^2\,D^3}$

M = torque, ρ = density, V = forward speed, D = diameter of propeller.

Valensi number, $Va = \dfrac{\omega\,l^2}{\nu}$

ω = pulsation of an oscillating body in a zero-viscosity fluid, l = length scale, ν = kinematic viscosity.

Weber number, $We = \dfrac{\rho\,l\,V^2}{\sigma}$

ρ = density, l = length scale, V = speed, σ = surface tension. It is the ratio (*Inertia force / Surface tension force*).

Weber number (rotating), $We_R = \dfrac{D^3\,\omega^2\,\rho}{\sigma}$

D = diameter, ω = angular velocity, ρ = density, σ = surface tension.

Womerseley number, $\alpha = (\omega\,\rho/\mu)^{1/2}\,r \equiv 2\,\pi\,Re \cdot Sr$

ρ = density of the fluid, μ = dynamic viscosity, r = radius of the pipe, ω = pulsation of the flow rate.

Transport parametr, $\dfrac{q_s}{\sqrt{g\,d^3\,(s-1)}}$

q_s = sediment volumetric flow rate per unit width, g = acceleration of gravity, d = diameter of sediments, s = relative specific gravity of sediments.

Glossary

Asymptotic independence of turbulence: the property of turbulent flow fields to be independent of the Reynolds number (and hence fluid viscosity) for a Reynolds number that tends to infinity.

Axiom: a general principle that is self-evident in itself, which does not need to be demonstrated or discussed and which can serve as the premise of a theory.

Coefficient: a number or known dimensional quantity that multiplies an algebraic quantity. This is defined because it concurs with the algebraic quantity to define a single product.

Coordinate system: a system for representing the position of a point in space. Not to be confused with the reference system.

Corollary: a proposition that follows by logical consequence from another proposition that has already been proven.

Dimensional equation: an equation involving only the physical dimensions of the variables.

Dimensional matrix: a matrix in which the variables are shown in the column and the fundamental quantities in the row. Each element of the matrix represents the exponent with which the fundamental quantity of the corresponding row appears in the dimensional expression of the variable in the corresponding column.

Equation: an expression or proposition asserting equality between two members and involving one or more variables. The symbol for equality between the two members is =.

Extensive quantity: a quantity whose measurement depends on the size of the system (e.g., mass, volume, area). The ratio of two extensive quantities is an intensive quantity if the two extensive quantities refer to the same dimension of the system. A function of extensive quantities is homogeneous in the 1st degree with respect to the quantities and satisfies Euler's Theorem on Homogeneous Functions (see Sect. A).

Fractal: a geometrical object that repeats itself in its structure in the same way on different scales, i.e., that does not change its appearance even when viewed through a magnifying glass. It has internal homothety.

S. G. Longo, *Principles and Applications of Dimensional Analysis and Similarity*, Mathematical Engineering, https://doi.org/10.1007/978-3-030-79217-6

Governed variable: a variable representing the response of a physical process.

Governing variable: a variable controlling a physical process.

Identity: an equation satisfied for any value of the variable(s), which can also be defined as a tautologically satisfied equation. The symbol of identity between the two members is $\equiv$.

Inertial reference system: a system in which the first law of dynamics $F = ma$ holds. Not to be confused with the coordinate system.

Intensive quantity: a quantity whose measure does not depend on the size of the system.

Jacobian: the matrix of all first-order partial derivatives of a function having domain and co-domain in a Euclidean space.

Linear proportionality: any power function having the size of a length calculated on the basis of the linear proportionality criterion.

Linear proportionality—criterion: a criterion for reducing the variables of a physical process to a smaller set of power functions having the size of a length (Barr, 1969).

Material objectivity: a principle according to which the laws governing the internal conditions of a physical system and the interactions between the various components must be independent of the reference system, be it inertial or non-inertial.

Measure: the assignment of a range of values to a property or characteristic of a material entity.

Measuring: the set of theoretical and practical operations used in carrying out a particular measure.

Monomial: a function containing the product of a number, called a coefficient (possibly unitary), and one or more variables raised to constant nonnegative integers.

Order of magnitude: the position of a quantity in a scale where each class contains values in a defined ratio to the preceding class. The most commonly used ratio is 10.

Physical process: any sequence of changes in a real object that can be observed on the basis of the scientific method. In a physical process, one or more physical variables can be identified, some governing, some governed.

Postulate: an unproved proposition that is accepted as the basis of a proof.

Power function: a function containing the product of a real number, called a coefficient (possibly unitary), and a variable raised to a constant real exponent. By extension, we also consider power functions to be the product between a coefficient and several variables, each of which is raised to a fixed real exponent.

Principle: a highly general or fundamental scientific law from which other laws are derived.

Prototype: it is the object or physical process to be modelled. Not necessarily coinciding with a physical process occurring in nature.

Rank of a matrix: the dimension of the vector space generated by its columns; computed as the maximum order of the square submatrices extracted from the matrix.

Scales: constant proportions between variables or between parameters occurring in the model and prototype. These are pure numbers.

Scale effects: the differences in the corresponding values of the variables involved in two similarity processes. They are caused by the practical impossibility of achieving complete similarity.

Similarity: equivalence condition of a process associated to two different models, or to a model and a prototype (some authors distinguish between similitude and similarity, the former being rigorous and complete, the latter only approximate).

Similarity criteria: the scale relationships between the variables involved in two processes in similarity. They can be strictly mathematically based or semi-empirical in nature. Their applicability is within the limits of the mathematical (or physical) model from which they originate.

Self-similarity: the property of a mathematical (or physical) object such that it is exactly or approximately equal to a part of itself.

Surfactant: a substance that is able to lower the surface tension of a liquid.

Tensor: an algebraic object independent of the coordinate system, defined intrinsically from a vector space. In its intrinsic definition, it does not need a base. From a mathematical point of view, the tensor generalises all algebraic structures from a vector space.

Theorem: a proposition that can be proven using axioms, postulates or theorems that have been proven previously.

True value of a quantity: the ideal measure of quantity, without uncertainty. The true value is inaccessible and is always replaced by an estimate.

Typical equation: an equation that, in symbolic form, expresses a functional relationship between the variables involved in a physical process.

General Bibliography

Antonets, V. A., Antonets, M. A., & Shereshevsky, I. A. (1991). The statistical cluster dynamics in the dendroid transfer systems. *Fractals in the Fundamental and Applied Sciences, 59–71.*

Bagnold, R. A. (1954). Experiments on a gravity-free dispersion of large solid spheres in a Newtonian fluid under shear. *Proceedings of the Royal Society London A, 225*(1160), 49–63.

Bairrao, R., & Vaz, C. (2000). Shaking table testing of civil engineering structures–the LNEC 3D simulator experience. In *Proceedings of the 12th World Conference on Earthquake Engineering. Auckland, New Zealand* (p. 2129).

Ball, T. V., & Huppert, H. E. (2019). Similarity solutions and viscous gravity current adjustment times. *Journal of Fluid Mechanics, 874,* 285–298.

Ball, T. V., Huppert, H. E., Lister, J., & Neufeld, J. (2017). The relaxation time for viscous and porous gravity currents following a change in flux. *Journal of Fluid Mechanics, 821,* 330–342.

Barenblatt, G. I. (1996). *Scaling, self-similarity, and intermediate asymptotics.* Cambridge University Press.

Barenblatt, G. I. (2003). *Scaling.* Cambridge University Press.

Barenblatt, G. I. (2014). *Flow, deformation and fracture: lectures on fluid mechanics and the mechanics of deformable solids for mathematicians and physicists* (Vol. 49). Cambridge University Press.

Barenblatt, G. I., Entov, V. M., & Ryzhik, V. M. (1989). *Theory of fluid flows through natural rocks.* Kluwer Academic Publishers.

Barenblatt, G. I., & Sivashinskii, G. I. (1969). Self-similar solutions of the second kind in nonlinear filtration. *Journal of Applied Mathematics and Mechanics, 33*(5), 836–845.

Barenblatt, G. I., & Zel'Dovich, Y. B. (1972). Self-similar solutions as intermediate asymptotics. *Annual Review of Fluid Mechanics, 4*(1), 285–312.

Barlow, J. B., Rae, W. H., & Pope, A. (1999). *Low-speed wind tunnel testing.* Wiley.

Barr, D. I. H. (1969). Method of synthesis-basic procedures for the new approach to similitude. *Water Power, 21*(4), 148–153.

Bear, J. (1972). *Dynamics of fluids in porous media.* Dover.

Benbow, J. J. (1960). Cone cracks in fused silica. *Proceedings of the Physical Society, 75*(5), 697.

Bertrand, J. (1848). *Note sur la similitude en Mécanique* (pp. 189–197). Cahier XX XII: Journal de l'École Polythechnique XIX.

Bhushan, B., & Nosonovsky, M. (2004). Scale effects in dry and wet friction, wear, and interface temperature. *Nanotechnology, 15*(7), 749.

Birkhoff, G. (1950). *Hydrodynamics: A study in logic, fact and similitude.* Dover.

S. G. Longo, *Principles and Applications of Dimensional Analysis and Similarity,* Mathematical Engineering, https://doi.org/10.1007/978-3-030-79217-6

Bluman, G. W., & Kumei, S. (1989). *Symmetries and differential equations*. Springer.

Booth, E., Collier, D., & Miles, J. (1983). Impact scalability of plated steel structures. In N. Jones & T. Wierzbicki (Eds.), *Structural crashworthiness* (136–174).

Bouc, R. (1967). Forced vibrations of mechanical systems with hysteresis. In *Proceedings of the Fourth Conference on Nonlinear Oscillations, Prague, 1967*.

Bridgman, P. W. (1922). *Dimensional Analysis*. Yale University Press.

Buckingham, E. (1914). On physically similar systems; illustrations of the use of dimensional equations. *Physical Review, 4*(4), 345–376.

Bucky, P. (1931). The use of models for the study of mining problems. *Technical Publication 425*, Class A, Mining methods no. 44. American Institute of Mining and Metallurgical Engineers.

Calladine, C. R., & English, R. W. (1984). Strain-rate and inertia effects in the collapse of two types of energy-absorbing structure. *International Journal of Mechanical Sciences, 26*(11–12), 689–701.

Carpinteri, A., & Corrado, M. (2010). Dimensional analysis approach to the plastic rotation capacity of over-reinforced concrete beams. *Engineering Fracture Mechanics, 77*(7), 1091–1100.

Carvallo, E. (1892). Sur une similitude dans les fonctions des machines. *Journal de Physique Théorique et Appliquée, 1*(1), 209–212.

Chen, Y., & Cheng, P. (2002). Heat transfer and pressure drop in fractal tree-like microchannel nets. *International Journal of Heat and Mass Transfer, 45*(13), 2643–2648.

Ciriello, V., Longo, S., Chiapponi, L., & Di Federico, V. (2016). Porous gravity currents: A survey to determine the joint influence of fluid rheology and variations of medium properties. *Advances in Water Resources, 92*, 105–115.

Coastal Engineering Research Center (CERC, US). (1984). *Shore Protection Manual (Vol. I-II)*: United States Army Corps of Engineers.

Corrsin, S. (1951). A simple geometrical proof of Buckingham's Π-theorem. *American Journal of Physics, 19*(3), 180–181.

Corti, G., Bonini, M., Mazzarini, F., Boccaletti, M., Innocenti, F., Manetti, P., Mulugeta, G., & Sokoutis, D. (2002). Magma-induced strain localization in centrifuge models of transfer zones. *Tectonophysics, 348*(4), 205–218.

Cranz, K. J. (1910). *Lehrbuch der ballistik* (Vol. 1). B. G: Teubner, Leipzig.

Dalrymple, R. A. (1989). Physical modelling of littoral processes. In R. Martins (Ed.), *Recent Advances in Hydraulic Physical Modelling* (pp. 567–588). Springer.

Damgaard, J. S., & Dong, P. (2004). Soft cliff recession under oblique waves: physical model tests. *Journal of Waterway, Port, Coastal, and Ocean Engineering, 130*(5), 234–242.

Di Federico, V., Longo, S., King, S. E., Chiapponi, L., Petrolo, D., & Ciriello, V. (2017). Gravity-driven flow of Herschel-Bulkley fluid in a fracture and in a 2D porous medium. *Journal of Fluid Mechanics, 821*, 59–84.

Dittus, F. W., & Boelter, L. M. K. (1985). Heat transfer in automobile radiators of the tubular type. *International Communications in Heat and Mass Transfer, 12*(1), 3–22.

Duncan, W. J. (1953). *Physical Similarity and Dimensional Analysis*. Edward Arnold & Co.

Ehrenfest-Afanassjewa, T. (1915). Der Dimensionsbegriff und der analytischen Base physikalischer Gleichungen. *Mathematische Annalen, 77*, 259–276.

Einstein, H. A., & Barbarossa, H. L. (1952). River channel roughness. *Transactions of the American Society of Civil Engineers, 117*(1), 1121–1132.

Ercole, G. (2011). *Vascelli e fregate della Serenissima: navi di linea della Marina veneziana 1652–1797 (in Italian)*. Gruppo modellistico trentino di studio e ricerca storica.

Euler, L. (1765). *Theoria motus corporum solidorum seu rigidorum*. A. F. Röse.

Exner, F. M. (1925). Über Wechselwirkung zwischen Wasser und Geschiebe in Flüssen (On the interaction between water and sediment in streams). *Akademie der Wissenschaften in Wien, Mathematisch-Naturwissenschaftliche Klasse, 134*(2a), 165–204.

Fleckstein, G. O. (1957). *Leonhardi Euleri opera omnia*, vol. 5(II). Orell Füssli, Lausanne.

Focken, C. M. (1953). *Dimensional methods and their applications*. Arnold.

Forchheimer, P. (1901). Wasserbewegung durch boden. *Zeitschrift des Vereins deutscher Ingenieure, 45*, 1782–1788.

Fourier, J. B. (1822). *Theorie analytique de la chaleur, par M* (p. 2009). Fourier: Chez Firmin Didot, père et fils, reprinted by Cambridge University Press.

Fredsøe, J., & Deigaard, R. (1992). *Mechanics of coastal sediment transport* (Vol. 3). World Scientific Publishing Co., Pte. Ltd.

Galilei, G. (1638). *Discorsi e dimostrazioni matematiche intorno a due nuove scienze attinenti la meccanica e i movimenti locali (only first 4 days + appendix, in Italian and in Latin).* Elzeviri.

Galilei, G. (1687). *De Motu Antiquiora* (1589–1592).

Galilei, G. (1718). *Discorsi e dimostrazioni matematiche intorno a due nuove scienze attinenti la meccanica e i movimenti locali (fifth and sixth days, in Italian and in Latin).* Tartini and Franchi.

Gibbings, J. C. (1980). On dimensional analysis. *Journal of Physics A: Mathematical and General, 13*(1), 75.

Gibbings, J. C. (1981). Directional attributes of length in dimensional analysis. *International Journal of Mechanical Engineering Education, 8*(3), 263–272.

Gibbings, J. C. (2011). *Dimensional analysis.* Springer Science & Business Media.

Giuliani, F., Petrolo, D., Chiapponi, L., Zanini, A., & Longo, S. (2021). Advancement in measuring the hydraulic conductivity of porous asphalt pavements. *Construction and Building Materials, 300*, 124110.

Goodridge, C. L., Shi, W. T., Hentschel, H. G. E., & Lathrop, D. P. (1997). Viscous effects in droplet-ejecting capillary waves. *Physical Review E, 56*(1), 472.

Gratton, J., & Minotti, F. (1990). Self-similar viscous gravity currents: Phase-plane formalism. *Journal of Fluid Mechanics, 210*, 155–182.

Grossel, S. S. (1996). Guidelines for evaluating the characteristics of vapour cloud explosions, flash fires and BLEVEs. *Journal of Loss Prevention in the Process Industries, 3*(9), 247.

Harris, H. G., Pahl, P. J., & Sharma, S. D. (1962). *Dynamic studies of structures by means of models.* MIT.

Hepburn, B. S. (2007). *Equilibrium and explanation in 18th century mechanics.* PhD thesis, University of Pittsburgh.

Hopkinson, B. (1915). UK Ordnance Board Minutes 13565. *Public records office, London Vol Sup6/187.*

Hornung, H. G. (2006). *Dimensional analysis: Examples of the use of symmetry.* Dover Publications Inc.

Hughes, S. A. (1993). *Physical models and laboratory techniques in coastal engineering* (vol. 7). World Scientific Publishing Co. Pte. Ltd.

Ipsen, D. C. (1960). *Units, Dimensions, and Dimensionless Numbers.* McGraw-Hill Book Co.

Iverson, R. M. (1997). The physics of debris flows. *Reviews of Geophysics, 35*(3), 245–296.

Ivicsics, L. (1980). *Hydraulic models.* Water Resources Publications.

Jackson, K. E., Kellas, S., & Morton, J. (1992). Scale effects in the response and failure of fiber reinforced composite laminates loaded in tension and in flexure. *Journal of Composite Materials, 26*(18), 2674–2705.

Jiménez-Portaz, M., Chiapponi, L., Clavero, M., & Losada, M. A. (2020). Air flow quality analysis of an open-circuit boundary layer wind tunnel and comparison with a closed-circuit wind tunnel. *Physics of Fluids, 32*(12), 125120.

Jiménez-Portaz, M., Clavero, M., & Losada, M. A. (2021). A New Methodology for Assessing the Interaction between the Mediterranean Olive Agro-Forest and the Atmospheric Surface Boundary Layer. *Atmosphere, 12*, 658.

Johnson, W. (1983). *Impact strength of materials.* Edward Arnold.

Julien, P. Y. (2002). *River mechanics.* Cambridge University Press.

Kang, J. H., Lee, K.-J., Yu, S. H., Nam, J. H., & Kim, C.-J. (2010). Demonstration of water management role of microporous layer by similarity model experiments. *International Journal of Hydrogen Energy, 35*(9), 4264–4269.

Keulegan, G. H. (1938). Laws of turbulent flow in open channels. *Journal of Research of the National Bureau of Standard, 21*(RP1151), 708–741.

Kleiber, M. (1947). Body size and metabolic rate. *Physiological Reviews, 27*, 511–541.

Langhaar, H. L. (1951). *Dimensional Analysis and Theory of Models*. Wiley.

Longo, S. (2009). Vorticity and intermittency within the pre-breaking region of spilling breakers. *Coastal Engineering, 56*(3), 285–296.

Longo, S., Chiapponi, L., & Liang, D. (2013). Analytical study of the water surface fluctuations induced by grid-stirred turbulence. *Applied Mathematical Modelling, 37*(12–13), 7206–7222.

Longo, S., Di Federico, V., & Chiapponi, L. (2015). A dipole solution for power-law gravity currents in porous formations. *Journal of Fluid Mechanics, 778*, 534–551.

Longo, S., & Lamberti, A. (2000). Granular streams rheology and mechanics. *Physics and Chemistry of the Earth, Part B: Hydrology, Oceans and Atmosphere, 25*(4), 375–380.

Longo, S., & Petti, M. (2006). *Misure e controlli idraulici*. McGrawHill.

Longo, S., Chiapponi, L., Petrolo, S., Lenci, A., & Di Federico, V. (2021). Converging gravity currents of power-law fluid. *Journal of Fluid Mechanics, 918*(A5), 1–30.

Ludwig, G. R., & Skinner, G. T. (1976). *Wind tunnel modeling study of the dispersion of sulfur dioxide in southern Allegheny County, Pennsylvania. Final report*. Calspan Corporation.

Macagno, E. O. (1971). Historico-critical review of dimensional analysis. *Journal of the Franklin Institute, 292*(6), 391–402.

Mandelbrot, B. B. (1982). *The Fractal Geometry of Nature* (Vol. 2). Freeman & Co. Ltd.

Massey, B. S. (1971). *Units, dimensional analysis and physical similarity*. Van Nostrand Reinhold Company.

Massey, B. S. (1978). Directional analysis? *International Journal of Mechanical Engineering Education, 6*(1), 33–36.

Massey, B. S. (1986). *Measures in science and engineering: Their expression, relation and interpretation*. Ellis Horwood Ltd.

Maxwell, C. J. (1869). Remarks on the mathematical classification of physical quantities. *Proceedings of the London Mathematical Society, 1*(1), 224–233.

Maxwell, C. J. (1878). III. On stresses in rarefied gases arising from inequalities of temperature. *Proceedings of the Royal Society London, 27*, 304–308.

Mignosa, P., Chiapponi, L., D'Oria, M., & Anelli, L. (2017). *Realizzazione del modello fisico delle pile del ponte sul fiume Po ex S.S 413 'Romana' (Physical model tests of the bridge piles in the riverbed of Po River ex S.S 413 'Romana')(in Italian)*. DICATeA, University of Parma.

Mignosa, P., Giuffredi, F., Danese, D., La Rocca, M., Longo, S., Chiapponi, L., et al. (2008). *Prove su modello fisico del manufatto regolatore della cassa di espansione sul Torrente Parma (Physical model tests of the control systems for the detention basin of the Parma Torrent) (in Italian)*. DICATeA, University of Parma, and AIPo.

Mignosa, P., Longo, S., Chiapponi, L., D'Oria, M., & Mammí, O. (2010). *Prove su modello fisico della vasca di dissipazione al termine della galleria di bypass del Lago d'Idro (Physical model tests for the stilling basin at the end of the Lake Idro bypass tunnel) (in Italian)*. DICATeA: University of Parma.

Newton, I. (1687). *Philosophiae naturalis principia mathematica (in Latin)*. Societatis Regiae.

Noda, E. K. (1972). Equilibrium beach profile scale-model. *Journal of Waterways, Harbors and Coastal Engineering Division, 98*(4), 511–528.

Novak, P., & Čábelka, J. (1981). *Models in hydraulic engineering: Physical principles and design applications* (Vol. 4). Pitman Publishing Ltd.

Nusselt, W. (1916). Die Oberflachenkondensation des Wasserdampfes (in German). *Zeitschrift Vereines Deutscher Ingenieure, 60*, 541.

Papageorgiou, D. T. (1995). On the breakup of viscous liquid threads. *Physics of Fluids, 7*(7), 1529–1544.

Pérez-Romero, D. M., Ortega-Sánchez, M., Moñino, A., & Losada, M. A. (2009). Characteristic friction coefficient and scale effects in oscillatory porous flow. *Coastal Engineering, 56*(9), 931–939.

Ramberg, H., & Stephansson, O. (1965). Note on centrifuged models of excavations in rocks. *Tectonophysics, 2*(4), 281–298.

Raszillier, H., & Durst, F. (1991). Coriolis-effect in mass flow metering. *Archive of Applied Mechanics, 61*(3), 192–214.

Raszillier, H., & Raszillier, V. (1991). Dimensional and symmetry analysis of Coriolis mass flowmeters. *Flow Measurement and Instrumentation, 2*(3), 180–184.

Reech, F. (1852). *Cours de Mécanique d'après la Nature Généralement Flexible et Elastique des Corps.* Carilian-Goeury.

Riabouchinsky, D. (1911). Methode des variables de dimension zéro, et son application en aérodynamique. *L'Aérophile, 1,* 407–408.

Rouse, H., & Ince, S. (1963). *History of Hydraulics.* Dover.

Ruark, A. E. (1935). Inspectional analysis: A method which supplements dimensional analysis. *Journal of the Elisha Mitchell Scientific Society, 51*(1), 127–133.

Sabnis, G. M., Harris, H. G., White, R. N., & Mirza, R. N. (1983). *Structural Modeling and Experimental Techniques.* Prentice-Hall Inc.

Schlichting, H. (1955). *Boundary layer theory.* McGraw-Hill Inc.

Sedov, L. I. (1959). *Similarity and Dimensional Methods in Mechanics.* Academic Press.

Sharp, J. J., Deb, A., & Deb, M. K. (1992). Applications of matrix manipulation in dimensional analysis involving large numbers of variables. *Marine Structures, 5*(4), 333–348.

Skinner, G. T., & Ludwig, G. R. (1978). Physical modelling of dispersion in the atmospheric boundary layer. *Technical report 201.* Calspan Corporation.

Smeaton, J. (1759). XVIII. An experimental enquiry concerning the natural powers of water and wind to turn mills, and other machines, depending on a circular motion. *Philosophical Transactions of the Royal Society of London, 51,* 100–174.

Sonin, A. A. (2004). A generalization of the Π-theorem and dimensional analysis. *Proceedings of the National Academy of Sciences, 101*(23), 8525–8526.

Sterrett, S. G. (2017). Physically Similar Systems-A History of the Concept. In Bertolotti Magnani (Ed.), *Springer Handbook of model-based science* (pp. 377–411). Springer.

Strutt, J. W. (Lord Rayleigh) (1877). *The theory of sound.* Macmillan Publishers Ltd.

Strutt, J. W., & (Lord Rayleigh),. (1915). The principle of similitude. *Nature, 95,* 66.

Szirtes, T. (2007). *Applied dimensional analysis and modeling.* Butterworth-Heinemann.

Takahashi, S., Tanimoto, K., & Miyanaga, S. (1985). Uplift wave forces due to compression of enclosed air layer and their similitude law. *Coastal Engineering in Japan, 28*(1), 191–206.

Tan, Q.-M. (2011). *Dimensional Analysis: with case studies in mechanics.* Springer Science & Business Media.

Tanner, R. I. (2000). *Engineering Rheology.* Oxford University Press.

Taylor, G. I. (1950). The formation of a blast wave by a very intense explosion I. Theoretical discussion. *Proceedings of the Royal Society London A 201*(1065), 159–174.

Taylor, R. N. (1995). Centrifuges in modelling: Principles and scale effects. In R. N. Taylor (Ed.), *Geotechnical Centrifuge Technology* (pp. 19–33). Blackie Academic and Professional.

Teixeira, M. A. C., & Belcher, S. E. (2006). On the initiation of surface waves by turbulent shear flow. *Dynamics of Atmospheres and Oceans, 41*(1), 1–27.

Tennekes, H., & Lumley, J. L. (1972). *A first course in turbulence.* The MIT Press.

Terzaghi, K. (1943). *Theoretical Soil Mechanics.* Wiley.

Thomson, W. (Lord Kelvin) (1883). Electrical Units of Measurement. *Popular Lectures and Addresses, 1*(73).

US Department of the Army. (1990). Structures to resist the effects of accidental explosions. *Technical Manual TM5-1300.* Picatinny Arsenal.

Van Driest, E. R. (1946). On dimensional analysis and the presentation of data in fluid-flow problems. *Journal of Applied Mechanics-Transactions of the ASME, 13*(1), A34–A40.

Various authors (1956). *I modelli nella tecnica. Vol. I–II.* Accademia Nazionale dei Lincei (in Italian).

Vaschy, A. (1892). Sur les lois de similitude en physique. *Annales Télégraphiques, 19,* 25–28.

Vaschy, A. (1896). *Théorie de l'électricité: Exposé des phénomènes électriques et magnétiques fondé uniquement sur l'expérience et le raisonnement.* Baudry et cie: Librairie Polytechnique.

Wang, Z. Y., Plate, E. J., Rau, M., & Keiser, R. (1996). Scale effects in wind tunnel modelling. *Journal of Wind Engineering and Industrial Aerodynamics, 61*(2–3), 113–130.

Warner, F. W. (2013). *Foundations of Differentiable Manifolds and Lie Groups* (Vol. 94). Springer Science & Business Media.

Weijermars, R., & Schmeling, H. (1986). Scaling of Newtonian and non-Newtonian fluid dynamics without inertia for quantitative modelling of rock flow due to gravity (including the concept of rheological similarity). *Physics of the Earth and Planetary Interiors, 43*(4), 316–330.

Wen, Y.-K. (1976). Method for random vibration of hysteretic systems. *Journal of the Engineering Mechanics Division, 102*(2), 249–263.

West, G. B. (1999). The origin of universal scaling laws in biology. *Physica A: Statistical Mechanics and its Applications, 263*(1), 104–113. In *Proceedings of the 20th IUPAP International Conference on Statistical Physics*.

West, G. B., Brown, J. H., & Enquist, B. J. (1997). A general model for the origin of allometric scaling laws in biology. *Science, 276*(5309), 122–126.

Westergaard, H. M. (1926). Stresses in concrete pavements computed by theoretical analysis. *Public Roads, 7*, 25–35.

Woisin, G. (1992). On J. J. Sharp et al. *Application of Matrix Manipulation in Dimensional Analysis Involving Large Numbers of Variables, 5*(4), 333–348; *Marine Structures 5*(4), 349–356.

Xuan, Y., & Li, Q. (2003). Investigation on convective heat transfer and flow features of nanofluids. *Journal of Heat Transfer, 125*(1), 151–155.

Yalin, M. S. (1971). *Theory of hydraulic models.* Macmillan Publishers Ltd.

Zhang, J., & Tang, Y. (2008). Dimensional analysis of soil-foundation-structure system subjected to near fault ground motions. In *Geotechnical Earthquake Engineering and Soil Dynamics Congress IV* (1–10).

Zhao, Y.-P. (1998). Prediction of structural dynamic plastic shear failure by Johnson's damage number. *Forschung im Ingenieurwesen, 63*(11–12), 349–352.

Author Index

Subject Index

A

Archimedes, 165
  number, 181, 220, 223, 388, 394

B

Bagnold
  number, 211, 301–303
Bingham
  number, 182
Boltzmann
  constant, 11, 90
Bond
  number, 184
British Imperial system, 6, 14, 15, 136
Buckingham - method or Theorem of, 25,
  29, 31, 34, 36, 37, 51

C

Capillarity
  number, 184
Capillary length scale, 184
Carvallo
  method, xvii
Cauchy
  number, 182
  similarity, 277
Cayley-Hamilton Theorem, 157
Centrifuge, 278, 293
  contaminant transport models, 285
  scale effects in the models, 283
  similarity conditions, 280
  similarity in dynamic models, 289
  similarity in tectonic processes, 291

Class
  of systems of units, 7, 16, 17, 30
Cliff
  recession, 306
Coefficient, 5
Coriolis
  acceleration, 284
  parameter, 321
Creep, 136, 155

D

Darcy
  equation, 202
  friction factor, 331
  number, 302
Dean
  number, 181
Deborah
  number, 182
Debris flow, 300, 303–305
  dimensional Analysis, 300
Dimensional
  quantity, 4
Dimensional basis, 31
Dimensional calculus
  arithmetics, 23
Dimensional homogeneity, 161
  principle, 9, 10, 20, 21, 23, 24, 86
Dimensional matrix, 58, 59, 61, 160, 236,
  291, 294, 296, 301
Dimensionless
  quantity, 4
Dimensionless groups
  non-power function, 96

S. G. Longo, *Principles and Applications of Dimensional Analysis and Similarity*,
Mathematical Engineering, https://doi.org/10.1007/978-3-030-79217-6

Printed in the United States
by Baker & Taylor Publisher Services